Osiris

**A RESEARCH JOURNAL DEVOTED
TO THE HISTORY OF SCIENCE
AND ITS CULTURAL INFLUENCES**

EDITOR
ARNOLD THACKRAY

MANAGING EDITOR
FRANCES COULBORN KOHLER

EDITORIAL OFFICE

NSYLVANIA

SUITE 460B

PHILADE   D1275540   )4-6228, USA

SUGGESTIONS FOR CONTRIBUTORS TO OSIRIS

OSIRIS is devoted to thematic issues, often conceived and compiled by guest editors.

1. Manuscripts should be **typewritten** or processed on a **letter-quality** printer and **double-spaced** throughout, including quotations and notes, on paper of standard size or weight. Margins should be wider than usual to allow space for instructions to the typesetter. The right-hand margin should be left ragged (not justified) to maintain even spacing and readability.

2. Bibliographic information should be given in **footnotes** (not parenthetically in the text), typed separately from the main body of the manuscript, **double-** or even **triple-spaced,** numbered consecutively throughout the article, and keyed to reference numbers typed above the line in the text.

   a. References to **books** should include author's full name; complete title of the book, underlined (italics); place of publication and publisher's name for books published after 1900; date of publication, including the original date when a reprint is being cited; page numbers cited. *Example*:

   [1]Joseph Needham, *Science and Civilisation in China,* 5 vols., Vol. I: *Introductory Orientations* (Cambridge: Cambridge Univ. Press, 1954), p. 7.

   b. References to articles in **periodicals** should include author's name; title of article, in quotes; title of periodical, underlined; year; volume number, Arabic and underlined; number of issue if pagination requires it; page numbers of article; number of particular page cited. Journal titles are spelled out in full on first citation and abbreviated subsequently. *Example*:

   [2]John C. Greene, "Reflections on the Progress of Darwin Studies," *Journal of the History of Biology,* 1975, *8:*243–272, on p. 270; and Dov Ospovat, "God and Natural Selection: The Darwinian Idea of Design," *J. Hist. Biol.,* 1980, *13:*169–174, on p. 171.

   c. When first citing a reference, please give the title in full. For succeeding citations, please use an abbreviated version of the title with the author's last name. *Example*:

   [3]Greene, "Reflections" (cit. n. 2), p. 250.

3. Please mark clearly for the typesetter all unusual alphabets, special characters, mathematics, and chemical formulae, and include all diacritical marks.

4. A small number of **figures** may be used to illustrate an article. Line drawings should be directly reproducible; glossy prints should be furnished for all halftone illustrations.

5. Manuscripts should be submitted to OSIRIS with the understanding that upon publication **copyright** will be transferred to the History of Science Society. That understanding precludes OSIRIS from considering material that has been submitted or accepted for publication elsewhere.

OSIRIS (SSN 0369-7827) is published once a year.

Subscriptions are $39 (hardcover) and $25 (paperback).

Address subscriptions, single issue orders, claims for missing issues, and advertising inquiries to *Osiris,* The University of Chicago Press, Journals Division, P.O. Box 37005, Chicago, Illinois 60637.

Postmaster: Send address changes to *Osiris,* The University of Chicago Press, Journals Division, P.O. Box 37005, Chicago, Illinois 60637.

Osiris is indexed in major scientific and historical indexing services, including *Biological Abstracts, Current Contexts, Historical Abstracts,* and *America: History and Life.*

Hardcover edition, ISBN 0-226-79376-1
Paperback edition, ISBN 0-226-79378-1

Constructing Knowledge in the History of Science

Edited by Arnold Thackray

A RESEARCH JOURNAL

DEVOTED TO THE HISTORY OF SCIENCE

AND ITS CULTURAL INFLUENCES

SECOND SERIES   VOLUME 10   1995

CONSTRUCTING KNOWLEDGE IN THE HISTORY OF SCIENCE

Cover: Earlier constructors of knowledge.
See page vi.

Scholars who attended the ten-day symposium "Critical Problems in the History of Science," held in Madison, Wisconsin, in 1957. Mentioned in the pages of this book are Marshall Clagett (49), Father Joseph Clark (31), Alistair C. Crombie (34), E. J. Dijksterhuis (5), Hunter Dupree (43), Charles Gillispie (41), Edward Grant (46), A. Rupert Hall (57), Thomas S. Kuhn (60), Dirk J. Struik (8), and L. Pearce Williams (58). Courtesy David C. Lindberg.

Preface

FORTY YEARS AGO the history of science was a new and fledgling field, enjoying its first glimpses of recognition in the American academy. Ironically, the death in 1956 of George Sarton—who had by then spent forty years as lone American prophet of his chosen discipline—coincided with the passing of the discipline itself into that promised land of which Sarton only dreamed. Sarton died in March 1956. In May of that year quite separate plans were officially launched for holding an ambitious "History of Science Institute" at the University of Wisconsin. The word "institute" had a Sartonian ring, but the plans in question owed little to Sarton.

In 1956 the exigencies of the cold war were driving the actions of the federal government. These actions included a steady growth in funding for the new National Science Foundation, and the NSF's decision to place the history of science at the leading edge of its cautious ventures into—or perhaps more accurately, its skirting of—the social sciences. Marshall Clagett and his colleagues in the new University of Wisconsin Department of the History of Science were thus taking advantage of the times when they applied for and received a munificent NSF grant to bring a small cadre of scholars to Madison in 1957, for the ten-day symposium "Critical Problems in the History of Science."

The roughly two dozen individuals who acted as speakers and commentators at the Madison meeting came close to constituting the totality of professional historians of science in the English-speaking world of their day. Their contributions, published in 1959 as *Critical Problems in the History of Science,* provided a "best practice" casebook of what the new, post-Sartonian, professional history of science aspired to be. If *Critical Problems* contained echoes of the debates of the 1930s, as in the seventy-five pages devoted to "The Scholar and the Craftsman," it also adumbrated the soon-to-be-dominant internalism of the adolescent profession, as in E. J. Dijksterhuis's paper on classical mechanics and Thomas S. Kuhn's grappling with "energy conservation as an example of simultaneous discovery."

Today, *Critical Problems* has itself passed into history, and the history of science as a discipline has passed beyond its adolescence.

The 1991 Madison "Conference on Critical Problems and Research Frontiers in History of Science and History of Technology" was thus far different from its 1957 predecessor. NSF funding was far less munificent. Those attending numbered in the hundreds. The conference was simultaneously part of the annual meetings of both the History of Science Society and the Society for History of Technology. Scholars had perforce to choose among competing sessions in scattered buildings. Individual participants jetted in and jetted out according to schedules of their own. There was little sense of community or coherence. Instead the meeting reflected the noisy, fractured, robust reality that is today's world of scholarship.

Constructing Knowledge in the History of Science offers only an imperfect glimpse into that world, for how could it be otherwise? Nonetheless, the eleven papers published here may serve as a useful starting point for those who wish an entry to present understanding and "best practice" in the history of science.

For help in producing this volume I am indebted to the officers of HSS and SHOT and especially to David C. Lindberg. Copy-editing and production services were once again ably supplied by Frances Coulborn Kohler.

With this book the revived *Osiris* reaches its tenth volume, and my editorship concludes. It has been a privilege to contribute to constructing knowledge in the history of science.

Arnold Thackray
Ash Wednesday 1995

KNOWLEDGE AND VALUES

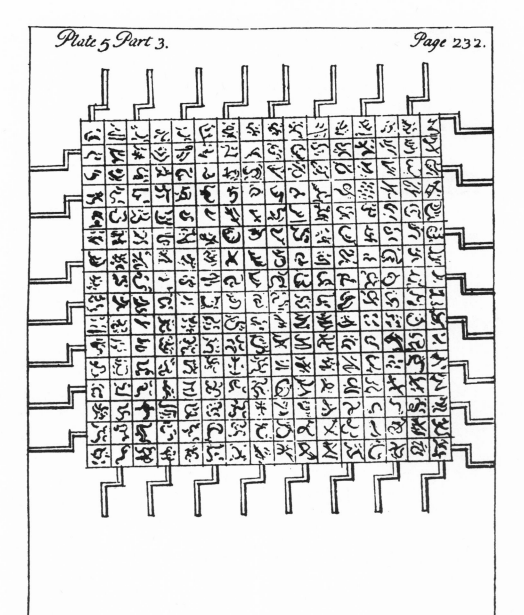

What is a moral economy good for? *Quantification (see p. 8): A mock version of Leibniz's "universal characteristic" was allegedly used by the Academy of Laputa to "write Books in Philosophy, Politicks, Law, Mathematicks, or Theology, without the least Assistance from Genius or Study"—the cranks were turned to create innumerable sentences. From Jonathan Swift,* Gulliver's Travels, *Part 3:* A Voyage to Laputa *(London, 1726).*

The Moral Economy of Science

By Lorraine Daston[*]

I. WHAT IS A MORAL ECONOMY?

The mind lives on the heart
Like any parasite.
—Emily Dickinson, "The Mind Lives
on the Heart" (c. 1876)

W E ARE HEIRS to an ancient tradition that opposes the life of the mind to
the life of the heart, and to a more recent one that opposes facts to values.
Because science in our culture has come to exemplify rationality and facticity, to
suggest that science depends in essential ways upon highly specific constellations
of emotions and values has the air of proposing a paradox. Emotions may fuel scientific work by supplying motivation, values may infiltrate scientific products as ideology or sustain them as institutionalized norms, but neither emotions nor values in-
trude upon the core of science—such are the boundaries that these habitual
oppositions would seem to dictate. The ideal of scientific objectivity, as currently
avowed, insists upon the existence and impenetrability of these boundaries. I will
nonetheless claim that not only does science have what I will call a moral economy
(indeed, several); these moral economies are moreover constitutive of those features
conventionally (and, to my mind, correctly) deemed most characteristic of science
as a way of knowing. Put more sharply and specifically: certain forms of empiricism,
quantification, and objectivity itself are not simply compatible with moral economies; they require moral economies.[1]

What exactly is a moral economy? Although several recent studies in the history
of science testify to the existence and significance of moral economies, such studies
have yet to crystallize around a common rubric, much less to rally around a common
standard.[2] Part of my work here will be to extrapolate implications and tendencies

---

[*] Department of History, University of Chicago, 1126 E. 59th Street, Chicago, Illinois 60637.

[1] I am grateful to John Carson and Nathan Reingold for pointing out that my use of the term *moral
economy* diverges significantly from E. P. Thompson's in "The Moral Economy of the English Crowd
in the Eighteenth Century," *Past and Present,* Feb. 1971, No. 50, pp. 76–136, reprinted in Thompson,
Customs in Common: Studies in Traditional Popular Culture (New York: New Press, 1991),
pp. 185–258, along with replies to critics and later reflections, "The Moral Economy Reviewed,"
pp. 259–351. My appeal here to "economies" of affects and values has little to do with Thompson's
accounts of corn markets and the tradition of "setting the price" by persuasion or riot, although it
does appeal to a broader sense of "legitimizing notion."

[2] See, e.g. (this list is by no means exhaustive), Owen Hannaway, "Laboratory Design and the Aim
of Science: Andreas Libavius versus Tycho Brahe," *Isis,* 1986, *77*:585–610; Theodore M. Porter,
"Objectivity as Standardization: The Rhetoric of Impersonality in Measurement, Statistics, and Cost-
Benefit Analysis," *Annals of Scholarship,* 1992, *9*:19–60; Porter, "Quantification and the Accounting

that seem to me to unite these scattered studies, and to clarify their contributions to a nascent investigation of moral economies in science.[3] What I mean by a moral economy is a web of affect-saturated values that stand and function in well-defined relationship to one another. In this usage, "moral" carries its full complement of eighteenth- and nineteenth-century resonances: it refers at once to the psychological and to the normative. As Gaston Bachelard decades ago remarked, to imbue objects or actions with emotion is almost always thereby to valorize them, and vice versa.[4] Here *economy* also has a deliberately old-fashioned ring: it refers not to money, markets, labor, production, and distribution of material resources, but rather to an organized system that displays certain regularities, regularities that are explicable but not always predictable in their details. A moral economy is a balanced system of emotional forces, with equilibrium points and constraints. Although it is a contingent, malleable thing of no necessity, a moral economy has a certain logic to its composition and operations. Not all conceivable combinations of affects and values are in fact possible. Much of the stability and integrity of a moral economy derives from its ties to activities, such as precision measurement or collaborative empiricism, which anchor and entrench but do not determine it.

It may help to etch the outlines of the notion of moral economy more crisply to point out what it is *not*. It is not a matter of individual psychology. Whatever and however vehement their other confessional differences, historians, sociologists, and philosophers of science share a certain horror of the psychological, properly so-called, and I confess I am no exception to this general hostility. The historians glare in distrust at the purported eternal verities of the mind, just because they are alleged to be eternal; the sociologists, recalling the warnings of Auguste Comte and Emile Durkheim, bare their teeth at the isolated individualism of much current psychology, including that labeled "social psychology"; the philosophers, post-Frege, take the word "psychological" into their mouths only as an epithet, as *ipso facto* proof that the problem or explanation at hand has nothing to do with genuine philosophy.[5] Although moral economies are about mental states, these are the mental states of

---

Ideal in Science," *Social Studies of Science,* 1992, 22:633–652; Simon Schaffer, "Astronomers Mark Time: Discipline and the Personal Equation," *Science in Context,* 1988, 2:115–145; Schaffer, "A Manufactory of Ohms: The Integrity of Victorian Values," in *Invisible Connections: Instruments, Institutions, and Science,* ed. Robert Bud and Susan Cozzens (Bellingham, Wash.: SPIE Press, 1992), pp. 23–56; Steven Shapin, "The House of Experiment in Seventeenth-Century England," *Isis,* 1988, 79:373–404; Shapin, " 'The Mind in Its Own Place': Science and Solitude in Seventeenth-Century England," *Sci. Context,* 1991, 4:191–217; Lorraine Daston and Peter Galison, "The Image of Objectivity," *Representations,* 1992, 40:81–128; and Daston, "Objectivity and the Escape from Perspective," *Soc. Stud. Sci.,* 1992, 22:597–618. I am grateful to authors who made prepublication versions of this recent work available to me when I was preparing this essay.

   [3] I undertake this task with considerable diffidence, given the obvious risks of misinterpretation and misappropriation. The studies I shall treat here have served as inspiration for my analysis, but their authors are wholly innocent of any responsibility for that analysis.

   [4] See, e.g., Gaston Bachelard, "Libido et connaissance objective," *La Formation de l'esprit scientifique* (1938), 14th ed. (Paris: Vrin, 1989), pp. 183–209.

   [5] Among the notable exceptions to this general disdain of the psychological within the history of science is Ryan D. Tweney, "Faraday's Discovery of Induction: A Cognitive Approach," in *Faraday Rediscovered: Essays on the Life and Work of Michael Faraday, 1791–1867,* ed. David Gooding and Frank James (London: Macmillan, 1985), pp. 189–209. The recent surge of interest among philosophers of science in approaches imported from cognitive science does not contradict their general dislike of psychology. Philosophers were repelled earlier not only because psychology smacked of the "irrational," but also because the psychological perspective seemed to them oozily invertebrate,

collectives, in this case collectives of scientists, not of lone individuals. To extend Ludwik Fleck's terminology, what is meant here is a *Gefühls-* as well as a *Denkkol-lektiv.*[6] Apprenticeship into a science schools the neophyte into ways of feeling as well as into ways of seeing, manipulating, and understanding. This is a psychology at the level of whole cultures, or at least subcultures, one that takes root within and is shaped by quite particular historical circumstances.[7] I hope that the collectivity and particularity of mental economies will go some way in assuaging the suspicions of the sociologists and historians, respectively; I shall return to the worries of the philosophers in the conclusion.

Nor is a moral economy confined to the level of motivation, whether in spurring individuals into scientific careers or in persuading society that science is worthy of encouragement and support. Since motivations of both sorts have been one of the principal loci for the discussion of values in science studies, it may be tempting to assimilate moral economies to them. However, this temptation should be resisted. The classical studies of how values, predominantly religious values, motivate both individuals and societies to pursue science grant such values at best a neutral and at worst a negative role with respect to the forms and content of scientific methods and assertions. Robert Merton's pioneering study *Science, Technology and Society in Seventeenth-Century England,* as well as his subsequent work in the sociology of science, advances the neutral alternative: the fervent desire of a seventeenth-century Englishman to glorify God through the investigation of His works might steer him toward a career in natural philosophy; the equally fervent desire of his twentieth-century counterpart to win the good opinion of a select circle of peers might propel him toward a scientific career. At the macrosocial level, utilitarianism, piety, or other cultural values may bolster the prestige and funding of science and even elevate some kinds of science above others. But none of these values impinges on scientific ways of knowing. As Steven Shapin points out in a lucid recent essay on the reception of the Merton thesis: "For Merton, the explanandum was emphatically not scientific method or scientific knowledge: it was the dynamics and social standing of a scientific enterprise that was itself conceived of as a black box."[8] Moral economies belong to the interior of Merton's black box. The outstanding example of the negative alternative also treats religion and science in seventeenth-century England, and it provides what are still some of the most exquisitely sensitive readings of how the Christian virtuoso's frame of mind and soul inclined him toward natural philosophy. R. S. Westfall was concerned, however, not only with the shading of religious

---

lacking all coherent structure. Although cognitive science has done little to rehabilitate the mind's rationality, it offers structures aplenty.

[6] Fleck in fact emphasized that emotions as well as concepts were shared by members of scientific thought collectives, and suggested that it was just this unanimity of feeling that created the illusion of freedom from emotions: Ludwik Fleck, *Genesis and Development of a Scientific Fact* (1935); ed. Thaddeus Trenn and Robert K. Merton; trans. Fred Bradley and Thaddeus Trenn (Chicago/London: Univ. Chicago Press, 1979), p. 49.

[7] I have in mind a gradual shaping of a collective personality akin (and, as will be seen below, sometimes identical) to Norbert Elias's "civilizing process": see Elias, "Synopsis: Towards a Theory of Civilizing Processes," in *Power and Civility,* Vol. II of *The Civilizing Process* (1939), trans. Edmund Jephcott (New York: Pantheon, 1982), pp. 229–333.

[8] Steven Shapin, "Understanding the Merton Thesis," *Isis,* 1988, *79:*594–605, on p. 595. Merton's work was originally published in *Osiris,* 1938, *4:*360–632; rpt. (New York: Harper Torchbooks, 1970).

reverence into scientific dedication, but also with the interaction of theological with natural philosophical doctrines. When these doctrines clashed, as Westfall believed they did on the topics of miracles and providences, the only role that values could play was to veil contradictions and foment inconsistencies.[9] Values could mix with scientific knowledge, but only as a contaminant. Moral economies, in contrast, are integral to science: to its sources of inspiration, its choice of subject matter and procedures, its sifting of evidence, and its standards of explanation.

Much the same might be said apropos of the relationship between moral economies and ideology in science. This is the other classical locus of how and why values enter science, this time opening the black box of scientific assumptions and assertions, and treating it very much as Pandora's box.[10] Whereas moral economies moralize scientists, ideologies moralize nature in the service of social interests. The numerous case studies in this genre run the gamut from piecemeal attempts to unmask this or that scientific claim as a piece of political interest tricked out as neutral fact, to more systematic exposés of all of science as a "social construction," laboriously if clandestinely built up out of interests, resources, and negotiations. Because it in principle encompasses all science, not just this or that ideologically tainted claim, the social-constructionist program comes closest to acknowledging the integral role of values in scientific work and its products: values do not distort science; they are science. This is why the annals of this program supply some of the most intriguing insights for the study of the moral economy of science.

However, because social constructionism focuses primarily on interests, be they political, social, or economic, and on (hidden) labor, it retains some of the muckraking character of more conventional revelations about ideology in science.[11] In contrast, to examine a moral economy of science may render familiar scientific procedures such as quantification strange, but seldom devious. Insofar as the study of moral economies in science is about power, it is power of the microscopic, internalized Foucauldian sort, rather than of the political (or martial), externalized kind.[12] In other words, the moral economy of science is more about self-discipline than coercion. Moreover, because social constructionists generally understand values as veiled interests, they are seldom concerned to explore the links between values and affects, unless these affects have an overtly societal character or influence. The stressed "social" in the social constructionist program refers not only to the dis-

[9] Richard S. Westfall, *Science and Religion in Seventeenth-Century England* (New Haven: Yale Univ. Press, 1958), pp. 90 et passim; for the sensitive readings see, e.g., pp. 27–28.

[10] Perhaps the most challenging of the current wave of ideology-and-science studies are those which address scientific accounts of gender: see, e.g., G.E.R. Lloyd, *Science, Folklore, and Ideology* (Cambridge: Cambridge Univ. Press, 1983); Ludmilla Jordanova, *Sexual Visions: Images of Gender in Science and Medicine between the Eighteenth and Twentieth Centuries* (Madison: Univ. Wisconsin Press, 1989); Londa Schiebinger, *The Mind Has No Sex? Women in the Origins of Modern Science* (Cambridge, Mass.: Harvard Univ. Press, 1989); and Cynthia Eagle Russet, *Sexual Science: The Victorian Construction of Womanhood* (Cambridge, Mass.: Harvard Univ. Press, 1989). [See also the articles by Evelyn Fox Keller (especially) and Sally Gregory Kohlstedt in this volume, with the relevant citations.—Eds.]

[11] For a more sanguine view of how ideology-laden or socially constructed science can sometimes count as an intellectual achievement rather than as a distortion, see M. Norton Wise, "Mediating Machines," *Sci. Context,* 1988, 2:77–113.

[12] See Michel Foucault, *Discipline and Punish: The Birth of the Prison* (1975), trans. A. Sheridan (New York: Pantheon, 1977); cf. Elias, "Civilizing Processes" (cit. n. 7), pp. 240–242. For the coercive, indeed bellicose view of power in science see Bruno Latour, *Science in Action: How to Follow Scientists and Engineers through Society* (Cambridge, Mass.: Harvard Univ. Press, 1987).

guised social components of which science is purportedly assembled, but also to the social uses to which science is put. Traffic flows in both directions across the science-society divide. Moral economies, however, tend to be one-way affairs. Although moral economies in science draw routinely and liberally upon the values and affects of ambient culture, the reworking that results usually becomes the peculiar property of scientists. Traces of the original cultural models—for example, the simplicity, dedication, and humility of Christian saints or the unworldly innocence of the pastoral idyll—lie ready to hand, and can be evoked by the spokesmen of science to win public approval and support. But the ultimate forms that moral economies assume within science, and the functions that they serve, are science's own.[13]

Finally, moral economies are not Mertonian norms, although here again there is a certain fleeting resemblance. Merton defines the "ethos of science" as "that affectively toned complex of values and norms which is held to be binding on the man of science. . . . These imperatives, transmitted by precept and example and reenforced by sanctions, are in varying degrees internalized by the scientist, thus fashioning his scientific conscience or, if one prefers the latter-day phrase, his super-ego."[14] However, the well-known norms of universalism, communism, disinterestedness, and organized skepticism, although "procedurally efficient," represent for Merton "one limited aspect of science as an institution," as carefully cordoned off from the "characteristic methods" and the "stock of accumulated knowledge" of science as motivations had been in his historical work. Moreover, these norms were, once established, immune to the vagaries of history and the pressures of context, for they were ultimately enforced not by human conscience but by nature. Scientists might violate the norms of universalism or communism, but only at their peril, for they were underwritten not simply by human sanctions but also by uniform, inexorable natural laws. Despite this alleged metaphysical grounding, a handful of scientific exposés, followed by a generation's worth of contextual studies in the history of science, apparently presented empirical refutation of Merton's norms, for here was candid testimony that violations could produce science of the first magnitude.[15] In contrast to Mertonian norms, moral economies are historically created, modified, and destroyed; enforced by culture rather than nature and therefore both mutable and violable; and integral to scientific ways of knowing.

To define an entity either directly or by contradistinction as I have tried to do above offers little proof that such entities exist, much less of their significance. Do

[13] On the evocation of these models in French academic éloges see Dorinda Outram, "The Language of Natural Power: The Eloges of Georges Cuvier and the Public Language of Nineteenth-Century Science," *History of Science*, 1978, *16:*153–178; see also Suzanne Delorme, "La vie scientifique à l'époque de Fontenelle après les éloges des savants," *Archeion*, 1937, *19:*217–235; and Charles B. Paul, *Science and Immortality: The Eloges of the Paris Academy of Sciences (1699–1791)* (Berkeley: Univ. California Press, 1980), esp. pp. 90–94.

[14] Robert K. Merton, "The Normative Structure of Science," rpt. in *The Sociology of Science: Theoretical and Empirical Investigations,* ed. Norman Storer (Chicago/London: Univ. Chicago Press, 1973), pp. 267–278, on pp. 268–269. Originally published as "Science and Technology in a Democratic Order," *Journal of Legal and Political Sociology,* 1942, *1:*115–126.

[15] The most spectacular of these exposés was James D. Watson, *The Double Helix: A Personal Account of the Discovery of the Structure of DNA* (New York: New American Library, 1968); for reactions see Watson, *The Double Helix*, including text, commentary, reviews, original papers, ed. Gunther Stent (New York: W. W. Norton, 1980), pp. 161–234. See also Nico Stehr, "The Ethos of Science Revisited: Social and Cognitive Norms," in *The Sociology of Science: Problems, Approaches, and Research;* ed. Jerry Gaston (San Francisco: Jossey-Bass, 1978), pp. 172–196.

moral economies of science really exist, and if so, what are they good for? These are challenges that can be met only by instantiation, not definition. In the next section I examine three examples of how moral economies have structured key aspects of how scientists come to know: quantification, empiricism, and objectivity.

II. WHAT IS A MORAL ECONOMY GOOD FOR? QUANTIFICATION

> L'excès de précision dans le règne de la quantité, correspond très exactement à l'excès du pittoresque dans le règne de la qualité.
> —Gaston Bachelard, *La formation de l'esprit scientifique* (1938)

Quantification is a portmanteau term that holds a multitude of meanings. It is part of our number fetishism that we seldom distinguish among them. Historians of science routinely use it to refer to abstract mathematical models that may or may not be tethered to measurements or even observations (e.g., Nicole Oresme's doctrine of the latitude of forms, or Jakob Bernoulli's probabilities of legal evidence); measurements that may or may not connect to a mathematical model of the phenomena under scrutiny (e.g., the physiological researches of Stephen Hales); straightforward counting (e.g., almost all of descriptive statistics); estimates grounded neither in measurement nor theory (e.g., many of William Petty's figures in his political arithmetic); methods of data representation and analysis (e.g., graphs and tables or the method of least squares); and the creation of new entities (e.g., index numbers such as the gross national product). The common denominator (so to speak) of all of these usages is not even numbers, for many historical instances of quantification in the sciences have been purely geometrical: when Galileo claimed in *Il Saggiatore* (1623) that the book of nature was written in the characters of "triangles, circles, and other geometric figures," he probably meant it quite narrowly.

Amidst this plurality of forms that scientific quantification has assumed, only some have aspired to accuracy, that is, to a close fit between mathematics and a select set of phenomena, although this is the virtue most heeded and praised by historians. Other mathematical virtues touted by quantifiers of various stripes have included precision, communicability, and impartiality, all of which can be cleanly detached from accuracy. For example, when in 1699 the English mathematician John Craig calculated the date of the millennium (A.D. 3150, when the credibility of the New Testament decays completely) on the basis of assumptions about the probability of human testimony, or when G. W. Leibniz proclaimed (with breathtaking optimism) that it would take a team of scholars less than five years to construct a Universal Characteristic by matching numbers to ideas and arithmetic operations to thought processes, they aimed primarily at precise knowledge, and only secondarily at accurate knowledge.[16] Accuracy concerns the fit of numbers or geometrical magnitudes to some part of the world and presupposes that a mathematical model can be anchored in measurement; precision concerns the clarity, distinctness, and intelligibility of concepts, and, by itself, stipulates nothing about whether and how those concepts match the world. Although striving for precision as a goal in and of itself is distinctive of much early modern quantification, in part because of a largely psycho-

---

[16] John Craig, *Theologiae christianae principia mathematica* (London, 1699); and Gottfried Wilhelm Leibniz, "Towards a Universal Characteristic" (1677), in *Leibniz Selections,* ed. Philip Wiener (New York: Scribners, 1951), pp. 22–23.

logical account of the grounds for mathematical certainty, it is by no means extinct among latter-day quantifiers.[17]

The cults of communicability and impartiality—again, with or without accuracy—also have an almost unbroken history in the sciences as well as in public life from the seventeenth century to the present. These quantifying virtues have often worked in tandem, usually to the end of damping controversy and compelling consensus. Even when neither measurements nor statistics were available, quantifiers of, say, the productivity of Holland or of the efficacy of smallpox inoculation pleaded for the superior clarity and communicability of numbers, favorably contrasted to "only comparative and superlative Words, and intellectual Arguments."[18] Leibniz contended that lack of clarity was at the root of almost all controversy and could therefore be cured by a goodly dose of numbers: "We need not be surprised then that most disputes arise from the lack of clarity in things, that is, from the failure to reduce them to numbers."[19] Although these attempts to silence dissent through quantification were (and still are) occasionally parasitic upon the vaunted certainty of mathematical demonstrations and operations, their dominant appeal was to consensus achieved through communication and thereby shared understanding, rather than through the necessity of demonstration. Even when the truth of the matter was not to be had, numbers could be invented, dispersed to correspondents at home and abroad, and, above all, mentally shared: you and I may disagree about the accuracy and the implications of a set of numbers, but we understand the same thing by them.

The moral economy of this form of quantification is sociable but intolerant of deviations, and it is not surprising that it flourishes under conditions of weak or confused authority—for example, the contested intellectual authority of sixteenth- and seventeenth-century natural philosophy, or, as Theodore Porter has recently argued, the contested political authority of twentieth-century pluralistic democracies.[20] In both cases the aim of quantification is not to secure individual conviction, but rather to secure the acquiescence of a diverse and scattered constituency. That is, the scientific polity that cherishes quantification is not only a collective, but also one whose members may differ from one another in nationality, skill, training, assumptions, or material resources such as laboratory equipment or statistical bureaus. It is quite possible to imagine, and to instantiate historically, scientific ideals and practices that preferred the solitary sage to the collective, or a more local and homogeneous collective that need not resort to the minimalist, information-losing techniques of quantification in order to communicate and persuade. For quantification, no matter how thorough and detailed, is necessarily a sieve: if it did not filter out local knowledge such as individual skill and experience, and local conditions such as this brand of instrument or that degree of humidity, it would lose its portability.[21] The moral commitment to a certain form of sociability among colleagues who may

---

[17] See, e.g., René Thom, "Mathématique et théorisation scientifique," in *Penser les mathématiques,* ed. Francois Guénard and Gilbert Lelièvre (Paris: Editions de Seuil, 1982), pp. 252–273.

[18] William Petty, *Political Arithmetick* (London, 1690), preface.

[19] Leibniz, "Universal Characteristic" (cit. n. 16), p. 24; cf. his plans for a language with "no equivocations or amphibolies," "Preface to the General Science" (1677), in *Leibniz Selections,* ed. Wiener (cit n. 16), p. 16.

[20] Porter, "Objectivity as Standardization" (cit. n. 2).

[21] *Ibid.* The same is true, *mutatis mutandis,* of the conditions for replicating empirical results: on the discord that ensues when aspects of local knowledge (e.g., a certain kind of glass prism) are not

never meet face to face must be strong in order to countenance the loss of so much hard-won detail.[22] It is in part the systematic erasure of these details in the service of extended sociability that creates the impression of the uniformity of nature: to turn Merton on his head, the uniformity of nature presupposes universalism among scientists, rather than the reverse.

Among the preconditions for this far-flung sociability are the oft-remarked impartiality and impersonality of quantified results and procedures. These qualities may flourish even in the absence of accuracy, and are indeed all the more highly valued when accuracy seems unattainable. Impartiality is first and foremost a judicial rather than a scientific virtue, and at most a prerequisite for rather than a guarantee of the truth of a verdict. Similarly, there is no a priori reason to believe that the elimination of all that is idiosyncratic will clear a path to the "really real": if the idiosyncrasy in question is skill, one might expect just the opposite. The point here is that impersonality and impartiality are cultivated by quantifiers as much for moral as for functional reasons. It is proverbial that both require dutiful self-abnegation so as to repress individuality and interest, and neither accrues automatically to quantified procedures and results. "Faceless numbers" fairly radiate personality in the hands of numerologists and cabalists; the chicaneries practiced with statistics are all too familiar. Abstraction alone never eliminates all traces of individuality and interest, and the history of applied mathematics, particularly social mathematics, is strewn with examples of partial impartiality.[23] Impersonality and impartiality in quantification might be better conceived as a continuum, more or less achieved by an effort of self-imposed restraint, rather than as properties inherent in the numbers themselves. To practice the form of quantification that breaches the boundaries of language, confession, nationality, and theoretical allegiance demands that the quantifiers voluntarily restrict their sphere of discretion. They must also sacrifice some of the meanings attached to numbers and techniques: Johannes Kepler's successors stripped his "laws" of their Pythagorean halo; Adolphe Quetelet's successors jettisoned his normative understanding of the normal curve. In other words, the choice of an extended form of scientific sociability incurs certain forms of moral obligation and discipline: the reining in of judgment, the submission to rules, the reduction of meanings—what Bachelard once called "that asceticism that is abstract thought."[24]

---

omitted see Simon Schaffer, "Glass Works: Newton's Prism and the Uses of Experiment," in *The Uses of Experiment: Studies in the Natural Sciences,* ed. David Gooding, Trevor Pinch, and Simon Schaffer (Cambridge: Cambridge Univ. Press, 1989), pp. 67–104. On the ideal of the solitary intellectual see Steven Shapin, "Mind in Its Own Place" (cit. n. 2); see also Martin Warnke, "Das Bild des Gelehrten im 17. Jahrhundert," in *Res publica litteraria: Die Institutionen der Gelehrsamkeit in der frühen Neuzeit,* ed. Sebastien Neumeister and Conrad Wiedemann (Wolfenbüttler Arbeiten zur Barockforschung, 14) (Wiesbaden: Otto Harrasowitz, 1987), Part I, pp. 1–34. For one collective that valued details over replicability or communicability see Lorraine Daston, "The Cold Light of Facts and the Facts of Cold Light: Luminescence and the Transformation of the Scientific Fact, 1600–1750," *Early Modern France,* in press; see also Steven Shapin, "Robert Boyle and Mathematics: Reality, Representation, and Experimental Practice," *Sci. Context,* 1988, 2:23–58.

[22] On the origins of this form of sociability among European intellectuals see Lorraine Daston, "The Ideal and Reality of the Republic of Letters in the Enlightenment," *Sci. Context,* 1991, *4:* 367–386.

[23] See, e.g., Donald A. Mackenzie, *Statistics in Britain, 1865–1930: The Social Construction of Scientific Knowledge* (Edinburgh: Edinburgh Univ. Press, 1981).

[24] Gaston Bachelard, "Les obstacles de la connaissance quantitative," *La formation de l'esprit scientifique* (cit. n. 4), pp. 211–238, on p. 237.

The affinities and arguably the origins of this ethos are bureaucratic, appealing to the rigid rationality of rules, conscientiously blind to variations of person or situation.[25] This is one moral economy of the several varieties of quantification.

When concerns for precision and accuracy combine in the enterprise of precision measurement, the moral economy takes another form. Whereas the quantification of precision alone aims at impersonality in the service of a collectivity, the quantification of precision measurement aims at integrity, sometimes in defiance of the collectivity. The more precise the measurement, the more it stands as a solitary achievement of the measurer, rather than as the replicable common property of the group. Not all scientific measurement aspires to precision: Robert Hooke, for example, recommended mathematics to the natural philosopher because it "accustoms the Mind to a more strict way of Reasoning, to a more nice and exact way of examining, and to a much more accurate way of inquiring into the Nature of things." But he did not require "Mathematical Exactness" of his measurements, "for we find that Nature it self does not so exactly determine its operations, but allows a Latitude almost to all its Workings, though . . . it seems to be restrain'd within certain Limits." The belief in the sharp-edged determinacy of nature grew slowly, and the scientific cult of precision measurement, with its rites of instrument making and error analysis, emerged only in the nineteenth century.[26]

With precision measurement emerged a quite different moral economy of quantification, one just as stern in its call for self-discipline, but self-discipline channeled to different ends. This is the self-discipline of caution and fastidious attention to detail, the painstaking prudence of the account ledger. In her fine recent study of Franz Neumann's physics seminar (established 1834) at Königsberg, Kathryn Olesko shows how the "ethos of exactitude" evolved in German astronomy, geodesy, and experimental physics, and how it was inculcated by the practices, particularly that of error analysis, taught in Neumann's seminar. The initiates of Königsberg scrupled to graph their measurements, for they distrusted the unobserved interpolated values. They warily sifted the results of colleagues, according to the known diligence and care of the experimenter. They balked at theoretical generalizations, unpersuaded that the data had been sufficiently purged of errors. In contrast to the moral economy of precision *tout court,* that of precision measurement cultivated certain personal idiosyncrasies, namely those of skill and, especially, the character traits of diligence, fastidiousness, thoroughness, and caution. Nor did scientific sociability figure prominently in their creed. Although the devotees of precision measurement never meant to withdraw from the scientific community, the rigor of their faith effectively isolated them even from other experimentalists, not to mention theorists, for all measurements were in principle subject to revision, correction,

[25] Gerd Gigerenzer *et al., The Empire of Chance: How Probability Changed Science and Everyday Life* (Cambridge: Cambridge Univ. Press, 1989), pp. 236–237; see also Porter, "Objectivity as Standardization" (cit. n. 2).
[26] Robert Hooke, "A General Scheme of the Present State of Natural Philosophy, and How its Defects may be Remedied By a Methodical Proceeding in the making Experiments and collecting Observations," in *The Posthumous Works of Robert Hooke* (1705), ed. Richard Waller, with an introduction by Richard S. Westfall (New York/London: Johnson Reprint, 1969), pp. 19, 38. On the instrumental preconditions for, and relative indifference to, precision measurement in eighteenth-century science see Maurice Daumas, "Precision of Measurement and Physical and Chemical Research in the Eighteenth Century," in *Scientific Change,* ed. A. C. Crombie (London: Heinemann, 1963), pp. 418–430.

improvement. To pursue the "duty" of perfecting precision led to the perceived in-commensurability of experimental results.[27]

This is perhaps a pathological expression of the moral economy of precision mea-surement, but like so many pathologies, simply an exaggeration of the same values and affects that sustained precision measurement under more normal conditions. Olesko correctly identifies integrity as the cardinal virtue of precision measurement, simultaneously applied to the character of the measurers and to the quality of the measurements.[28] Yet paradoxically integrity sometimes teetered on the edge of disin-tegration: the disintegration of a smooth curve into discrete data points, the disinte-gration of a set of apparently uniform measurements, the disintegration of the bonds between experiment and theory, the disintegration of the scientific collectivity.

III. WHAT IS A MORAL ECONOMY GOOD FOR? EMPIRICISM

> "Next vnto *Arui* there are two riuers *Atoica* and *Caora,* and on that braunch which is called *Caora* are a nation of people, whose heades appeare not aboue their shoulders, which though it may be thought a meere fable, yet for mine owne parte I am resolued it is true, because euery child in the prouinces of *Arro-maia* and *Canuri* affirme the same."
> —Sir Walter Raleigh, *The Discoverie of the large and bewtiful Empire of Guiana* (1596)

> "As it happened to a Dutch ambassador, who entertaining the king of Siam with the particularities of Holland, which he was inquisitive after, amongst other things told him, that the water in his country would sometimes, in cold weather, be so hard, that men walked upon it, and that it would bear an elephant, if he were there. To which the King replied, *Hitherto I have be-lieved the strange things you have told me, because I took you for a sober fair man, but now I am sure you lie."*
> —John Locke, *Essay Concerning Human Understanding* (1690), Book IV, Ch. 15, Sect. 5.

Empiricism is at least as multifarious as quantification, and correspondingly fertile in moral economies. Here I shall restrict myself to three distinctive aspects of the empiricism of seventeenth-century natural philosophy: testimony, facticity, and nov-elty. Each relied crucially upon intertwined values and affects: testimony upon trust, selectively extended; facticity upon academic civility; novelty upon the rehabilita-tion and transformation of curiosity. Critical for all three was the emergence of a new understanding of experience in natural philosophy in the middle decades of the seventeenth century.

Aristotelian natural philosophy aspired to causal knowledge, formulated in dem-onstrations about universals. It is history that deals with particulars, and this is why history is inferior not only to philosophy but also to poetry. As Aristotle explains in

[27] Kathryn M. Olesko, *Physics as a Calling: Discipline and Practice in the Königsberg Seminar for Physics* (Ithaca,N.Y./London: Cornell Univ. Press, 1991), pp. 250–252, 287, 392–393, 378–386. On precision measurement as a matter of character see Schaffer, "Astronomers Mark Time" (cit. n. 2). On the constraints practical and economic considerations could place upon the atomizing ten-dencies of precision measurement see Crosbie Smith and M. Norton Wise, *Energy and Empire: A Biographical Study of Lord Kelvin* (Cambridge: Cambridge Univ. Press, 1989), pp. 684–722.
[28] On the integrity of values, in both senses of both words, see Schaffer, "Manufactory of Ohms" (cit. n. 2).

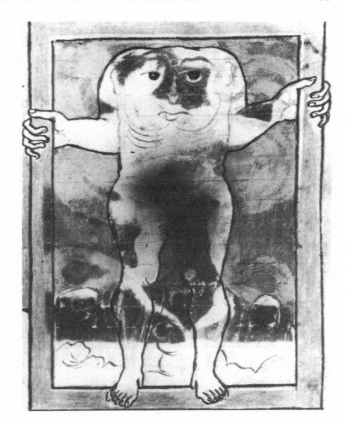

Empiricism: This blemmye from a medieval manuscript on the monstrous races of the East closely resembles Sir Walter Raleigh's secondhand description of the people of Caora. From MS Cotton Tiberius B.v., folio 82a; courtesy of the British Library, London.

the introduction to *On the Parts of Animals,* particulars occupy the philosopher only as stepping stones to generalizations and the discovery of causes. It is a happy fact that our minds are so constructed as to synthesize universals out of particulars: "Sense perception must be concerned with particulars, whereas knowledge depends on the recognition of the universal."[29] This does not imply that Aristotle's natural philosophy was not empirical, for his natural philosophical treatises reveal him to have been a sharp-eyed and indefatigable observer of an astonishing variety of phenomena. However, Aristotle's experience was common experience, "that which is always or that which is for the most part."[30] Medieval scholasticism was too long-lived and varied a set of doctrines to admit of any monolithic generalization concerning the meaning of experience in natural philosophy; it can nevertheless be cautiously asserted that most observations cited were indeed about what happened always or for the most part, and that these universals of experience served as the axioms for scholastic demonstrations.

Peter Dear has argued that in the course of the seventeenth century natural philosophers gradually abandoned universals in favor of particulars as the primary form of

[29] Aristotle, *Posterior Analytics,* 100a10–15, 87b37–39, translation from *The Complete Works of Aristotle,* ed. Jonathan Barnes, 2 vols. (Princeton: Princeton Univ. Press, 1984). See also *Poetics,* 1451b1–7; and *On the Parts of Animals,* 639a13–640a10.

[30] Aristotle, *Metaphysics,* 1027a20–27.

scientific experience.[31] Specific events, described in first-person singular historical narratives, with details of who saw what when and where, replaced universal generalizations. These particulars were increasingly published as short, semiepistolary articles in fledgling scientific journals such as the *Journal des Savants, Philosophical Transactions of the Royal Society of London,* the *Histoires et Mémoires de l'Académie des Sciences de Paris,* and the *Acta Eruditorum* rather than as long, systematic treatises. Just because natural philosophical experience had been transformed into discrete, specific events, previously deemed the stuff of history, their credibility was to be assessed by the traditional historical (and legal) means: that is, by a cloud of witnesses, each testimony carefully weighted by degree of credibility according to familiar criteria, including sex, age, character, and social standing.[32]

The new-style natural philosophical experience had at least one striking disadvantage vis-à-vis the old-fashioned scholastic sort: whereas universals and commonplaces are by definition accessible to all, specific events, particularly those produced by experiment with finicky, expensive equipment, were not. As the large medieval and early modern literature on "secrets" testifies, some kinds of knowledge about nature, particularly alchemy and artisanal techniques, had been at least partially shrouded from public view.[33] But as Owen Hannaway and Steven Shapin show, this retreat to the privacy of the monastic cell or the alchemist's den clashed with the reformed natural philosophy's pointed critique of both forms of intellectual aloofness.[34] For reasons we still do not fully understand, seventeenth-century natural philosophers envisioned themselves as members of an international collective ("the ingenious of Europe," as their title pages had it), and empiricism as a collaborative enterprise. In order to honor simultaneously the ideals of particulate experience and publicity, natural philosophers had necessary recourse to witnessing, both actual and "virtual," and to testimony.[35]

Yet not all testimony was trustworthy: how to assess both the credibility of the witness and the plausibility of the event became a central epistemological problem in the latter half of the seventeenth century.[36] As the dilemmas of Sir Walter Raleigh and John Locke's king of Siam show, it was all too easy to err on the side of either excessive credulity or excessive skepticism. The dilemma was sharpened by the dis-

[31] Peter Dear, *"Totius in verba:* Rhetoric and Authority in the Early Royal Society," *Isis,* 1985, *76:* 145–161; and Dear, "Jesuit Mathematical Science and the Reconstitution of Experience in the Early Seventeenth Century," *Studies in History and Philosophy of Science,* 1987, *18:*133–175.

[32] See, e.g., Francis Bacon's lawyerly instructions for sifting evidence for a "Natural and Experimental History": Bacon, *Description of a Natural and Experimental History* (1620), in *The Works of Francis Bacon,* ed. J. Spedding, D. Heath, and R. L. Ellis, 15 vols. (London, 1870), Vol. I, p. 401, Aphorism VIII.

[33] William Eamon, "Arcana Disclosed: The Advent of Printing, the Books of Secrets Tradition, and the Development of Experimental Science in the Sixteenth Century," *Hist. Sci.,* 1984, *22:*111–150; and B. J. T. Dobbs, "From the Secrecy of Alchemy to the Openness of Chemistry," in *Solomon's House Revisited: The Organization and Institutionalization of Science,* ed. Tore Frängsmyr (Canton, Mass.: Science History Publications, 1990), pp. 75–94.

[34] Hannaway, "Laboratory Design"; and Shapin, "House of Experiment" (both cit. n. 2).

[35] On virtual witnessing and its relationship to the establishment of "matters of fact" in Robert Boyle's experimental philosophy see the seminal study by Steven Shapin and Simon Schaffer, *Leviathan and the Air-Pump: Hobbes, Boyle, and the Experimental Life* (Princeton: Princeton Univ. Press, 1985).

[36] On philosophical and mathematical treatments of the credibility of testimony see Lorraine Daston, *Classical Probability in the Enlightenment* (Princeton: Princeton Univ. Press, 1988), pp. 306–342. For the English legal background see Barbara J. Shapiro, *Probability and Certainty in Seventeenth-Century England* (Princeton: Princeton Univ. Press, 1983), pp. 163–180.

tinctly odd character of many of the particulars retailed in the annals of the natural philosophers, a point to which I shall return shortly. Shapin suggests that trust among natural philosophers, as well as access to the places where experimental particulars were produced, was extended to gentlemen, following codes of honor and courtesy that sanctified a gentleman's word, however implausible his report, and opened his home (where most experiments took place) to other gentlemen, however inconvenient their visits.[37] There is considerable evidence that these conventions applied to natural philosophers of established reputation, as well as to the titled and well-born. When, for example, the Paris Académie des Sciences abandoned in frustration its attempts to replicate Johann Bernoulli's glowing barometers, Perpetual Secretary Bernard de Fontenelle reassured Bernoulli that "the confidence one has in his [Bernoulli's] word" made the academicians more willing to believe in protean nature than to doubt his unverifiable account.[38] Trust, rather than replicability, made the collaborative empiricism of particulars possible among natural philosophers. Belief in natural regularities wavered before belief in the testimony of trusted witnesses.

This trust was all the more sorely tried by the kind of particulars that often exercised early modern natural philosophers. Raleigh's tribe with "their eyes in their shoulders, and their mouths in the middle of their breasts"[39] could have taken their place alongside the numerous reports of anatomical anomalies such as monstrous births, celestial apparitions such as armies battling in the clouds, odd weather such as cyclones or rains of blood, and other strange phenomena in the *Journal des Savants, Philosophical Transactions,* and *Histoire et Mémoires.* The experience that replaced Aristotelian universals with particulars also replaced Aristotelian commonplaces with rarities and singularities. Many heeded Francis Bacon's charge that the axioms of Aristotelian natural philosophy were abstracted from too scanty a collection of particulars, and that experience of nature's ordinary course alone was inadequate to reveal the rules and species of nature. Scholastic commonplaces must be supplemented and corrected by a collection of "Deviating Instances, that is, errors, vagaries, and prodigies of nature, wherein nature deviates and turns aside from her ordinary course."[40] This "history of pretergenerations," or of "nature out of course," was meant to stand as warning and reproach to premature generalization and theorizing in natural philosophy, a collection of exceptions to all rules. The impact of these instances of "nature erring" was indeed chastening: amongst the numerous reports of strange phenomena published in the natural philosophical journals, very few offered an explanation or attempted to subsume the anomaly under a theory.

These strange facts had several obvious disadvantages. Rare unto marvelous, they strained even the ample trust of the new-style empiricism nearly to the breaking point. Observing them was a matter of luck, and they were even less amenable to public witnessing and sustained study than the most capricious of experimental phenomena. Moreover, the ban on premature theorizing could be used to cover a mute

---

[37] Shapin, "House of Experiment" (cit. n. 2). Shapin points out that the taboo against doubting a gentleman's word was so strong that skepticism concerning testimony was all but unknown in natural philosophy (p. 398); see also Shapin, "O Henry" (essay review), *Isis,* 1987, 78:417–424.

[38] Bernard de Fontenelle, "Sur le phosphore," *Histoire de l'Académie Royale des Sciences,* Année 1701 (Paris, 1743), pp. 1–8.

[39] Sir Walter Raleigh, *The Discoverie of the large and bewtiful Empire of Guiana* (1596), ed. V. T. Harlow (London: Argonaut Press, 1928), p. 56.

[40] Francis Bacon, *Novum organum* (1620), Aphorisms 1.25, 2.29, translation from *Works,* ed. Spedding, Heath, and Ellis (cit. n. 32).

and uninquisitive wonder, which hindered the causal explanations of uncommon as well as of common experience that Bacon had sought. Yet strange phenomena had the virtues of their vices, at least within the context of the seventeenth-century scientific academies that avidly pursued them. These academies self-consciously distinguished themselves from what they perceived to be the pedantry and pugnacity of university scholastics by insisting upon civility in their discussions.[41] The rivalries that proved most divisive were theoretical ones, and the most explosive of these conflicts pitted one member's pet theory against another's. Hence the pronounced preference among academicians for strange phenomena, which baffled theories on all sides. As Thomas Sprat commented apropos of discussions of experiments at the Royal Society, "There was no room left, for any to attempt, to heat their own, or others minds, beyond a due temper; where they were not allow'd to expatiate, or amplifie, or connect specious arguments together." This applied doubly to strange phenomena, the immediate effect of which was to paralyze speculation, and therefore, it was hoped, to pacify discussion.[42]

Therein lay the otherwise obscure attraction of strange phenomena, which became the archetypes of the first scientific facts. Facticity in science has a history, and these early facts resemble those honored by later generations only in part. There is considerable historical variability in what kind of phenomena can, in principle, become facts. Whole domains of experience—dreams, electrophosphorescence, musical harmonies—have drifted in and out of facticity since the seventeenth century. There is also historical variability in the virtues becoming to a fact. The facts of strange phenomena were neither reliable nor robust—they could not be produced at will, much less against one's will. These facts were stubborn not because they would not go away (the problem was to make them stay), but rather because they resisted explanation by any and all available theories. In contrast to the inductive and statistical facts of the eighteenth and nineteenth centuries, seventeenth-century matters of fact were neither mundane, repetitive, homogeneous, nor countable. Rather, they were rare, heteroclite, and singular. They qualify as facts because they were the first form of empiricism within natural philosophy to pulverize the continuum of experience into discrete particulars and to sever radically the link between a datum of experience and the inferences and conjectures founded upon it. Part and parcel of the moral economy of scientific civility, strange phenomena shaped seventeenth-century empiricism—its standards of evidence, its (very) peculiar objects, its model of facticity.[43]

For those familiar with the empiricism of scholastic natural philosophy, seventeenth-century natural philosophy presents yet another striking novelty—namely, novelty itself. While scholastic natural philosophy was far from static, its appetite

---

[41] On the "moral conventions" used to regulate dispute see Shapin and Schaffer, *Leviathan and the Air-Pump* (cit. n. 35), pp. 72–76; and Shapin, "House of Experiment" (cit. n. 2). Although Shapin and Schaffer situate the irenicism of Boyle and other early Fellows of the Royal Society within the context of the Restoration and Clarendon Code, the academic call for civility was a pan-European phenomenon. See, e.g., the prohibition against name-calling in *Histoire de l'Académie Royale des Sciences*, Année 1699 (Paris, 1718), p. 7.

[42] Thomas Sprat, *The History of the Royal Society of London* (London, 1667), p. 91. On academic civility and the facts of strange phenomena see Lorraine Daston, "Baconian Facts, Academic Civility, and the Prehistory of Objectivity," *Annals of Scholarship*, 1992, 8:337–363.

[43] On the nature and origins of this brand of facticity see Lorraine Daston, "The Factual Sensibility" (essay review), *Isis*, 1988, 79:452–470.

for novelties seldom embraced whole new vistas of experience. Instead, scholastic natural philosophers—like philosophers to this day—worked at the ever more subtle and penetrating analysis of a largely fixed stock of examples. In contrast, the appetite of their seventeenth-century counterparts for empirical novelties (it is not clear that they were so enamored of theoretical novelties) was gluttonous. Their journals and books were crammed with observations of new objects—the moons of Jupiter, a hurricane in Bermuda, a brilliant artificial phosphor—and old objects seen in a new way—a flea enlarged under the microscope, a lark suffocated in a bell jar by the air pump. The performing of experiments was in the first instance a method of manufacturing novelties of experience, which were served up with much the same breathless terseness in natural philosophical articles as the latest tidings of battles or heinous crimes were in the broadsides of the day. We are so accustomed to the pell-mell, headlong pace of scientific novelties that we are hard put to recognize its oddity: why crave new experiences before the old ones have been duly digested?

The answer lies in part in the sensibility and the epistemology of curiosity that distinguish much early modern science. In the course of the sixteenth and seventeenth centuries curiosity was not only elevated from grave vice to peccadillo to outright virtue. It was also transformed through a realignment in the field of vices and virtues, passions and interests: briefly, curiosity wandered from the pole of lust and pride to that of greed and avarice.[44] Whereas Augustine and a host of medieval commentators had criticized curiosity as a form of incontinence and passivity, early modern writers associated it with self-disciplined activity, all faculties marshaled and bent to the quest. Newly distinguished from the desires of the body, which could be lulled and sated, curiosity was insatiable, pure *conatus* or endeavor, and marked, as Thomas Hobbes said, "by a perseverance of delight in the continuall and indefatigable generation of Knowledge, [which] exceedeth the short vehemence of any carnall Pleasure." Marin Mersenne also mused upon the restless acquisitiveness of curiosity: "And thus we always desire to go beyond, such that acquired truths only serve as means to arrive at others: this is why we take no more stock of those we have than a miser does of the treasures in his coffers."[45]

Early modern curiosity had become a subspecies of consumerism, and its dynamics mirrored those of the trade in luxuries. Both curiosity and the luxury market thrived on novelty, for today's luxuries—tea, shoes, white bread—were tomorrow's necessities, and today's knowledge staled just as quickly for voracious curiosity. Like the market in luxuries, curiosity had become open-ended and insatiable, and this structural affinity was decisive for the chosen objects of curiosity in early modern science. All that was small, intricate, and, especially, hidden exerted a particular fascination for early modern investigations of nature, for the most suitable objects

[44] On the elevation see Hans Blumenberg, *Der Prozeß der theoretischen Neugierde* (Frankfurt am Main: Suhrkamp, 1988); Jean Céard, ed., *La curiosité à la Renaissance* (Paris: Société d'Edition d'Enseignement Supérieur, 1986); and Carlo Ginzburg, "High and Low: The Theme of Forbidden Knowledge in the Sixteenth and Seventeenth Centuries," *Past and Present*, Nov. 1976, No. 73, pp. 28–41. On the transformation and its impact on early modern science see Lorraine Daston, "Neugierde als Empfindung und Epistemologie in der frühmodernen Wissenschaft," in *Macrocosmos im Microcosmos: Die Welt in der Stube: Zur Geschichte des Sammelns, 1450–1800*, ed. Andreas Grote (Opladen: Leske & Budrich, 1994), pp. 35–60.

[45] Thomas Hobbes, *Leviathan* (1651), Book I, Ch. 6; and Marin Mersenne, *Les questions theologiques, physiques, morales et mathematiques* (1634), reprinted in Mersenne, *Questions inouyes . . .*, ed. André Pessel (Paris: Fayard, 1985), Quest. 23, p. 302 (my translation).

of curiosity were those that matched insatiable desire of the Hobbesian sort to inexhaustible detail, allowing the eye of the body or the eye of the mind to wander from one minute part or labyrinthine convolution to another, never at rest yet still fixed upon the same object in all its multiplicity. This power to awaken, hold, and even deepen attention made curiosity an indispensable part of the militant empiricism of seventeenth-century natural philosophy, with its abiding distrust of Aristotelian generalizations and natural kinds. Curiosity was most easily ignited by "Things strange and rare," but by means of a habit of estrangement it could also be marshaled to study more prosaic objects: "In the making of all kinds of Observations or Experiments there ought to be a huge deal of Circumspection, to take notice of every least perceivable Circumstance . . . And an Observer should endeavour to look upon such Experiments and Observations that are more common, and to which he has been more accustom'd, as if they were the greatest Rarity, and to imagine himself a Person of some other Country or Calling, that he had never heard of, or seen the like before."[46] The curious sensibility singled out objects, subjects, and stance: strange objects (or common ones estranged) studied with every-nerve-strained attention by people often united only in their taste for such objects and their cultivation of that stance. It also impressed the prestissimo pace of novelty tumbling after novelty upon seventeenth-century scientific empiricism.

Trust, civility, and curiosity were thus three moments of the moral economy of seventeenth-century empiricism. It was a moral economy that set evidentiary standards, stipulated the forms of facticity, selected certain objects as worthy of inquiry, and accelerated the rate of that inquiry. Its models lay in gentlemanly codes of honor, civic humanism, and an emotional mutation that mimicked the dynamic of consumerism. However, the whole, once assimilated by the natural philosophers, was more than the sum of these parts, for trust, civility, and curiosity meshed into an economy heretofore unknown within early modern culture. Civility privileged the facts of strange phenomena; trust in testimony expanded to correspondingly generous proportions; curiosity initially excited by rarities and oddities could also be schooled to examine more ordinary objects. This is only one possible moral economy of empiricism, and a rather short-lived one at that, but it was of great significance for the reformed natural philosophy of the seventeenth century.

IV. WHAT IS A MORAL ECONOMY GOOD FOR? OBJECTIVITY

> For that part of the scientific world whose opinion is of most weight, is generally so unreasonable, as to neglect altogether the observations of those in whom they have, on any occasion, discovered traces of the artist. In fact, the character of an observer, as of a woman, if doubted is destroyed.
> —Charles Babbage, *Reflections on the Decline of Science in England and on Some of its Causes* (1830)

The various forms of quantification and empiricism have their distinctive moral economies; objectivity however *is* a moral economy. The philosopher Thomas Nagel writes: "Objectivity is a method of understanding. It is beliefs and attitudes that are objective in the primary sense. Only derivatively do we call objective the truths that can be arrived at in this way."[47] As in the case of quantification and empiricism, it

---

[46] Hooke, "General Scheme" (cit. n. 26), pp. 61–62.
[47] Thomas Nagel, *The View from Nowhere* (Oxford: Oxford Univ. Press, 1986), p. 4.

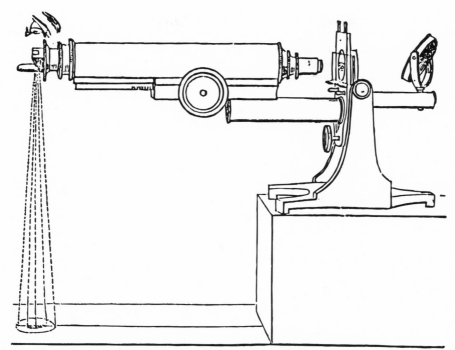

*Objectivity: The camera lucida, invented in 1807 by W. H. Wollaston as a
mechanical means of reproducing objects in perspective, is here adapted for use
with a microscope. From William B. Carpenter,* The Microscope and Its
Revelations *(London, 1868).*

would be more accurate to speak of objectivities, in the plural, for because scientific
objectivity has a history, it too displays diversity. Here I can only sketch two of its
most important variants, both of nineteenth-century vintage: mechanical objectivity
and aperspectival objectivity.

Mechanical objectivity is objectivity in the postlapsarian mode.[48] It is grounded
simultaneously in an epistemology of authenticity—in a yearning for "data" in its
root sense of "givens," bestowed with the effortlessness of grace—and also in the
guilty conviction that fallen humans, left unsupervised, can only meddle with the
givens of nature. This is the form of objectivity that strives to eliminate all forms of
human intervention in the observation of nature, either by using machines, such as
self-inscription devices or the camera, or by mechanizing scientific procedures, as
in deploying statistical techniques to choose the best of a set of observations.[49] Most
forms of objectivity share an enmity towards the personal, but which aspect of the
personal—judgment, inarticulable skill, an intense aesthetic response to nature, partiality towards one's own pet ideas—depends on the particular kind of objectivity.
In contrast to aperspectival objectivity, which combats idiosyncrasies of individuals,

[48] On the nature and history of mechanical objectivity see Daston and Galison, "Image of Objectivity" (cit. n. 2).

[49] Zeno G. Swijtink, "The Objectification of Observation: Measurement and Statistical Methods
in the Nineteenth Century," in *The Probabilistic Revolution,* ed. Lorenz Krüger *et al.,* 2 vols. (Cambridge, Mass./London: MIT Press, 1987), Vol. I: *Ideas in History,* pp. 261–286.

mechanical objectivity battles the general, all-too-human tendencies to aestheticize, anthropomorphize, judge, interpret, or in any other way "tamper" with the givens of nature. Goethe gave voice to the worries that impel mechanical objectivity when he preached caution in interpreting experimental results: "For here at this pass, this transition from empirical evidence to judgment, cognition to application, all the inner enemies of man lie in wait: imagination, which sweeps him away on its wings before he knows his feet have left the ground; impatience; haste; self-satisfaction; rigidity; formalistic thought; prejudice; ease; frivolity; fickleness—this whole throng and its retinue. Here they lie in ambush and surprise not only the active observer but also the contemplative one who appears safe from all passion."[50]

Mechanical objectivity found its characteristic expression in successive waves of first enthusiasm for and then disappointment in devices that seemed to promise an escape from this inner ambush, from the automatic registration of observations to the photograph. The latter became emblematic of mechanical objectivity and reveals its essentially moral, rather than epistemological core: the advantages of the photograph for art as well as science were not necessarily those of verisimilitude—naturalistic drawings in color often excelled photographs in creating a copy of what could be seen—but rather those of authenticity. By its very automatism the photograph created the illusion of an unmediated image, free of human intervention, if not visually faithful to its original.[51] It is this craving for authenticity that also explains why some scientists toyed with the idea of employing uneducated assistants. Not only was their labor cheap;[52] they were, it was thought, nearly as much tabula rasa as machines were, and therefore more fit to record observations than the all-too-well-informed and expectant scientist. For this reason Claude Bernard divided the labor of the experiment between the design, to be entrusted to the scientist's mind, fully equipped with hypotheses, and the execution, carried out by the senses "that observe and note," or even by an "uneducated man . . . knowing nothing of theory" and therefore able to see "new facts unperceived by a man preoccupied with an exclusive theory." Whereas Bernard was hesitant "to raise ignorance to a principle," the Astronomer Royal John Pond was blunter: "But to carry on such investigations, I want indefatigable, hard-working, and above all obedient drudges (for so I must call them, although they are drudges of a superior order), men who will be contented to pass half their day in using their eyes and hands in the mechanical act of observing, and the remainder of it in the dull process of calculation."[53]

These attempts to divide and thereby mechanize at least a part of the scientific labor of observation are redolent of the attitudes of contemporary manufacturers, for whom the workers and the machines in their factories were interchangeable, and

---

[50] Johann Wolfgang Goethe, "The Experiment as Mediator between Object and Subject" (1792, publ. 1823), in *Goethe: Scientific Studies,* ed. and trans. Douglas Miller (New York: Suhrkamp, 1988), p. 14.

[51] Charles Rosen and Henri Zerner, *Romanticism and Realism: The Mythology of Nineteenth-Century Art* (New York: Viking Press, 1984), pp. 107–108.

[52] Charles Babbage, *On the Economy of Machinery and Manufacturers,* 4th ed. (London: Charles Knight, 1835), p. 195.

[53] Claude Bernard, *An Introduction to the Study of Experimental Medicine* (1865), trans. Henry Copley Greene (New York: Dover, 1957), pp. 23, 38; and Pond in minutes of the Council of the Royal Society, 6 April 1826, quoted in Charles Babbage, *Reflections on the Decline of Science in England and on Some of its Causes* (1830), *The Works of Charles Babbage,* ed. Martin Campbell-Kelly (London: William Pickering, 1989), Vol. VII, p. 64n.

for whom the word *mechanical* still carried the derogatory lower-class associations of "rude mechanical." Although scientists often policed the character of their observer-drudges, the more striking moral overtones in mechanical objectivity were those of self-discipline, confronting Goethe's "inner enemies" on their own territory. Here scientists attempted to hold their own predilection to judge, interpret, aestheticize, and anthropomorphize in check, and their language of self-restraint sometimes echoes that of Christian asceticism.[54] Charles Gillispie caught this note of manful self-denial, of speculation crushed and beguiling illusions willfully destroyed, in his evocations of the "cruel edge of objectivity."[55] It is in the nineteenth century that stubborn facts also turn "nasty," "ugly," and "contrary." By the turn of the twentieth century, to "face the facts" always implies some unpleasantness, and therefore objectivity—some measure of resolve and self-command.[56]

If the emblem of mechanical objectivity is the photograph, the motto of aperspectival objectivity could be "the view from nowhere," in Nagel's brilliant oxymoron. Whereas mechanical objectivity is about suppressing the universal human propensity to judge, interpret, and aestheticize, aperspectival objectivity is about eliminating the idiosyncrasies of particular observers or research groups. Although all idiosyncrasies came to be tarred with the brush of subjectivity in the nineteenth century, they are by no means always handicaps: The ability to split a double star with the naked eye is as much an idiosyncrasy as a sluggish reaction time. Charles Babbage thus spoke for the ideal of aperspectival objectivity when he insisted that "genius marks its tract, not by the observation of quantities inappreciable to any but the acutest senses, but by placing nature in such circumstances, that she is forced to record her minutest variations on so magnified a scale, that an observer, possessing ordinary faculties, shall find them legibly written."[57]

Aperspectival objectivity serves scientific sociability and therefore enlists the various techniques of standardization, both quantitative and experimental. Transcendence of individual viewpoints seemed a precondition for a coherent scientific community to many nineteenth-century scientists and philosophers. The existence of such a community, stretching over time and space, in turn seemed to some a precondition—or even an eventual guarantee—for reaching scientific truth. Charles Sanders Peirce conceived of this necessarily communal form of truth-seeking as proceeding by a kind of symmetric cancellation of individual errors: "The individual may not live to reach the truth; there is a residuum of error in every individual's opinions. No matter, it remains that there is a definite opinion to which the mind of man is, on the whole and in the long run, tending. . . . This final opinion, then, is independent, not indeed of thought in general, but of all that is arbitrary or individual in

[54] See Daston and Galison, "Image of Objectivity" (cit. n. 2). On policing drudges see Schaffer, "Astronomers Mark Time" (cit. n. 2).

[55] Charles C. Gillispie, *The Edge of Objectivity: An Essay in the History of Scientific Ideas* (Princeton: Princeton Univ. Press, 1960), pp. 44–45; cf. pp. 73, 107, 156, 241.

[56] See, e.g., Max Weber's typical reflections on the link between "wissenschaftliche Objektivität" and "unbequeme Tatsachen": Weber, "Die 'Objektivität' sozialwissenschaftlicher und sozialpolitischer Erkenntnis" (1904), *Gesammelte Aufsätze zur Wissenschaftslehre,* ed. Johannes Winckelmann, 3rd ed. (Tübingen: J. C. B. Mohr, 1968), pp. 154–155.

[57] Babbage, *Reflections* (cit. n. 53), p. 86. On aperspectival objectivity and its origins see Daston, "Objectivity and the Escape" (cit. n. 2). On Weber's views on the prospects for objectivity in the social sciences see Robert N. Proctor, *Value-Free Science? Purity and Power in Modern Knowledge* (Cambridge, Mass.: Harvard Univ. Press, 1991), pp. 134–154.

thought; it is quite independent of how you, or I, or any number of men think."[58] Peirce here gives philosophical voice to changing scientific ideals, ideals rooted in changing scientific practices in the middle decades of the nineteenth century. International scientific congresses and collaborations in demography, astronomy, geodesy, and meteorology multiplied, creating networks of coordinated observers intent on capturing phenomena so vast that they were invisible to the individual observer.[59] It is no accident that Peirce himself contributed to several of these international efforts during his stint at the United States Coast and Geodetic Survey.

Within these global collaborations aperspectival objectivity became a scientific creed, the ideal that corresponded to the practices of well-nigh constant, coordinated communication: articles were circulated across oceans and continents, measurements exchanged, observations tallied, instruments calibrated, units and categories standardized. Aperspectival objectivity was the ethos of the interchangeable and therefore featureless observer—unmarked by nationality, by sensory dullness or acuity, by training or tradition, by quirky (even if superior) apparatus, or by any other idiosyncrasy that might interfere with communication, comparison, and accumulation of results. It should come as no surprise that proponents of this ideal frowned upon carrying precision measurement to hair-splitting extremes: "The extreme accuracy required in some of our modern enquiries has, in some respects, had an unfortunate influence, by favouring the opinion, that no experiments are valuable, unless measures are most minute, and the accordance among them most perfect."[60] Scientists paid homage to the ideal of aperspectival objectivity by contrasting the individualism of the artist with the self-effacing cooperation of scientists, who no longer came in the singular—"L'art, c'est moi; la science, c'est nous," as Claude Bernard neatly put it. For those like Bernard who celebrated the ideal of aperspectival objectivity, there was a certain nobility in the abandonment of the personal, a sacrifice of the self for the collective—if not for the collective good, then at least for the collective comprehension. Norbert Elias argued that all complex coordinations of human activity partake of the "civilizing process," that is, the "permanent self-control" demanded by "the lengthening of the chains of social action and interdependence," and science is no exception here. The self-control and detachment required of scientists by aperspectival objectivity was strenuous: scientists must not only wait to be recognized for their efforts; they must give up recognition altogether. Bernard exhorted scientists to bury their pride and vanity in order "to unite our efforts, instead of dividing them or nullifying them by personal disputes," for all scientists are ultimately equal in their anonymity: "In this fusion [of particular truths into general truths], the names of promoters of science disappear little by little, and the further science advances, the more it takes an impersonal form and detaches itself from the past."[61]

No doubt the manifestos of both mechanical and aperspectival objectivity

---

[58] Charles Sanders Peirce, "A Critical View of Berkeley's Idealism" (1871), *Values in a Universe of Chance: Selected Writings of C. S. Peirce (1839–1914)*, ed. Philip Wiener (New York: Doubleday, 1958), pp. 81–83.

[59] See Susan Faye Cannon, *Science in Culture: The Early Victorian Period* (Folkstone: Dawson; New York: Science History Publications, 1978), Ch. 3, on "Humboldtian science."

[60] Babbage, *Reflections* (cit. n. 53), p. 85.

[61] Bernard, *Introduction* (cit. n. 53), pp. 43, 39, 42. See Elias, *Power and Civility* (cit. n. 7), pp. 230–248, 273–274.

reflected a high-minded ideal rather than a sociological reality. Yet even values honored erratically are nonetheless genuine values, reflecting choices and revealing attitudes. Moreover, the values and affects of both mechanical and aperspectival objectivity left visible traces in the conduct of scientists, in their ever stronger preference for mechanized observation and methods, in their ever more refined division of scientific labor, in their preference for the authentic as opposed to the verisimilar or clear-cut image, and in their ever more exclusive focus on the communicable. The self-restraining and self-effacing counsels of mechanical and aperspectival objectivity reverberate with the stern voice of moral duty: the self-command required in both cases to suppress the merely personal is indeed the very essence of the moral. This does not mean we must admire these ideals, for the annals of cultural anthropology are full of moral conventions enforced by the most stringent self-command which we nonetheless abhor. But we cannot ignore them, particularly since they are interwoven into those key aspects of science we somewhat grossly term quantification, empiricism, and objectivity.

V. CONCLUSION: DISTINCTIONS WITHOUT PRIVILEGE

The few facets of quantification, empiricism, and objectivity that I have so briefly explored here by no means exhaust those aspects of science that are in whole or in part constituted by a moral economy. Many other practices and commitments might, I believe, be profitably so studied. One thinks, for example, of the kinship of form (short, private dated entries) and coincidence of timing (sixteenth and seventeenth centuries) of the diary and the notebook of experimental and field observations.[62] The *Verinnerlichung* and construction of self by the more expansive diary writers seem to parallel the construction of what F. L. Holmes has described as the construction of "larger units of meaning" in the reports experimentalists write up from their lab notebooks.[63] The checkered history of scientific attitudes toward secrecy also cries out for investigation *qua* moral economy.

However, I do not mean to suggest that moral economies in science are the Universal Key to all Mythologies. As I hope will be clear from the above examples, they tell us very little about the detailed contents of scientific articles and treatises, and equally little about the institutional framework of science. Rather, they answer old questions and pose new ones about how scientists at a given time and place dignify some objects of study at the expense of a great many others, trust some kinds of evidence and reject other sorts, and cultivate certain mental habits, methods of investigation, and even characters of a distinctive stamp. Above all, they focus our attention on the distinctions between and the history of the heterogeneous pursuits we are accustomed to lump together under the headings of "quantification" or "empiricism" or "objectivity." And because moral economies are part of what one might

[62] On the history of the diary see Elisabeth Bourcier, *Les journaux privés en Angleterre de 1600 à 1660* (Paris: Publications de la Sorbonne/Imprimerie Nationale, 1976); and Albert Gräser, *Das literarische Tagebuch* (Saarbrücken: West-Ost Verlag, 1955), pp. 19–38. Literature on the origins of the practice of keeping a laboratory notebook is harder to come by, but see W. E. Knowles Middleton, *The Experimenters: A Study of the Accademia del Cimento* (Baltimore/London: Johns Hopkins Univ. Press, 1971), pp. 359–382; and M. J. van Lieburg, "Isaac Beeckman and His Diary-Notes on William Harvey's Theory of Blood Circulation (1633–1634)," *Janus*, 1982, *69*:161–183, for some suggestive early examples.

[63] F. L. Holmes, "Scientific Writing and Scientific Discovery," *Isis*, 1987, *78*:220–235, on p. 235.

call historical epistemology—a history of the categories of facticity, evidence, objectivity, and so forth—they should interest the philosopher as well as the historian.

As for the sociologist, an investigation of how intellectual work is saturated with moral, emotional, and aesthetic elements at a collective, not just a biographical, level opens perspectives into psychology that go beyond the calculus of self-interest, strategically deployed to the ends of discipline- or career-building. Here there is ample room for a revival of Weberian-style historical sociology that carries the study of the cultural origins and character of scientific rationality beyond a few mournful sighs over the disenchantment of the world. There is excellent evidence that the moral economies of science derive both their forms and their emotional force from the culture in which they are embedded—gentlemanly honor, Protestant introspection, bourgeois punctiliousness—and here are promising grounds for an alliance of science studies with the new cultural history and its anthropological affiliates. However, there is also evidence that once these cultural forms have been uprooted and combined within a moral economy of science, they become naturalized to that milieu. Honor among scientists is not quite what it was among gentlemen, asceticism among scientists is not quite what it was among the devout. Swimming against the stream of contextual studies of science, moral economies reassert rather than dissolve the boundaries that separate the mentalities and sensibilities of scientists from those of ambient society. Science is not thereby privileged—an analysis of moral economies sometimes has the power to render the practices of the tribe of scientists as bizarre as those of any other tribe—but it is distinct. By examining in a new light just those ways of knowing once thought to exempt science from the realm of emotions and values, a study of moral economies may illuminate the nature of the rationality that seemed to exclude them.

KNOWLEDGE AND GENDER

Knowledge and gender

Marie Curie is the name that first comes to mind at the words "woman scientist." New research is not only uncovering the work of less famous women but also redefining the term to include those who computed, illustrated, and otherwise contributed to the scientific enterprise (see page 39). Courtesy the Chemical Heritage Foundation.

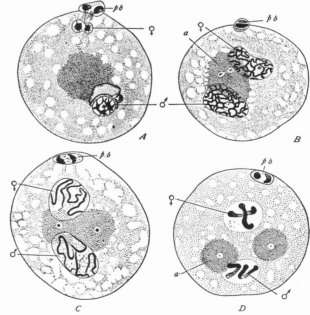

The issue of gender and science also encompasses the effect of social expectations on representations of nature—even, for example, the fertilization of the egg (see p. 34). Figure 65 from E. B. Wilson, The Cell in Development and Inheritance *(New York, 1896).*

Gender and Science:
Origin, History, and Politics

By Evelyn Fox Keller[*]

HISTORIANS OF SCIENCE may be tempted to treat gender and science as a subspecialty within the history of science—indeed, to judge from new course titles, bibliographies, session headings in history of science conferences, one of the fastest growing subspecialities in the field. But to make sense of this subspecialty, and especially to understand the extraordinary interest the term (and the subject) have attracted in recent years both in this country and abroad, it is necessary to begin by locating it first, not within the history of science, but in relation to contemporary feminist theory—in other words, in the context out of which it in fact emerged.

A feminist historian, Mary Poovey, introduced the term *border cases* to denote historical phenomena that, by virtue of their location on the "border between two defining alternatives," constitute privileged sites for examining the ideological work of gender. "Border cases," she wrote, "mark the limits of ideological certainty." Peter Galison, a historian of science, introduced the kindred notion of *trading zones* to call attention to the extensive traffic across borders (in his case, between experiment, instruments, and theory) and hence to the impossibility of clear demarcations.[1] Both notions might be invoked to highlight the cultural and historical specificity of disciplinary perspectives in general and, at the same time, to problematize the very concept of disciplinary boundaries. The issue of gender and science, I suggest, is a border case par excellence. It sits not on one border, but on multiple borders—indeed, on the borders between feminist theory and all the scientific and metascientific disciplines. It is also a trading zone, a domain of cross talk, exchange, and struggle. By its very existence, it calls into question the borders of all these disciplines.

Given my personal involvement with the history of both the term and the subject, I have found it difficult to review that history as a disinterested observer. What follows, therefore, is a frankly first-person perspective on the history and problems of gender and science.

I. GENDER AND SCIENCE

As far as I can tell, the phrase *gender and science* first made its appearance in an article I published under that title in 1978, not, as it happened, in a history of science

[*] Program in Science and Technology Studies, E51–228, Massachusetts Institute of Technology, Cambridge, Massachusetts 02139.

[1] Mary Poovey, *Uneven Developments: The Ideological Work of Gender in Mid-Victorian England* (Chicago: Univ. Chicago Press, 1989), p. 12. Cf. Peter Galison, "The Trading Zone: Coordinating

journal but in a psychoanalytic journal. By that title I sought to rouse readers out of a certain habitual complacency, explaining my concerns as follows:

> The historically pervasive association between masculine and objective, more specifi-
> cally between masculine and scientific, is a topic that academic critics resist taking
> seriously. Why is that? Is it not odd that an association so familiar and so deeply en-
> trenched is a topic only for informal discourse, literary allusion, and popular criticism?
> How is it that formal criticism in the philosophy and sociology of science has failed to
> see here a topic requiring analysis? . . .
>   The survival of [such] mythlike beliefs in our thinking about science . . . ought, it
> would seem, to invite our curiosity and demand investigation. Unexamined myths . . .
> have a subterranean potency; they affect our thinking in ways we are not aware of, and
> to the extent that we lack awareness, our capacity to resist their influence is undermined.
> The presence of the mythical in science seems particularly inappropriate. What is it
> doing there? From where does it come? And how does it influence our conceptions of
> science, of objectivity, or, for that matter, of gender?[2]

Although I was technically the author of this article, I did not then, and I do not now, regard the ideas behind it as originating with me. In the introduction to the book that the article eventually became, I tried to emphasize the ways in which these questions derived from the logic of a collective endeavor we had just begun to call "feminist theory," and in one of the early drafts I even tried to spell out the sense in which they were not mine at all, but rather, "ours." Feminists have recently become as suspicious of the first person plural as they were earlier of the impersonal pro-noun, so I need to try to be quite clear about my notion of a collective "we." Who, and what, did I have in mind? Certainly not historians of science—I am not sure I even knew any at that time. Rather, what I had in mind was a very local collectivity of women academics who had actively participated in what has been called "the women's movement" (recall that 1975 was the designated "International Year of Women"), who called themselves feminists, who were involved in "consciousness-raising" groups, and who had begun to deploy their heightened consciousness in radical theoretical critiques of the disciplines (and the worlds) from which they had come. By the mid 1970s works in "feminist theory," the name we gave to this collec-tive endeavor, began to appear in anthropology and sociology, history, literature, and psychoanalysis—though not yet in any discipline relating to the natural sciences.[3]

Action and Belief," Ch. 9 in *Image and Logic: The Material Culture of Modern Physics* (Chicago: Univ. Chicago Press, forthcoming).

[2] Evelyn Fox Keller, "Gender and Science," *Psychoanalysis and Contemporary Thought,* 1978, *1*(3):409–433.

[3] Anthropology and sociology: Michelle Zimbalist Rosaldo and Louise Lamphere, eds., *Woman, Culture, and Society* (Palo Alto, Calif.: Stanford Univ. Press, 1974); Sherry B. Ortner, "Is Female to Male as Nature Is to Culture?", *ibid.,* pp. 67–87; Rayna Rapp Reiter, ed., *Toward an Anthropology of Women* (New York: Monthly Review, 1975); Gayle Rubin, "The Traffic in Women: Notes on the Political Economy of Sex," *ibid.,* pp. 157–210; and Dorothy Smith, "Women's Perspective as a Radical Critique of Sociology," *Sociological Inquiry,* 1974, *44:*7–13. History: Joan Kelly-Godol, "The Social Relations of the Sexes: Methodological Implications of Women's History," *Signs,* 1976, *1*(4):809–823. Literature: Kate Millett, *Sexual Politics* (New York: Doubleday, 1970); Elaine Showalter, *A Literature of Their Own: British Women Novelists from Bronte to Lessing* (Princeton: Princeton Univ. Press, 1977); and Sandra M. Gilbert and Susan Gubar, *The Madwoman in the Attic: The Woman Writer and the Nineteenth-Century Imagination* (New Haven: Yale Univ. Press, 1979). Psychoanaly-sis: Nancy Chodorow, "Family Structure and Feminine Personality," in *Woman, Culture, and Society,*

The first step was to appropriate the term *gender* to underscore and elaborate Simone de Beauvoir's dictum that "one is not born a woman." In a classic and self-conscious deployment of naming as a form of political action, they (we) redefined *gender,* in contradistinction to sex, to demarcate the social and political, hence variable, meanings of *masculinity* and *femininity* from the biological or presumably fixed categories of *male* and *female*. The function of this redefinition was to redirect attention away *from* the meaning of sexual difference and *to* the question of how such meanings are deployed. To quote Donna Haraway, "Gender is a concept developed in order to contest the naturalization of sexual difference."[4] Very quickly feminists began to see, and as quickly to exploit, the analytic power of this distinction for exploring the force of gender and gender norms, not only in the making of men and women, but also as silent organizers of the cognitive and discursive maps of the social and natural worlds that we, as humans, simultaneously inhabit and construct—even of those worlds that women rarely enter.

Even, that is, the world of the natural sciences. It was just a matter of time before feminists who were involved in these conversations and reading these papers and who knew something about the natural sciences would take on, as they say, the "hard" case. I may have been the first to use the term *gender and science,* but I was hardly alone in recognizing the kind of questions now brought into view, which, once in view, demanded analysis. By the late 1970s a generation of feminists from a range of different disciplines, who brought with them varying senses of a collective "we," were taking due note of the traditional naming of the scientific mind as "masculine" and the collateral naming of nature as "feminine," and accordingly calling for an examination of the meaning and consequences of these historical connotations.[5] We hoped by that route both to undermine these traditional dichotomies and to pave the way for a restoration or relegitimation of just those values that had been excluded from science by virtue of being labeled "feminine." All of us were variously fueled, and to various degrees, by what Donna Haraway calls "paranoid fantasies and academic resentments,"[6] by what I might call "utopian aspirations," as well as by more straightforward recognitions of intellectual (and soon, even academic) opportunities. For some a suspicion that "objectivity" might be a code word for "domination" went hand in hand with the fantasy that we had hold of a lever with which we could not only liberate women, but also turn our disciplines upside down—perhaps, even change the world. (Some of us were humbler: I, for one, merely thought of changing science.) In other words, it was a pretty heady time.

The first book-length response came from Carolyn Merchant, writing as a historian of science, as a Marxist, as a feminist, and as a committed environmentalist. In *The Death of Nature* she focused squarely on the significance of the metaphor of nature as woman—for science, for capitalism, for women, and for nature—in the displacement of organicist by mechanist world views. By arguing that this

---

ed. Rosaldo and Lamphere, pp. 43–66; Jean Baker Miller, *Toward a New Psychology of Women* (Boston: Beacon Press, 1976).

[4] Donna Haraway, *Simians, Cyborgs, and Women: The Reinvention of Nature* (New York: Routledge, 1991), p. 131.

[5] I might mention Elizabeth Fee, Donna Haraway, Sandra Harding, Hilde Hein, Leigh Star, Carolyn Merchant, and Helen Longino.

[6] Haraway, *Simians* (cit. n. 4), p. 183.

displacement implied, at least to the users of that language, a symbolic act of violence both against nature-as-woman and against woman-as-nature, her work played a major role in mobilizing, at least briefly, a coalition between feminists and environmentalists.[7]

By the time my own book on gender and science came out, I too had discovered history, and I sought to tie such shifts in metaphors of nature (and simultaneously of mind and knowledge) to changing conceptions of individuality, selfhood, and masculinity—changes that were themselves neither epiphenomenal nor causal but deeply enmeshed in the social, economic, and political changes of the time. Especially, I attempted to argue for the confluence of new definitions of masculinity and new conceptions of what constituted a "proper" and epistemologically "productive" relation between "mind" and "nature." I sought, in sum, to locate the popular equation between "masculine" and "objective" in a particular historical transition. But to understand the relationship between "objectivity" and domination, I returned to the psychodynamics of individual development, adding to my earlier foray an analysis of the relations between love, knowledge, and power.[8]

By the early 1980s quite a bit of feminist literature about science had already been written, out of widely varying conceptions of feminism. Besides the work growing directly out of feminist theory, another literature, on the history of women in science, was growing with equal rapidity directly out of the history of science—most notably, Margaret Rossiter's book on women scientists in America (my own book on Barbara McClintock might also belong here).[9] There was also a body of work, mostly by biologists, devoted to a critical examination of scientific constructions of "woman." Perhaps the earliest of these was the edited volume *Genes and Gender,* published by Ruth Hubbard and Marian Lowe in 1979. Ruth Bleier's *Science and Gender* appeared in 1981, and Anne Fausto-Sterling's *Myths of Gender* in 1985.[10] By grouping these different literatures together, one could now begin to offer courses under the rubric "gender and science"—provided, that is, one expanded the meaning of the term *gender* to include women and sex. Such an expansion-cum-coalition was strategic; it seemed reasonable enough at the time, especially since other problems were more pressing. One such problem was academic shelter, as it were. Where would one put such a course, composed of so many different disciplinary and intellectual agendas? And as became increasingly clear over the 1980s, even of different political agendas? I was fortunate; I had access to a science and technology studies program that was hospitable. Others tried women's studies; eventually, some even tried history of science programs. And indeed it was with the entry of this catchall label of gender and science into the history of science—as titles for courses, bibliographies, and conference proceedings—that I originally attempted to begin this paper. In other words, at precisely the point where the use of the rubric

---

[7] Carolyn Merchant, *The Death of Nature: Women, Ecology, and the Scientific Revolution* (New York: Harper & Row, 1980).

[8] Evelyn Fox Keller, *Reflections on Gender and Science* (New Haven: Yale Univ. Press, 1985).

[9] Margaret Rossiter, *Women Scientists in America: Struggles and Strategies to 1940* (Baltimore: Johns Hopkins Univ. Press, 1982); and Evelyn Fox Keller, *A Feeling for the Organism: The Life and Work of Barbara McClintock* (San Francisco: Freeman, 1983).

[10] Ruth Hubbard and Marian Lowe, eds., *Genes and Gender* (Staten Island, N.Y.: Gordian Press, 1979); Ruth Bleier, *Science and Gender: A Critique of Biology and Its Themes on Women* (New York: Pergamon, 1984); and Anne Fausto-Sterling, *Myths of Gender: Biological Theories about Men and Women* (New York: Basic Books, 1985).

to cover a loose coalition of assorted works on women, sex, gender, and science threatened to break down altogether. Let me explain.

II. WOMEN, SEX, AND GENDER

Even at the very beginning, the slippage between women and gender had been a source of discomfort for feminists. For one thing, and most trivially, the equation of *women* with *gender* is a logical error—in fact, as Donna Haraway points out, exactly the same kind of error involved in equating *race* with people of color.[11] Whatever the term is taken to mean, when we use it to apply to people, strictly speaking, we mean it to apply to at least most if not all people. In fact, my own interest in gender and science focused neither on women nor on "femininity," but on men and conceptions of "masculinity." What, then, invites the elision, and why would feminists permit such an elision? One answer to the first question is immediately made clear by the analogy with race: women are culturally and historically marked by their sex or gender in a way that men are not, much as people of color are marked by their race. But the principal reasons that feminists at least initially tolerated and to some extent even supported the slippage between women and gender were twofold: first, the primary concern with which most of us had begun was with the force of gender on women's lives; second, even when our concerns moved outward, as they conspicuously did in feminist theory, that slippage was endured out of simple expedience. More recently, however, at least in most parts of the academy, it has widely come to be seen necessary to mark explicitly the distinction between women and gender: witness the renaming of a number of feminist research programs as "program for the study of women and gender." I want to suggest that this distinction has now become especially necessary in the study of gender and science.

A decade ago a loose coalition of works on women, sex, gender, and science could be held together by the common denominator of feminism—by a commitment to the betterment of women's lives. Differences in other commitments, differences, say, in disciplinary, theoretical, or political perspectives, might well pale by comparison—as long, that is, as that common denominator remained primary, as long as the women whose lives we sought to improve seemed to have common needs, and as long as we continued to see ourselves as a subversive force operating from the margins and interstices of academic life. In the intervening years, however, in part because of the very successes of contemporary feminism, all these differences became both more visible and more pressing as we, and our concerns, began to move into established academic and disciplinary niches. Today it has become conspicuously evident that not all women have the same interests or needs; so too has it become evident that not all scholars who call themselves "feminist" have common or even reconcilable theoretical and disciplinary agendas.

These remarks surely pertain to feminist scholarship in general, but I would like to try to spell out their particular relevance for the present status of the catchall "gender and science" in the U.S. academy, and especially in history of science. First, insofar as our interest in the history of women in science was initially motivated by a protest against a history of exclusion and by a political quest for equity, the dramatic changes that have occurred in the participation of at least some women in science

---

[11] Haraway, *Simians* (cit. n. 4), p. 243.

over these fifteen years need to be noted. I do not mean to suggest that women have achieved equity in the sciences, but rather that where fifteen years ago gender appeared as the principal axis of exclusion, today a glance at the racial and ethnic profile of the increased numbers of women who have entered scientific professions over these years suggests that it no longer does. Furthermore, from the perspective of those who have broken through the gender barrier and have now forged alliances within the scientific enterprise, it is not at all obvious how a continued focus on gender, especially given its emphasis on criticism, might be in their interest. These, I think, are some of the issues that Londa Schiebinger was referring to when she stressed the importance of context in what she somewhat elliptically calls "arguments for and against gender differences"; they are also the issues contributing to the increasing difficulty experienced by so many historians of science who attempt to teach the disparate subjects of women, sex, and gender in science together under one rubric.[12]

But if the primacy of gender as an occupational barrier in the sciences has receded and its utility as a critical wedge been blunted by occupational success, recognition of both its cultural and analytic importance has in other areas only increased—attesting, once again, to the successes of contemporary feminism. It is alas true that feminist theory has not proven itself powerful enough to change the world, as I once hoped it might. It has, however, radically, and I think irrevocably, changed the landscape of a number of academic disciplines. Some more so than others. Notably not the natural sciences, nor—at least not yet—the philosophy of science; hardly at all the social studies of science, and only some areas of the history of science.

For feminist theory to realize more fully its analytic promise in the history of science, obviously historians need to start reading its literature. But also, I suggest, we need a new taxonomy: "gender and science" needs to be disaggregated into its component parts. Schematically, these might be described as those studies examining the history of (1) women in science; (2) scientific constructions of sexual difference; and (3) the uses of scientific constructions of subjects and objects that lie both beneath and beyond the human skin (or skeleton). Each of these subjects has by now accumulated a rich literature in its own right and requires its own reconfiguration into new kinds of "trading zones." Here I will focus on the third of these, "gender *in* science"—trading not between assorted studies of women, sex, and gender in science, but between historical studies of gender, language, and culture in the production of science. For such studies of gender *in* science, I want to make a particular plea for a consolidated, two-way, effort toward integrating a number of analytic perspectives that are currently (mis)perceived as disjoint.

It will be evident that in proposing a disaggregation of what has come to be known as gender and science and focusing only on gender in science, I have made my task considerably easier, for I have now bracketed most of the literature usually considered under this term. My aim, however, is not to privilege my own or any other particular political or intellectual agenda; rather, it is to provide room for all the different concerns these subcategories separately raise. I want especially to sidestep the question of whether one is for or against gender differences and to allow

---

[12] Londa Schiebinger, *The Mind Has No Sex? Women in the Origins of Modern Science* (Cambridge, Mass.: Harvard Univ. Press, 1989), p. 273. In the session on syllabi for courses in gender and science at the 1990 History Science Society meeting, such difficulty was widely attested.

for the possibility of being either both or neither. The principal point is that I take the role of gender ideology to be but one aspect of the constitutive role of language, culture, and ideology in the construction of science, and hence, though the roots of such analyses have been and must continue to lie in feminist theory, I take their place in the history of science proper to be just one part of that more general inquiry. I suggest that work in this area has not only raised novel kinds of questions for historians, but also offers some novel models of and sites for historiographic analysis that might even be of use to historians of science who are not women, who may not even be gendered, and, possibly, who do not necessarily think of themselves as feminists.

III. GENDER IN SCIENCE

Let me now illustrate this claim with a few examples of the kinds of questions feminists have raised about the implications of a gendered vocabulary in scientific discourse, proceeding from those that have relatively straightforward implications for the reading of scientific texts to those with rather more indirect implications. In all of these examples metaphors of gender can be seen to work, as social images in science invariably do, in two directions: they import social expectations into our representations of nature, and by so doing they simultaneously serve to reify (or naturalize) cultural beliefs and practices. Although the dynamics of these two processes are almost surely inextricable, many feminists focus on the latter, emphasizing their effects (usually negative) on women; here my focus will be on the former, on their influence on the course of scientific research.

I will start where concerns about women, sex, and gender are most likely to intersect, in analyses of past and current work in the biology of reproduction and development. Many of these analyses reduce to a common basic form, the identification of synecdochal (or part for whole) errors of the following sort: (1) the world of human bodies is divided into two kinds, male and female (i.e., by sex); (2) additional, extraphysical properties are culturally attributed to those bodies—active or passive; independent or dependent; primary or secondary (read gender); and (3) the same properties that have been ascribed to the whole are then attributed to the subcategories of, or processes associated with these bodies. Often, though not necessarily, these analyses are undertaken from the vantage point of present, presumably superior, knowledge.

Undoubtedly the most conspicuous examples of such synecdoches are found in the history of theories of generation. Nancy Tuana has sought to augment the existing literature on reproductive theories from Aristotle to the preformationists by focusing on the imposition of prevailing views of women (i.e., as passive, weak, and generally inferior) onto their roles in reproduction. Thomas Laqueur's more recent and more probing analysis, *The Making of Sex,* adds substantially to such an effort. And some authors have undertaken corresponding analyses of contemporary discussions of fertilization. Scott Gilbert and his students, for example, have traced the language of courtship rituals in standard treatments of fertilization in twentieth-century textbooks.[13]

---

[13] Nancy Tuana, ed., *Feminism and Science* (Bloomington: Indiana Univ. Press, 1989); Thomas Laqueur, *Making Sex: Body and Gender from the Greeks to Freud* (Cambridge, Mass.: Harvard

Emily Martin has continued this effort by tracking the "importation of cultural ideas about passive females and heroic males into the 'personalities' of gametes" in the most recent technical literature. This is how the argument goes. Conventionally, the sperm cell has been depicted as "active," "forceful," and "self-propelled," qualities that enable it to "burrow through the egg coat" and "penetrate" the egg, to which it "delivers" its genes and "activate[s] the developmental program." By contrast, the egg cell "is transported," "swept," or merely "drifts" along the fallopian tube until it is "assaulted," "penetrated," and fertilized by the sperm.[14] The technical details that elaborate this picture have, until the last few years, been remarkably consistent: they provide chemical and mechanical accounts for the motility of the sperm, their adhesion to the cell membrane, and their ability to effect membrane fusion. The activity of the egg, assumed nonexistent, requires no mechanism. Only recently has this picture shifted, and with that shift, so too has shifted our technical understanding of the molecular dynamics of fertilization. One early and self-conscious marking of this shift by two researchers in the field, Gerald and Helen Schatten, appeared in 1983:

> The classic account, current for centuries, has emphasized the sperm's performance and relegated to the egg the supporting role of Sleeping Beauty. . . . The egg is central to this drama, to be sure, but it is as passive a character as the Grimm brothers' princess. Now, it is becoming clear that the egg is not merely a large yolk-filled sphere into which the sperm burrows to endow new life. Rather, recent research suggests the almost heretical view that sperm and egg are mutually active partners.[15]

And indeed, the most current research on the subject routinely emphasizes the activity of the egg cell in producing the proteins or molecules necessary for adhesion and penetration. At least nominal equity (and who, in 1994, could ask for anything more?) seems even to have reached the most recent editions of *The Molecular Biology of the Cell,* where "fertilization" is defined as the process by which egg and sperm "find each other and fuse."[16]

For historians of science, this recapitulation may raise more questions than it supplies answers, but one question is critical: What *is* the relation between the shift in metaphor in these accounts, the development of new technical procedures for representing the mechanisms of fertilization, and the concurrent embrace of at least nominal gender equity in the culture at large? If nothing else, tracking the metaphors of gender in this literature has provided us with an ideal site in which, with more extensive analysis, we can better appreciate and perhaps even sort out the complex lines of influence and interactions between cultural norms, metaphor, and technical development. Such a task does not require us to embrace the most recent version as

---

Univ. Press, 1990); and Biology and Gender Study Group, "The Importance of Feminist Critique for Contemporary Cell Biology," in *Feminism,* ed. Tuana, pp. 172–187.

[14] Emily Martin, "The Egg and the Sperm: How Science has Constructed a Romance Based on Stereotypical Male-Female Roles," *Signs,* 1991, *16:*485–501, on pp. 500, 489–490.

[15] Gerald Schatten and Helen Schatten, "The Energetic Egg," *The Sciences,* 1983, *23*(5):28–34, on p. 29.

[16] Bruce Alberts *et al., Molecular Biology of the Cell* (New York/London: Garland Press, 1994), p. 868.

correct, either scientifically or politically; witness the work of Frederick Churchill and John Farley on nineteenth-century debates about sexual reproduction. It is merely necessary to register the extent to which gender ideologies are implicated in the construction of (at least some) scientific stories.[17]

Similar reviews could be provided of biological accounts of sex determination: Anne Fausto-Sterling, for example, has explored the language of presence and absence in recent discussions of the sex gene. Or of the relationship between cytoplasm and nucleus over the past hundred years, in which, far from coincidentally, the cytoplasm has at least tacitly been routinely figured as female, and the nucleus as male. Jan Sapp has given us an excellent account of the history of cytoplasmic inheritance, but he neglected to note the significant marks of gender in this history, some of which were noted by Scott Gilbert and his students.[18] Indeed, the history of depictions of cytoplasm and nucleus is remarkably parallel to that for the egg and sperm: like the reformulations of the gametes' roles, recent support of cytoplasmic inheritance and of an important role for cytoplasmic determinants in development coincides with the rise in the 1980s of an ideology of gender equality. There are even some rumblings indicating a similar theoretical shift in the most recent discussion of sex genes. I group these examples together because of the similarity of their structure and the simplicity of their morals.

Other examples, in which readings of gender are less closely tied to readings of biological sex, are correspondingly less straightforward in both their structure and their implications. One general area of the history of science that has attracted particular interest among feminist scientists is that of the relation between genetics and developmental biology, and here an undercurrent of resistance to genetic determinism and a corresponding championship of the organizing models of developmental biology can clearly be seen. The arguments of Lynda Birke, Ruth Bleier, Scott Gilbert's study group, and Ruth Hubbard especially come to mind.[19] Similar (or related) preferences can be seen in analyses of brain and behavior science, as carried out by Helen Longino and Ruth Doell; of mechanism and organicism, examined by Merchant; and even in certain analyses of models in physics, explored by Stephen Kellert.[20] The common denominator in these discussions might be described as a preference for interactionist, contextual, or global models over linear, causal, or "master

[17] Frederick Churchill, "Sex and the Single Organism: Biological Theories of Sexuality in Mid-Nineteenth Century," *Studies in the History of Biology,* 1979, *3:*139–177; and John Farley, *Gametes and Spores: Ideas about Sexual Reproduction, 1750–1914* (Baltimore: Johns Hopkins Univ. Press, 1982).

[18] Anne Fausto-Sterling, "Life in the XY Corral," *Women's Studies International Forum,* 1989, *12*(3):319–331; Jan Sapp, *Beyond the Gene: Cytoplasmic Inheritance and the Struggle for Authority in Genetics* (New York: Oxford Univ. Press, 1987); and Biology and Gender Study Group, "Feminist Critique" (cit. n. 13).

[19] See Lynda Birke and J. Silverton, eds., *More than the Parts: Biology and Politics* (London: Pluto Press, 1984); Bleier, *Science and Gender* (cit. n. 10); Biology and Gender Study Group, "Feminist Critique" (cit. n. 13); Hubbard and Lowe, eds. *Genes and Gender* (cit. n. 10); and Keller, *Feeling for the Organism* (cit. n. 9).

[20] Helen Longino and Ruth Doell, "Body, Bias, and Behavior," *Signs,* 1983, 9(2):206–227; Helen Longino, *Science as Social Knowledge: Values and Objectivity in Scientific Inquiry* (Princeton: Princeton Univ. Press, 1990); and Stephen H. Kellert, *In the Wake of Chaos: Unpredictable Order in Dynamical Systems* (Chicago: Univ. Chicago Press, 1993).

molecule" theories.[21] The question is, What does gender have to do either with these concerns or with these preferences?

One link to gender of particular relevance to discussions of genetics and development can be traced in the tacit coding, already suggested, of the cytoplasm (and more generally, of the body) as female and the corresponding coding of the nucleus or gene as male. Just as in the story about egg and sperm, such codings carry with them traces of social relations between male and female, inviting the suspicion that even when merely implicit, they have been silently working to support correspondingly hierarchical structures of control in biological debates. To date only fragmentary evidence has been brought to bear on this suspicion, but it provides sufficient support, I think, to demonstrate the need to attend to gender markings in future, more detailed, investigations of these subjects.[22]

But feminist concerns about "master molecule" theories do not necessarily depend on the allocation of gender labels to the constituent elements of debate. For some critics the concern is more explicitly political, based on the fear that such hierarchical structures in biology are themselves rationales for existing social hierarchies; or that they reflect values supporting not only social but also individual constraint. Such concerns led Helen Longino to develop her sophisticated philosophical analysis of the ways in which political and social values inevitably enter into theory choice.[23] Other scholars have employed (as I have) yet a different kind of argument, seeing in these hierarchical models the expression of a mind-set predicated on control and domination, unconsciously projecting its own sense of self-and-other onto representations of processes operating in the natural world. Today I find this argument by projection to be unduly limited—above all, by its failure to take into account the particular kinds of material consequences that models or metaphors of domination have and, accordingly, the particular kinds of material ambitions such models support. As I have argued elsewhere, master molecule theories are not only psychologically satisfying; they are also remarkably productive—productive, that is, in relation to particular kinds of aims.[24]

IV. STILL TO BE DONE

By focusing on mind-sets and metaphors, feminists have made visible the possibility of alternative mind-sets and metaphors—by itself no mean feat for our thinking about science. But now we need a further analysis of the ways in which metaphors work to bridge the gulf between representing and intervening, of how they help to organize and define research trajectories. Such studies in language and science might be said to constitute one of the most interesting new frontiers in the history

---

[21] The term *master molecule* was originally invoked by David Nanney in protest over the conception of genes as "dictatorial elements in the cellular economy": Nanney, "The Role of Cytoplasm in Development," in *The Chemical Basis of Heredity,* ed. W. D. McElroy and H. B. Glass (Baltimore: Johns Hopkins Univ. Press, 1957). Only later was it appropriated by feminists for a larger critique. In the late 1980s it was reappropriated by the NSF in celebration of the successes of DNA.

[22] For further discussion see Evelyn Fox Keller, "Rethinking the Meaning of Genetic Determinism," Tanner Lecture, University of Utah, 1993.

[23] Longino, *Science as Social Knowledge* (cit. n. 20); and Keller, *Reflections on Gender and Sex* (cit. n. 8).

[24] See Evelyn Fox Keller, "Critical Silences in Scientific Discourse," *Secrets of Life, Secrets of Death: Essays on Language, Gender, and Science* (New York: Routledge, 1992), pp. 73–92.

of science, and our understanding of the conceptual dynamics of gendered metaphors has been well assisted by the sophistication Gillian Beer, Ludmilla Jordanova, and Nancy Stepan have brought to their own work at this frontier.[25]

More generally, and of utmost importance for historians of science, is the need for deep and thoroughgoing contextualization of all the early work that was undertaken with such broad strokes. Feminist scholars may have been among the first in modern times to raise the meaning of objectivity as a central issue for investigation, but it has taken more mainstream historians of science to turn this question into a significant historiographic pursuit, as in the extremely interesting work on the history of objectivity by Lorraine Daston, Peter Dear, and Theodore Porter.[26] Now, however, it is time to integrate gender issues into this admirably careful and contextsensitive historiographic work, and perhaps even to acknowledge how the inclusion of gender can transform the very questions we ask.

Still, *context* is a big word, and it points in many different directions, a fact which the generic notion of historical specificity, even with the addition of gender, does not always capture. It may fail to do justice to the need to attend to specificities of local disciplinary or even subdisciplinary interests, or to the specificities of local social (national, ethnic, or racial) interests. Donna Haraway offers us some useful ways of talking here that, for many, resonate with current political and intellectual priorities simultaneously: she defines feminist objectivity as *situated knowledges* and stresses the need for "partial perspectives." And indeed, it is with Haraway's radical though controversial deployment of the method of partial perspectives that I will end this all too cursory review.

In *Primate Visions* Haraway attempts, in contrast to a subject-rooted approach, a subject-free unpacking of how "love, power, and science [are] entwined in the constructions of nature in the late twentieth century," and she pursues this aim through an insistent scrutiny of the politics of narratives. As Gregg Mitman writes, "Her craft is the art of storytelling; her model for the construction of scientific knowledge is 'contested narrative fields.' Science, and in this instance primatology, is a story about nature, a tale circumscribed by its narrator, but one constantly evolving as new storytellers enter."[27] Haraway vigorously eschews the idea that gender can be understood independently of the politics of race and class, and her subject is an ideal one for making this case. As she makes abundantly clear, constructions of "nature" in twentieth-century primatology, like constructions of gender, are profoundly implicated in twentieth-century politics of race and colonialism. The fact

[25] See, e.g., Gillian Beer, "Translation or Transformation? The Relations of Literature and Science," *Notes and Records of the Royal Society of London,* 1990, *44*:81–99; Ludmilla Jordanova, *Sexual Visions: Images of Gender in Science and Medicine between the Eighteenth and Nineteenth Centuries* (Madison: Univ. Wisconsin Press, 1989); and Nancy Stepan, "Race and Gender: The Role of Analogy in Science," *Isis,* 1986, 77:261–277.

[26] See the articles published in *Social Studies of Science,* 1992, 22 (originally presented at the History of Science Society Meeting, Seattle, 1990): Lorraine Daston, "Objectivity and the Escape from Perspective," pp. 597–618; Peter Dear, "From Truth to Disinterestedness in the Seventeenth Century," pp. 619–631; and Theodore M. Porter, "Quantification and the Accounting Ideal in Science," pp. 633–652. See also Daston and Peter Galison, "The Image of Objectivity," *Representations,* 1992, *40*:81–128; and Porter, "Objectivity as Standardization: The Rhetoric of Impersonality in Measurement, Statistics, and Cost-Benefit Analysis," *Annals of Scholarship,* 1992, 9:19–60.

[27] Donna Haraway, *Primate Visions: Gender, Race, and Nature in the World of Modern Science* (New York: Routledge, 1989), p. 1; and Gregg Mitman, review of *Primate Visions, Isis,* 1991, 82: 163–165, on p. 164.

that the (mostly white) women who entered the field in the 1970s and 1980s took the lead in restructuring traditional narratives does not, for Haraway, provide support for the idea of "a feminist science" (and with this I agree), but rather demonstrates once again the dependence of scientific narratives on their authors' historical "positioning in particular cognitive and political structures of science, race, and gender."[28] Haraway's very method precludes telling or even hoping for one coherent story, and many readers may be left feeling a bit too destabilized. But her reach toward a postmodern historiography has not only provided new models for working with "gender" in science; it also suggests new models for any politically oriented analysis of science.

Undoubtedly, my choice of examples is idiosyncratic, and I have omitted much work that has had an enormous impact on other disciplines.[29] But the moral of my account will, I hope, be clear. For the future of gender in science in the particular context of the history of science, I want to add now, complementing Haraway's emphasis on fractures and partiality, a plea for affiliation and integration. Fifteen years ago it took an organized effort on the part of feminists to rouse historians' attention to the marks and significance of gender. Working from the strengths of their political consciousness of gender and their irreverence for received boundaries operating both within and between disciplines, feminist theorists have brought home powerful lessons about the cognitive and institutional politics of gender that even historians of science can no longer ignore. But to move the analysis of gender in science forward in this discipline, I suggest that these strengths need now to be integrated with the strengths of more conventional historiographic scholarship. There needs to be a lot more traffic and two-way exchange in this trading zone if it is to do the work it is capable of doing.

[28] Haraway, *Primate Visions*, p. 303.
[29] One subject especially close to my own heart that I have not discussed is the impact that feminist analyses of the personal and subjective dimensions of science have had on other disciplines; nor have I discussed the potential value of these analyses for the history of science—were it not for the intellectual and political climate that currently prevails. Interestingly, it was the gender coding of these dimensions as "feminine" that led us to examine their exclusion from investigation in the first place; might, I wonder, the same gender coding still be operative?

Women in the History of Science:
An Ambiguous Place

By Sally Gregory Kohlstedt[*]

ISSUES OF EQUITY AND FAIRNESS are fundamental for most historians who explore the participation of women in science. The social and political circumstances of the last two decades helped spur the investigation of women in history even as other intellectual trends pushed an increasing number of scholars to investigate the history of science and technology, particularly in the modern period. Certainly the proliferation of women's studies courses across the disciplines stimulated contemporary scholarship that insists on making women and gender part of the investigation in the history of science as well. Professional circumstances for historians of science, discussed below, enabled scholars to share and debate their research results, to identify ways of incorporating new and complex issues into their classrooms, and to achieve recognition for their work. This essay will explore major results of that enterprise and comment briefly on the enabling circumstances as well. The somewhat arbitrary categories and distinctions employed here are not meant to diminish the rich variety and provocative analyses surrounding the "women, gender, and science question"; indeed, this paper is intended to encourage a dialogue among various clusters of researchers exploring the topic.[1]

In her pathbreaking history of women in American science to 1940, Margaret Rossiter found it useful to frame her commentary in economic terms, noting that the limited participation of women in modern science (or their lack of participation) had significant financial as well as social explanations and implications.[2] Other historians have chosen to focus on a more traditional intellectual history, looking at

[*] Institute of Technology, University of Minnesota, 107 Walter Library, 117 Pleasant Street S.E., Minneapolis, Minnesota 55455–0291.

My article has benefited from the thoughtful comments of Margaret W. Rossiter, James H. Capshew, Donald Opitz, and from the assistance of Mark Jorgensen, Natalie McIntire, and Mary Thomas. An abbreviated version was presented in Moscow for the conference on Science and Social Justice, organized by Loren Graham, September 1992. Like many others in our discipline and this volume, I am deeply indebted to Frances Coulborn Kohler for her recommendations and attention to detail.

[1] Such a dialogue is also the purpose of a conference, "The Women, Gender and Science Question: What Do Research on the History of Women in Science and on Science and Gender Have to Do with One Another?" planned for 10–14 May 1995 at the University of Minnesota in Minneapolis. See also Ludmilla Jordanova, "Gender and the Historiography of Science," *British Journal for the History of Science,* 1993, *26*:469–483. For Jordanova, "Gender is emphatically not another way of talking about women, nor is it a topic or subject area; it is an analytical category. . . . [I]t is simultaneously a way of ordering experience, a system of representation, and a metaphor for particular kinds of relationships" (p. 474).

[2] Margaret W. Rossiter, *Women Scientists in America: Struggles and Strategies to 1940* (Baltimore: Johns Hopkins Univ. Press, 1982). The sequel, entitled *Women Scientists in America: Officially Encouraged, Institutionally Discouraged, 1940–1970* (Baltimore: Johns Hopkins Univ. Press, forth-

women—a few well recognized but most hardly visible—whose participation resulted in acknowledged contributions to a particular discipline in science. What,
other scholars ask, are the ideological and social circumstances that led to the inclusion of women in or their exclusion from the communities in which scientific activity
flourished? Still other scholars seek to understand how scientific theory intersects
with the cultural understandings of gender roles and sexual identity. Investigators
of scientific practice, particularly those from sociology and philosophy, often use
historical cases in their effort to elaborate on the cultural circumstances that establish a context for (or even construct) scientific research and theorizing, although
surprisingly few deal with gender. This essay will review representative and recent
works that investigate and interpret the position of women in the history of science.
Because many of them address the last two centuries, a period when segregation
by race and class kept large parts of the population from the education requisite
for participating in science, there is a critical edge to some of this historical assessment.[3]

Woven into recent historical books and articles are two themes, sometimes explicit
and distinct but at other times muted or blurred by interconnections that are presumed but not clearly articulated. One theme concentrates on the activities of
women and uses biography, publication statistics, employment patterns, and the ac-
knowledgment of peers, as well as practices of exclusion as relevant data: *women
in the history of science.* The second theme teases out scientific estimates of women
(sometimes noted as "woman") viewed as an apparent essentialist group and comments on their physical and mental traits, and the use of gendered metaphors within
science: *science and gender.* This latter point is addressed in greater detail in the
essay in this volume by Evelyn Fox Keller, who focuses on the significant intersec-
tions between gender and science, as well as in a recent review by Ludmilla Jordanova. Both themes can be traced from the ancients. Learned Greek philosophers
admitted that women might have scientific understanding, but their generalizations
tended to establish binary explanations of male and female that permeated the life
sciences, medicine, and human sciences.[4] Increasingly today some feminist philosophers and scientists posit a relationship between participation of women in science
and scientific conclusions about women's bodies and minds, the nature of scientific

coming), argues that despite rising expectations among increasing numbers of women in the middle
of the twentieth century, relatively little actually changed in terms of opportunities.

[3] Many of the general readership journals on science have featured articles on women in science,
often combining historical assessments with contemporary issues. See, e.g., Marguerite Holloway,
"A Lab of Her Own," *Scientific American,* Nov. 1993, *269:*94–103; and Stephen G. Brush, "Women
in Science and Engineering," *American Scientist,* Sept. 1991, *79:*401–419. For other largely excluded
groups see Kenneth Manning's pioneering work on the participation of people of color and Jewish
heritage, "The Complexion of Science," *Technology Review,* Nov. 1991, *94:*60–69, originally presented as the Sarton Lecture at the meeting of the American Association for the Advancement of
Science in Washington, D.C., February 1991.

[4] "Gendered pairs . . . provided languages for alchemy, chemistry, and the study of electricity":
Jordanova, "Gender and the Historiography of Science" (cit. n. 1), p. 479. See also Nancy Tuana,
The Less Noble Sex: Scientific, Religious, and Philosophical Conceptions of Women's Nature
(Bloomington: Indiana Univ. Press, 1993); Natalie Harris Bluestone, *Women and the Ideal Society:
Plato's Republic and Modern Myths of Gender* (Amherst: Univ. Massachusetts Press, 1987); and
Rosemary Agonito, *History of Ideas on Women* (New York: Putnam, 1977).

work and sponsorship, and the language of science.[5] Just how this connection affects both women and science is only beginning to be analyzed by historians who look at how other socially prescribed conditions affect a given individual's intellectual work.[6]

Recently archaeologists, whose research depends heavily on artifacts rather than written historical records, have offered new interpretations of gender roles in prehistorical science and technology. Rethinking how skills and implements improved over time and unconvinced that "man the hunter" was solely responsible, these scholars argue that women have always played a role in such developments, and that women probably developed such fundamental devices as vessels for holding water and cooking foods as well as fishhooks for the protein that supplemented red meat supplies from hunters.[7] Egyptian women practiced medicine and studied chemistry and Greek women were known to teach and debate in the academies.[8] Women have thus been continuing and significant participants throughout the history of science. Historians are only beginning to explore the comparative dimensions of their activities across cultural as well as chronological boundaries; recent international congresses have particularly contributed to this discussion (see Section IV).[9] Current historical work on women in science, much of it concentrated on the nineteenth and twentieth centuries, has frequently utilized biography as a means first to identify and document women's participation and, second, to learn of their experiences in broader historical contexts that allow for comparative analysis.

Until very recently, the only women known, if rarely studied, were those whose lives stand out from the norm and whose contributions to mainstream scientific research can be established—although sometimes with considerable difficulty. The initial work on these women documented their important roles as formulators, educators, and translators of scientific expertise. Recognition of their contributions to the process of science led historians next to reconcile women's stories with the existing historical narratives of science, to show their relationship to that history. A third approach, much more recent and building on the other two, illuminates the often rather different experiences of women and men practitioners and raises issues about the nature and definition of the scientific enterprise; the gender sensibilities that influenced work relationships and outcomes when women worked in largely

[5] Mary Jacobus, Evelyn Fox Keller, and Sally Shuttleworth, eds., *Body/Politics: Women and Discourses of Science* (New York: Routledge, 1990); Brian Easlea, *Fathering the Unthinkable: Masculinity, Scientists, and the Nuclear Arms Race* (London: Pluto Press, 1983); and Marina Benjamin, ed., *Science and Sensibility: Gender and Scientific Inquiry, 1780–1945* (Oxford: Blackwell, 1991).

[6] Oddly, historians interested in the social construction of science seldom consider the issue of women or gender in any detail. See, e.g., the essays in Andrew Pickering, ed., *Science as Practice and Culture* (Chicago: Univ. Chicago Press, 1992).

[7] Joan M. Gero, David M. Lacy, and Michael L. Blakey, eds., *The Socio-politics of Archeology,* Research Report no. 23 (Amherst: Department of Anthropology, 1983).

[8] Marilyn Bailey Ogilvie, introduction, *Women in Science, Antiquity through the Nineteenth Century: A Biographical Dictionary with Annotated Bibliography* (Cambridge, Mass.: MIT Press, 1988), pp. 2–4.

[9] Historians of medicine are particularly interested in the role women have played as healers in rural communities, as well as among Southern white women, Native American women, Chicana women, and black women in the Americas. For relevant references see "Domestic Healing and Other Lay Practices," in Phyllis Holman Weisbard, ed., *The History of Women and Science, Health, and Technology: A Bibliographic Guide to the Professions and the Disciplines* (Madison: University of Wisconsin System Women's Studies Librarian, 1993).

male settings is only beginning to be explored. A still elusive goal is to integrate women and other groups within a coherent narrative that yet accounts for pluralism of experience.[10]

I. PATTERNS OF WOMEN'S PARTICIPATION IN SCIENCE

Important early agendas for those studying women in science were to document the lives of famous women and to discover others whose names deserved to be added to the list of such known luminaries as Hypatia, Mary Somerville, Sofia Kovalevskaia, and Marie Curie. Biographies, autobiographical notes, and eulogies demonstrated that women had the talent for, and, given opportunities, might pursue studies in astronomy, chemistry, natural history, mathematics, and related subjects.[11] This approach had its own history. A number of women authors, from Christine de Pisan in the fourteenth century through Margaret Fuller in the nineteenth century, illustrated women's capacity to pursue virtually any occupation or activity by recounting the lives of those engaged in a range of intellectual pursuits. In the latter nineteenth century exceptional, literate women extended this tradition by highlighting efforts of their peers working within science.[12]

The first comprehensive history in English to focus exclusively on women scientists was, perhaps unexpectedly, that of an American priest who wrote under the pseudonym H. J. Mozans in the early twentieth century. Father John Augustus Zahm of Notre Dame was an educational reformer and scientific enthusiast who compiled an account of women in science in order to challenge colleagues and contemporaries who, he said, assumed women had an "incurable mental weakness." This unusual priest, even as he compiled an impressive number of those who had made contributions, was remarkably prescient about the specific circumstances and larger cultural concerns that defined women and thus inhibited most from pursuing science. In 1943 Edna Yost, searching for historical women who were other than "saints and courtesans," compiled a series of twelve biographies for her book *American Women of Science*. In 1945 Helen Wright published a thoughtful biography of the astronomer Maria Mitchell. These biographical studies were researched and published ex-

---

[10] For some overviews of works on women in science see Carolyn Iltis Merchant, "Clio's Consciousness Raised," *Isis,* 1982, *73*:398–409; Londa Schiebinger, "The History and Philosophy of Women in Science: A Review Essay," *Signs,* 1987, *12*:305–332; and Dorinda Outram, "Essay Review: Fat, Gorillas, and Misogyny: Women's History in Science," *Brit. J. Hist. Sci.,* 1991, *24*:361–367.

[11] Eve Curie's *Madame Curie* (New York: Doubleday, Doran, 1937) was a best-seller in France and sold over a million copies in English translation. Elizabeth Chambers Patterson's *Mary Somerville and the Cultivation of Science, 1815–1840* (Boston: Kluwer, 1983) is heavily based on Somerville's own memoirs and manuscripts. See also Ann Hibner Koblitz, *A Convergence of Lives: Sofia Kovalevskaia, Scientist, Writer, Revolutionary,* rev. ed. (New Brunswick: Rutgers Univ. Press, 1993).

[12] Schiebinger, "The History and Philosophy of Women in Science," (cit. n. 10). See Christine de Pisan, *Book of the City of Ladies,* trans. E. J. Richards (New York: Persea Books, 1982); and Margaret Fuller, *Woman in the Nineteenth Century and Kindred Essays Relating to the Sphere, Condition, and Duties of Woman* (Boston: J. Jewell, 1855). Late nineteenth-century authors include friends and colleagues of Maria Mitchell like Matilda Joslyn Gage, *Woman as Inventor* (Fayetteville, N.Y.; 1870); and Phebe A. Hanaford, *Daughters of America; or, Women of the Century* (Augusta, Me.: True, 1882). Weisbard, ed., *History of Women* (cit. n. 9), pp. 88–89, highlights books and articles for older children. A very recent publication for children is *A Celebration of Women in Mathematics and Science* (Notre Dame, Ind.; St. Mary's College, 1992).

plicitly to assert claims about women's capacity for science and thus challenge negative assumptions by providing counterexamples.[13]

Activism and inquiry into women's history since the 1960s has built upon the painstaking labor of scholars and bibliographers who create finding aids for manuscripts and published sources. Their work contributes substantively to documentation of the presence of women in science and opens entire new areas for research. A comprehensive bibliography coordinated by Rima Apple as chair of the Committee on Women of the History of Science Society was published in 1987; recently it was updated as *The History of Women and Science, Health, and Technology: A Bibliographic Guide to the Professions and the Disciplines* (1993). It references over 2,500 published articles and books, nearly all of which, like the works cited in this essay, are in English and concentrate on Western Europe and North America.[14] Marilyn Bailey Ogilvie's biographical dictionary *Women in Science* includes over two hundred major women scientists active through the nineteenth century and is useful for its introductory notes and especially its annotated bibliography. Else Høyrup's *Women of Science, Technology, and Medicine* covers somewhat similar biographical territory but extends the list of non-American women.[15]

A difficulty for those who create biographical dictionaries is defining the term science and deciding which fields and what kinds of activity merit inclusion. Most of the projects to date have used the twentieth-century academic disciplines and sought evidence of employment in universities, academies, museums, public agencies, or sometimes corporations as a principal basis for the inclusion of an individual; publications and organizational affiliation are typically taken as the measure of achievement. Unfortunately, discriminatory practices and social constraints kept many women from such affiliations even as they served in adjunct roles that required considerable skills but gave little recognition, such as anonymous illustrator of scientific texts or "computer" in an astronomical observatory. Thus while using institutional measures is practical and the results are suggestive, dictionaries based on them better indicate where investigation has focused in the past than where future research needs to go.

Given the current interest in women in science, surprisingly few full scale biographies of women scientists have been written in the past decade or so. Evelyn Fox Keller's account of Barbara McClintock is widely cited, as is Ann Hibner Koblitz's study of Sofia Kovalevskaia. Women in medicine and nursing often have scientific training and interests; Alice Hamilton, as Barbara Sicherman demonstrates in a

---

[13] H. J. Mozans, *Woman in Science* (1913), rpt. with preface by Cynthia Russett (Notre Dame, Ind.: Univ. Notre Dame Press, 1991) was the first comprehensive history of women who participated in scientific activity. On Mozans see R. Scott Appleby, *Church and Age Unite! The Modernist Impulse in American Catholicism* (Notre Dame, Ind.: Univ. Notre Dame Press, 1992), pp. 27–52. Edna Yost, *American Women of Science* (Philadelphia: Frederic A. Stokes, 1943); see also Yost, *Women of Modern Science* (New York: Dodd, Mead, 1959). Helen Wright, *Sweeper in the Sky: The Life of Maria Mitchell, First Woman Astronomer in America* (New York: Macmillan, 1949).

[14] Weisbard, ed., *History of Women* (cit. n. 9). The first edition was produced by Susan Searing, with Rima Apple and others. Other early bibliographies were Audrey B. Davis, *Bibliography on Women, with Special Emphasis on Their Roles in Science and Society* (New York: Science History Publishers, 1974); Phyllis Zweig Chinn, *Women in Science and Mathematics: A Bibliography* (Arcata, Calif.: Humboldt State Univ., 1983); and Patricia Siegel and Kay Thomas Finley, *Women in the Scientific Search: An American Bio-bibliography* (Metuchen, N.J.: Scarecrow, 1985).

[15] Ogilvie, *Women in Science* (cit. n. 8); and Else Høyrup, *Women of Science, Technology, and Medicine: A Bibliography* (Roskilde, Denmark: Roskilde Universitetsbibliotek, 1987).

carefully edited volume of her letters, was both physician and scientist as well as founder of the field of industrial toxicology. More controversial and mysterious figures like Augusta Ada Lovelace, whose enthusiasm for machine computation influenced Charles Babbage, have also attracted recent biographers.[16] Some women worked literally on the frontier, as did Martha Maxwell, a naturalist and taxidermist who earned her living by identifying, collecting, and displaying specimens in Colorado, and Alice Fletcher, a pioneering woman anthropologist in the Southwest.[17] A number of other biographical studies are under way or recently completed, and some women, like Clemence Royer and Lise Meitner, have attracted more than one potential biographer.[18]

Exceptional women, whose careers and personal stories let them carve out a place within traditional scientific inquiry and institutions, demonstrate the fragility of their circumstances as well. Reading their life stories, historians begin to understand the variables that allowed a few women to engage in science. Scientific access has been acquired through social class, personal relationships, political experimentation, and the institutional configuration of education at any given time and place.[19] Historians studying exceptional women also uncover the problematic constraints on their scientific activity, including negative assessments of women's mental capacity and their family responsibilities, lack of financial power, and limited legal autonomy—restraints that vary with time and place and discipline but recur with relentless monotony. Even the most successful women scientists could be challenged, dismissed as eccentric, excluded from prestigious societies, and literally displaced from jobs and professional affiliations.

Some historians have used collective biography to explore and demonstrate some recurring themes about women in science and the elements of empowerment and discrimination that exist in particular times and places. Collective and comparative

[16] Evelyn Fox Keller, *A Feeling for the Organism: The Life and Work of Barbara McClintock* (New York: Freeman, 1983); Koblitz, *A Convergence of Lives* (cit. n. 11); Barbara Sicherman, *Alice Hamilton: A Life in Letters* (Cambridge, Mass.: Harvard Univ. Press, 1984); and Dorothy Stein, *Ada: A Life and a Legacy* (Cambridge, Mass.: MIT Press, 1985). A useful introduction to several pioneers in health and physiology is Toby A. Appel, "Physiology in American Women's Colleges: The Rise and Decline of a Female Subculture," *Isis*, 1994, *85*:26–56.

[17] Maxine Benson, *Martha Maxwell: Rocky Mountain Naturalist* (Lincoln: Univ. Nebraska Press, 1986); and Joan Mark, *A Stranger in Her Native Land: Alice Fletcher and the American Indians* (Lincoln: Univ. Nebraska Press, 1982). See also Marcia Myers Bonta, *Women in the Field: America's Pioneering Women Naturalists* (College Station: Texas A&M Press, 1991). Local historians are beginning to write about others whose life work had important regional influence, as Martha E. Hellander, *The Wild Gardener: The Life and Selected Writings of Eloise Butler* (St. Cloud, Minn.: North Star Press, 1992).

[18] On Royer see Sara Joan Miles, "Evolution and Natural Law in the Synthetic Science of Clemence Royer" (Ph.D. diss., Univ. Chicago, 1988); and Joy D. Harvey, "Races Specified, Evolution Transformed: The Social Context of Scientific Debates Originating the Société d'Anthropologie de Paris, 1859–1902" (Ph.D. diss., Harvard Univ., 1983). On Meitner see Patricia E. Rife, *Lise Meitner: Ein Leben für die Wissenschaft* (Dusseldorf: Claassen, 1990); and Fritz Krafft, *Lise Meitner: Eine Biographie* (Berlin: Hans-Meitner-Institut, 1988). Ruth Lewin Sime is also completing a biography of Meitner. See also Fay Ajzenberg-Felove, *A Matter of Choices: Memoirs of a Female Physicist* (New Brunswick, N.J.: Rutgers Univ. Press, 1994).

[19] See, e.g., Sharon Bertsch McGayne, *Nobel Women in Science: Their Lives, Struggles and Momentous Discoveries* (Secaucus, N.J.: Birch Lane Press, 1993); Mary R. S. Creese, "British Women of the Nineteenth and Early Twentieth Centuries Who Contributed to Research in the Chemical Sciences," *Brit. J. Hist. Sci.*, 1991, *24*:275–305; and Mary R. S. Creese and Thomas M. Creese, "British Women Who contributed to Research in the Geological Sciences in the Nineteenth Century," *Brit. J. Hist. Sci.*, 1994, *27*:23–54.

Women traditionally tended bees. Here a group of women engaged in systematic studies of bee culture, identifying parasites and documenting behavior, at the University of Minnesota, about 1917. Courtesy of the University of Minnesota Archives.

analysis helps contextualize the individual experience. Lois Arnold's *Four Lives in Science* traced the significant contributions of nineteenth-century North American women whose dependent status masked their achievements; like others, she uses the vocabulary of marginality that portrayed women on the periphery, the edge, the outside. Others sought to emphasize achievement, as Marianne G. Ainsley did in *"Despite the Odds": Essays on Canadian Women and Science.* More recently Pnina Abir-Am and Dorinda Outram prepared a well-reviewed volume, *Uneasy Careers and Intimate Lives,* on women whose personal relationships intersect with research careers in complex ways that at times have enabled and at times inhibited scientific achievement.[20] A sequel, *Creative Couples,* is being edited by Helena Pycior, Nancy Slack, and Pnina Abir-Am. Together these two sets of portraits reveal the mechanisms through which family situations—and indeed the roles of sisters, daughters, wives, and mothers—connect to intellectual accomplishments.[21] Helena Pycior's

[20] Lois Barber Arnold, *Four Lives in Science: Women's Education in the Nineteenth Century* (New York: Schocken, 1984); and Marianne G. Ainsley, *"Despite the Odds": Essays on Canadian Women and Science* (Montreal: Vehicule Press, 1990).

[21] Pnina Abir-Am and Dorinda Outram, eds., *Uneasy Careers and Intimate Lives: Women in Science, 1789–1979* (New Brunswick, N.J.: Rutgers Univ. Press, 1987); and Helena Pycior, Nancy Slack, and Pnina Abir-Am, eds., *Creative Couples in the Sciences* (New Brunswick, N.J.: Rutgers Univ. Press, forthcoming).

recent work on the Curie family suggests that the collaboration of Pierre and Marie Curie, although exceptionally successful, may provide a useful model of how productive and satisfying partnerships were accomplished.[22] *Sage,* a scholarly journal devoted to African American women, featured women in science in a special issue in 1989; the volume also provided a series of biographical and autobiographical sketches of twentieth-century achievers and discussed the particular circumstances that confronted them in their efforts to gain an education and employment in science.[23] Advocates eager to recruit women into science around the world have provided contemporary biographical and autobiographical materials, directed at adults, teachers, and even older children, for whom the details of struggle and survival might provide role models.[24] Read singly, such women's stories reveal the accommodations they had to make to engage in science; taken collectively, they expose the boundaries imposed more generally on practitioners of science even as they reveal the persistence and sometimes exceptional qualities that allowed some women to achieve.

Historical recognition is further broadened by studies that examine women who took the roles of patron, collector, assistant, editor, illustrator, teacher, librarian, and support staff. Like many men whose enthusiasms for science were avocational or viewed as supplementary to the "basic work" of science, these women figured in the lists of contributors to the major works written or edited by others, paid dues to societies that sponsored museum publications and expeditions, and spent long, often tedious hours gathering and compiling data to complete taxonomic descriptions or astronomic tables. While the names of such women contributors are often obscure, determined researchers like Elizabeth Wagner Reed and others have argued that their combined efforts advanced science and their local efforts invigorated communities of investigators.[25]

II. THE DOMAINS OF SCIENCE

Identification of women scientists, with their varied roles and affiliations, is well under way. Not surprisingly, the results do not integrate neatly into well-established

[22] Helena Pycior, "Reaping the Benefits of Collaboration While Avoiding Its Pitfalls: Marie Curie's Rise to Scientific Prominence," *Social Studies of Science,* 1993, *23*:301–323. Another example of a highly productive collaboration, Anna and John Comstock, is discussed by Pamela Henson in the forthcoming *Creative Couples* (cit. n. 21).

[23] *Sage,* Fall 1989, *4*(3):1–54. See also Shirley M. Malcom, Paula Q. Hall, and Janet Brown, eds., *The Double Bind: The Price of Being Minority Women in Science* (Washington, D.C.: American Association for the Advancement of Science, 1976).

[24] Examples include Vivian Gornick, *Women in Science: Portraits from a World in Transition* (New York: Simon & Schuster, 1983); and Harriet Zuckerman, Jonathan Cole, and John T. Bruer, eds., *The Outer Circle: Women in the Scientific Community* (New York: Norton, 1991). For a similar effort on Soviet women see Inna Kosheleva, *Women in Science* (Moscow: Progress Publishers, 1983). Women from a variety of Western and Western-colonized countries contributed sketches to Derek Richter, *Women Scientists: The Road to Liberation* (London: Macmillan, 1982). Australian women have recently been identified in Farley Kelly, ed., *On the Edge of Discovery: Australian Women in Science* (Melbourne: Text Pub. Co., 1993); and Julie Marcus, ed., *First in Their Field: Women and Australian Anthropology* (Melbourne: Melbourne Univ. Press, 1993).

[25] Elizabeth Wagner Reed, *Women in Science before the Civil War* (Minneapolis: privately printed, 1992) examined 24 mid-nineteenth-century publications and identified numerous contributions by women. She also calculated that 31 women contributed 175 years of observation to the Smithsonian Institution's meteorological survey between 1847 and 1861. The volume contains biographical sketches of 22 women. On the contributions of women in libraries and other occupations that contributed to the organization and communication of scientific knowledge, see Margaret W. Rossiter,

historical accounts. The new data require a reconsideration of traditional narratives and interpretations. Studying the lives of women in science forces us to reconsider what was formerly presumed: Who gets educated? What determines employment? Who selects and certifies work that is acknowledged and rewarded? Women's lives provide a fresh perspective on the process of producing and disseminating scientific research. Tracing out the experiences of women confronts familiar perceptions, events, institutions, and explanations about scientific activity in ways hardly anticipated by George Sarton.

As scholars investigate women's history, they have enlarged their focus to include places and activities where women's lives have been concentrated, particularly in the home, around community activities, and in educational settings. A number of historians in the 1970s reconsidered the ways in which women took advantage of home resources for their studies, the activities they promulgated in their communities, the schools they established and conducted, and the allies they found at home and at work. The fact that nearly all the prominent women scientists before 1850 and many thereafter gained access to books, equipment, and advice through family members has been increasingly emphasized.[26] The extension of domestic life into more public arenas, largely through voluntary agencies, helps explain the growth of women's natural history clubs and energetic support of conservation groups so evident in England and North America by the end of the nineteenth century.[27] Elizabeth Keeney carefully integrated women into her account of nineteenth-century botanizers, showing that even as isolated individuals they participated in correspondence networks and organized local natural history clubs. Gradually the most earnest amateurs joined regional and national organizations that published new discoveries; many also contributed to preservation movements.[28] Joan Burstyn studied those women who helped establish the scientific laboratory at Penikese Island, which later led to the Woods Hole Oceanographic Institute, as an example of how women could provide special facilities and financial support in order to advance career development of other women.[29] Women's colleges were essential in expanding women's opportunities, as Margaret Rossiter and others have made clear. It is perhaps ironic that American women, like Russian women, gained initial access to advanced education outside their own countries, at German and Swiss universities, and were thus

---

"Women and the History of Scientific Communication," *Journal of Library History*, 1986, *21*:39–59. On the leadership of women in some local midwestern academies of science, see Daniel Goldstein, "Midwestern Naturalists: Academies of Science in the Mississippi Valley, 1850–1900" (Ph.D. diss., Yale Univ., 1989), esp. pp. 188–213.

[26] Ann B. Shteir, "Linnaeus's Daughters, Woman and British Botany," in Abir-Am and Outram, *Uneasy Careers and Intimate Lives* (cit. n. 21); Deborah Jean Warner, *Graceanna Lewis: Scientist and Humanitarian* (Washington, D.C.: Smithsonian Institution Press, 1979); and Sally Gregory Kohlstedt, "Parlors, Primers, and Public Schooling: Education for Science in Nineteenth Century America," *Isis*, 1990, *81*:424–445.

[27] David Elliston Allen, *The Naturalist in Britain: A Social History* (London: A. Lane, 1976); Lynn Barber, *The Heyday of Natural History, 1820–1870* (New York: Doubleday, 1980); Sally Gregory Kohlstedt, "In from the Periphery: American Women in Science, 1830–1880," *Signs*, Autumn 1978, *4*:81–96; and Dolores Hayden, *The Grand Domestic Revolution: A History of Feminist Designs for American Homes, Neighborhoods, and Cities* (Cambridge, Mass.: MIT Press, 1981).

[28] Elizabeth B. Keeney, *The Botanizers: Amateur Scientists in Nineteenth-Century America* (Chapel Hill: Univ. North Carolina Press, 1992).

[29] Joan N. Burstyn, "Early Women in Education: The Role of the Anderson School of Natural History," *Journal of Education*, 1977, *59*:50–64.

enabled to embark on scientific careers. These women often returned to join the faculty of women's colleges and trained the generation of women who pressured all-male institutions in the United States into letting them pursue graduate degrees.[30]

A considerable effort has gone into tracing, often in chronological sequence, the particular women who worked in specific sciences, particularly botany, mathematics, and astronomy.[31] Articles about individual women "pioneers" are also increasingly published within specialized scientific and educational journals and even included in highlighted sidebars in scientific textbooks, but these accounts are typically more anecdotal than analytical.[32] Some simply account for women's contributions while others attend to the forces that shaped their lives. The recovery of women's lives in science, however, does make it clear that women have and continue to expand the domain of science. A few essays, like that by James Capshew and Alejandro Laszlo, show how demographics, ideology, and politics affected women psychologists during World War II, a time when the discipline was expanding and entering new areas of investigation.[33] And a recent set of bibliographical essays takes one more step toward "righting the record" of women's participation in several scientific fields.[34]

III. NETWORKS AND SURVIVAL STRATEGIES

Yet a third approach to the history of women in science also builds on biography, but it uses the detail from the arena of women's history more generally. In fact, uncovering women's networks and the continuities that existed for women across geographical and time boundaries—clubs, correspondence, correspondence schools, mentor chains—has led to an inside story of women and science. Those women whose experience or education led them to scientific inquiry viewed themselves as participating in science whether or not their work and efforts were so recognized; often they introduced scientific approaches into arenas where they had special concerns. Such activities as home economics, public health and medicine, physiology and physical therapy, and science education and nature study all show the confluence of women's issues and scientific expertise. The recovery of this his-

---

[30] Ann Hibner Koblitz, "Science, Women, and the Russian Intelligentsia: The Generation of the 1860s," *Isis*, 1988, *79*:208–226; and Rossiter, *Women Scientists in America* (cit. n. 2), esp. Chs. 1 and 2.

[31] Emanuel D. Rudolph, "Women in Nineteenth Century Botany," *American Journal of Botany*, 1982, *69*:1346–1355; Louise S. Grinstein and Paul S. Campbell, ed., *Women of Mathematics: A Bibliographical Sourcebook* (Westport, Conn.: Greenwood, 1987); Lynn M. Olsen, *Women in Mathematics* (Cambridge, Mass.: MIT Press, 1974); Louise S. Grinstein, Rose K. Rose, and Miriam H. Rafailovich, eds., *Women in Chemistry and Physics: A Biobibliographic Sourcebook* (Westport, Conn.: Greenwood, 1991); and John Lankford and Rickey Slavings, "Gender and Science: Women in American Astronomy, 1859–1940," *Physics Today*, March 1990, *43*(3):58–65.

[32] Some scientists become deeply interested in a historical figure and pursue that research seriously, see, e.g., Ruth Lewin Sime, "Lise Meitner and the Discovery of Fission," *Journal of Chemical Education*, 1989, *66*:373–376; and Sime, "Lise Meitner's Escape from Germany," *American Journal of Physics*, 1990, *58*:262–267.

[33] James H. Capshew and Alejandra C. Laszlo, "'We Would Not Take No for an Answer': Women Psychologists and Gender Politics during World War II," *Journal of Social Issues* (a special issue devoted to 50 years of psychology and social issues), 1986, *42*:157–180. See also Elizabeth Scarborough and Laurel Furumoto, *Untold Lives: The First Generation of American Women Psychologists* (New York: Columbia Univ. Press, 1987).

[34] G. Kass-Simon and Patricia Farnes, *Women of Science: Righting the Record* (Bloomington: Indiana Univ. Press, 1990).

Girls and women studied botany as an avocation throughout the nineteenth century, and many eagerly attended classes in horticulture at land grant colleges at the turn of the century. Courtesy of the University of Minnesota Archives.

tory reveals quite different dimensions to the history of science and, in some instances, challenges conventional accounts.

As in many other areas of women's history, the rediscovery of the women's schools and colleges together with women's role in their establishment has led to interest in curriculum development in individual institutions and school systems throughout the United States. Pamela Mack and Miriam Levin are examining peer and student-teacher dynamics in science at Mount Holyoke, one of the oldest women's colleges and one with a tradition of science in the curriculum. Their work should document the way in which colleges segregated by sex provided opportunities not only to students but to women faculty.[35] Essentially no historical work has been done to investigate the experiences of women at historically African-American institutions or predominantly Hispanic or Native American colleges. Existing quite apart from the history of science literature is a growing literature on history of women's education.

[35] Miriam R. Levin and Pamela E. Mack, "The Transformation of Science Education at Mount Holyoke in the Gilded Age," presented at the Joint Meeting of the American Historical Association and the History of Science Society, 28 Dec. 1988, and available from ERIC, No. ED309930. See also Barbara Sicherman, "Colleges and Careers: Historical Perspectives on the Lives and Work Patterns of Women College Graduates," in *Women and Higher Education in American History: Essays for the Mount Holyoke Sesquicentennial Symposia,* ed. John Mack Faragher and Florence Howe (New York: Norton, 1988), pp. 130–164.

It typically records, with evident surprise, the relatively substantial component of science in the curriculum. Margaret Rossiter brought some of these accounts together and extends them in her two-volume study of women in American science to 1970.[36]

Rossiter's first volume is exceptional in its comprehensiveness, shrewd in its insights and analysis, and perceptive in its orientation toward women scientists as workers whose experiences can be interpreted in terms of labor force economics. She documents the experimental steps, trials, and tests of those women whose struggles and strategies seemed so different from those of the men whose lives she had previously studied. Even before completing her research, this traditionally educated historian of science observed that she felt like Alice in Wonderland who had fallen into a rabbit hole and found a world turned topsy-turvy. Rossiter moved toward creating a history of women in science, an account generated and shaped by these women's activities. Her results offer a commentary on standard historical work. *The Physicists,* for example, Daniel Kevles's pioneering and clearly written account of post–Civil War science in America (frequently used as a textbook), describes well the professionalization process through which men created new organizations and drew up standards that defined their disciplines. In the process they also circumscribed their membership, typically excluding women or limiting their roles within the group. This positive account of professional achievement contrasts with Rossiter's discussion of women's struggles to remain visible and viable when excluded by definition of their sex from most scientific societies and academic departments throughout higher education, except for the enclave of women's colleges.[37]

Thus establishing a separate line of inquiry based on a feminist point of view explicitly challenges presentations that imply a certain inevitability about the arrival of the status quo. Like the classic essay by Joan Kelly that argued that women did not experience a renaissance of learning in the fifteenth century, Rossiter's book suggests that professionalization did not confer on women opportunities and recognition comparable to those accorded men in science.[38] Concentrating on protégée chains, women's organizations, and the often self-conscious support that women provided other women, Rossiter reveals experiences that were previously hidden and apart from traditional histories. She never, however, loses sight of that better-known history whose principal players and programs inevitably reverberate in women's lives. The chronologically comprehensive survey of Margaret Alic in *Hypatia's Heritage* is another effort to uncover women's experiences as distinct from, yet inevitably in the context of, that traditional history.[39]

As noted above, the fact of exclusion from science has led a number of historians to look at images and estimates of women's abilities, particularly of their intelligence

[36] Rossiter, *Women Scientists in America* (cit. n. 2). The pioneering work of Thomas Woody, *History of Women's Education in the United States* (New York: Science Press, 1929), emphasizes science curriculum in its text and appendixes.

[37] Rossiter, *Women Scientists in America,* esp. Ch. 2; and Daniel J. Kevles, *The Physicists: The History of a Scientific Community in Modern America* (New York: Knopf, 1978).

[38] Joan Kelly,"Did Women Have a Renaissance?" *Women, History, and Theory: The Essays of Joan Kelly* (Chicago: Univ. Chicago Press, 1984). This challenge to conventional ways of interpreting science is explicit in many of the essays in Nancy Tuana, ed., *Feminism and Science* (Bloomington: Indiana Univ. Press, 1989).

[39] The only recent comprehensive compilation is Margaret Alic, *Hypatia's Heritage: A History of Women in Science from Antiquity through the Nineteenth Century* (Boston: Beacon Press, 1986).

and capacity for science. At most times and places, women have expressed their own, often quite independent assessments. Opinions date from the ancient Greek philosophers.[40] For the Scientific Revolution and the Enlightenment, besides investigating attitudes toward women who themselves studied science, Carolyn Merchant, Londa Schiebinger, and others have explored the various formulas regarding women's abilities devised by literati, physicians, and some scientists of those times. Merchant laid open the fundamental questions of scientific (re)definition in her book *The Death of Nature* by arguing that science and nature became gender-identified in the period of the Scientific Revolution in a distinctively new way. Masculinity was correlated with themes of authority and control, especially in those sciences dominated by physical and mathematical concepts. These traits contrasted with feminine qualities ascribed to both nature and women themselves. Merchant's book is pathbreaking, and thus contested, in its use of sources as well as its argument, and her implicit lament for a lost organic cosmology has inspired many in the current movement gathered under the umbrella of ecofeminism.[41] Londa Schiebinger's *The Mind Has No Sex?* reassesses a somewhat later period and reaches a different although related conclusion about women's access to science, finding considerable variation among those who commented on women's capacities, including the women themselves.[42] In her important recent book *Nature's Body: Gender in the Making of Modern Science,* Schiebinger investigates the way in which sexual categories were insinuated into the animal kingdom, most notably but by no means exclusively in the work of Carl Linnaeus. She joins a growing number of scholars interested in definitions of sex and sexuality and the closely related issue of how science reads women's bodies.[43] Patricia Phillips takes yet another approach to women's experience in *The Scientific Lady: A Social History of Women's Scientific Interests,* arguing that leisured women of Britain and France manifested an unprecedented predilection for science in the seventeenth and eighteenth centuries.[44]

The fierce debate over "the woman question" at the end of the nineteenth century, which ranged from debates over cranial size in relationship to intelligence to the vulnerability of women's reproductive organs, has generated another important

---

[40] Thomas Laqueur, *Making Sex: Body and Gender from the Greeks to Freud* (Cambridge, Mass.: Harvard Univ. Press, 1990); Joan Cadden, *The Meanings of Sex Difference in the Middle Ages: Medicine, Natural Philosophy, and Culture* (Cambridge: Cambridge Univ. Press, 1993); see also David F. Noble, *A World without Women: The Christian Clerical Culture of Western Science* (New York: Alfred A. Knopf, 1992); and Bluestone, *Women and the Ideal Society* (cit. n. 4).

[41] Carolyn Merchant, *The Death of Nature: Women, Ecology, and the Scientific Revolution* (New York: Harper & Row, 1980); and Merchant, *Ecological Revolution: Gender, Nature, and Science in New England* (Chapel Hill: Univ. North Carolina Press, 1989). On women and the environment see also Vera Norwood, *Made from This Earth: American Women and Nature* (Chapel Hill: Univ. North Carolina Press, 1993).

[42] Londa Schiebinger, *The Mind Has No Sex? Women in the Origins of Modern Science* (Cambridge, Mass.: Harvard University Press, 1989); and Ludmilla Jordanova, *Sexual Visions: Images of Gender in Science and Medicine between the Eighteenth and Twentieth Centuries* (Madison: Univ. Wisconsin Press, 1989).

[43] Londa Schiebinger, *Nature's Body: Gender in the Making of Modern Science* (Boston: Beacon Press, 1993).

[44] Patricia Phillips, *The Scientific Lady: A Social History of Women's Scientific Interests* (New York: St. Martin's Press, 1990). For a similar argument see Bonnie Smith, *Ladies of the Leisured Class: The Bourgeoises of Northern France in the Nineteenth Century* (Princeton: Princeton Univ. Press, 1981). For a slightly different view of domestic science see James A. Secord, "Newton in the Nursery: Tom Telescope and the Philosophy of Tops and Balls, 1761–1838," *History of Science,* 1985, *23:*127–151.

literature.[45] One subset deals with the effects of evolutionary theories of species development and the new awareness of the evident biodiversity of the earth. Biology became destiny in new ways. Historians interested in women's topics have investigated the implications for science and for women of the ideas about the mechanism of sexual selection held by Darwin and the popularizers of his evolutionary theory.[46] Several scholars have looked at new definitions of sexuality and sexual identity, among them Louise Mitchell Newman, who scrutinized essays on women in *Popular Science Monthly* during the late nineteenth century and entitled her results *Men's Ideas/Women's Realities*. Men and women typically, although not inevitably, held quite different opinions of what natural selection and evolution meant for men and women of the species *Homo sapiens*.[47] Scientific assumptions about sex and gender became firmly planted in higher education and in public culture, especially in North America. A few women challenged the assumptions built into the theories promulgated in Darwin's *Descent of Man* and elaborated by his supporters.[48] Nineteenth-century anthropology was another science deeply influenced by evolutionary theory, with a consequent impact on the larger society—and another science in which women had to struggle for acceptance.[49]

Research on women trained in science has discovered how many of them were affected by such debates. Rosalind Rosenberg has elucidated the pioneering work of women psychologists like Helen Thompson Woolley at Chicago and Leta Stetter Hollinworth at Teachers College, Columbia, who focused energy on research that demonstrated that intelligence is sexblind.[50] Others gravitated to fields connected with nourishment and human comfort rather than economic power and military might.[51] Women in physiology paid considerable attention to women's comparable strengths and requisites for good health, according to Martha H. Verbrugge, and thus established yardsticks for women in athletics as well as physiology, health, and hygiene. Historical investigation on such areas of activity as pharmacy, nutrition,

[45] See, e.g., Elizabeth Fee, "Nineteenth Century Craniology: The Study of the Female Skull," *Bulletin of the History of Medicine*, 1979, *53*:415–433; Cynthia Eagle Russett, *Sexual Science: Victorian Construction of Womanhood* (Cambridge, Mass.: Harvard Univ. Press, 1989); Stephanie A. Shields, "The Variability Hypothesis: The History of a Biological Model of Sex Differences in Intelligence," *Signs*, 1982, *7*:769–785; and Evelleen Richards, "Darwin and the Descent of Woman," in *The Wider Domain of Evolutionary Thought*, ed. D. Oldroyd and I. Langham (Dordrecht: Reidel, 1983).

[46] Sara Delamont, "Prisoners of Progress: Women and Evolution," in *Nineteenth Century Woman: Her Cultural and Physical World*, ed. Delamont and Lorna Duffen (New York: Barnes & Noble, 1978); and Susan Sleeth Mosedale, "Science Corrupted: Victorian Biologists Consider 'The Woman Question,'" *Journal of the History of Biology*, 1978, *11*:1–55.

[47] Louise Mitchell Newman, *Men's Ideas/Women's Realities: Popular Science, 1870–1915* (New York: Pergamon Press, 1985).

[48] Sally Gregory Kohlstedt and Mark Jorgensen, "'The Irrepressible Woman Question': Women's Responses to Biological Determinism, 1871–1918," paper read at the Conference on Responses to Darwin, 11–13 May 1994, Dunedin, New Zealand.

[49] Elizabeth Fee, "The Sexual Politics of Victorian Anthropology," in *Clio's Consciousness Raised*, ed. Mary S. Hartman and Lois Banner (New York: Harper & Row, 1974); and Eveleen Richards, "Huxley and Women's Place in Science: The Woman Question and Control of Victorian Anthropology," in *History, Humanity, and Evolution: Essays in Honor of John C. Greene*, ed. James R. Moore (Cambridge: Cambridge Univ. Press, 1989), pp. 253–284.

[50] Rosalind Rosenberg, *Beyond Separate Spheres: Intellectual Roots of Modern Feminism* (New Haven: Yale Univ. Press, 1982).

[51] A conference organized by Joan Jacobs Brumberg at Cornell University, October 1991, "Women, Home Economics, and American Culture," has opened that field up to fresh investigation and interpretation. (A volume of proceedings will be published by Cornell University Press.)

journalism, and education may well indicate other ways in which women's participation and concerns influenced the direction and application of scientific research.[52]

Why did women develop these fields, and what difference did it make to the other sciences that so few women pursued them? That thorny issue is evident in a number of important works, either explicitly or by implication. For some scholars and activists, including those involved in the contemporary movement known as ecofeminism, the answer is that women have very different perspectives from men, which influences how they carry out their research and apply it.[53] Others argue that science should be free of sex bias simply because the exclusion of women not only harms aspiring women but perhaps limits the outcomes of scientific research in general.[54]

In the twentieth century modern biology, chemistry, and medicine have introduced additional ways of defining sexual identity and extending the definitions into the broader social context.[55] The biologists Ruth Hubbard and Ruth Bleier were among the first to challenge twentieth-century sociobiology from a feminist standpoint.[56] Much remains to be done in terms of both substantive historical research and theory. Only when we attend to sexual identity (along with race, class, and other prescribed characteristics) will historians be able to explain why a select number of women did astronomy so well in some scientific centers of seventeenth-century northern Europe, but not later or elsewhere at that time; or why women were presumed to be particularly capable of botany in the eighteenth and nineteenth centuries, but later excluded from botanical societies; or why women in the late nineteenth century set out so boldly to establish distinctive professional fields in home economics, social work, and nursing.[57]

Research on women in science requires that history be rewritten in order to challenge generalizations made by those who ignore issues of gender, race, and class. One of the most important recent books to engage explicitly in such a commentary

---

[52] Martha H. Verbrugge, *Able-Bodied Womanhood, Personal Health and Social Change in Nineteenth-Century Boston* (Cambridge, Mass.: Harvard Univ. Press, 1988). On scientific journalism see Marcel LaFollette, *Making Science Our Own: Public Images of Science, 1910–1950* (Chicago: Univ. Chicago Press, 1990); and LaFollette, "Eyes on the Stars: Images of Women Scientists in Popular Magazines," *Science, Technology, and Human Values,* 1988, *13*:262–275.

[53] An early explicit call for liberated science was Elizabeth Fee, "Is There a Feminist Science?" *Science and Nature,* 1981, *4*:46–57. See also Sue Rosser, *Biology and Feminism: A Dynamic Interaction* (New York: Twayne, 1992).

[54] Many readers interpret Barbara McClintock's success with hybrid corn, as depicted in Keller's *A Feeling for the Organism* (cit. n. 16), as related to her having a special sensibility about her subject. Keller, however, makes it clear that she does not believe there is a special "feminist" science; see Evelyn Fox Keller, *Reflections on Gender and Science* (New Haven: Yale Univ. Press, 1985).

[55] See esp. Diana Long Hall, "Biology, Sex Hormones, and Sexism in the 1920s," *Philosophical Forum,* 1973, *5*:81–96. See also Hall, "Physiological Identity of American Sex Researchers between the Two World Wars," *Physiology in the American Context, 1850–1940,* ed. Gerald L. Geison (Bethesda, Md.: American Physiological Society, 1987).

[56] Ruth Bleier, *Science and Gender: A Critique of Biology and Its Theories on Women* (New York: Pergamon, 1984); Bleier, ed., *Feminist Approaches to Science* (New York: Pergamon, 1986); Ruth Hubbard, *Women Look at Biology Looking at Women* (Cambridge: Schenkman, 1979); and Hubbard, *The Politics of Women's Biology* (New Brunswick, N.J.: Rutgers Univ. Press, 1990).

[57] Redefining science to include those subject areas developed by women has not been much attempted. For some discussion of the origins of home economics, applied social science research, library science, and baccalaureate nursing at one women's college during the Progressive period, see Sally Gregory Kohlstedt, "Single Sex Education and Leadership: The Early Years of Simmons College," in *Women and Educational Leadership: A Reader,* ed. Sari Knopp Biklen and Marilyn Brannigan (Boston: Lexington Press, 1979), pp. 93–112.

is Donna Haraway's pathbreaking and widely reviewed *Primate Visions*. Haraway studies the history of twentieth-century primatology in the United States and places women at the center of her account. In particular, she studies the women working in the field with apes; they discover, as Haraway and others have argued, different social patterns as well as unanticipated similarities between the lives of male and female primates. She is not satisfied with this gender revision alone, but points out that nearly all female and male primatologists have been white and middle-class; throughout the text she offers suggestions about the need for a genuinely new understanding of primatology (and, by inference, science) with more diverse participation and more imagination.[58]

Recent historical work on women in science runs along multiple paths, sometimes intersecting but often not. No grand theoretical program has emerged and even the emphasis on differences between men and women is problematic. If the earliest work took for granted distinctions between men and women and praised pioneers, later assessments insisted on pointing out the barriers and obstacles women commonly faced. Most recent work, while admitting cultural constraints, looks at the unconventionality and originality of those whose work seemed within the norms despite peculiar obstacles and those who chose paths to personal satisfaction and success in less visible ways. What persists within current feminist scholarship is the insistence on hearing women's voices, understanding their experiences, and listening to their definitions of science and their own goals for scientific work. This brings us, full circle, back to the necessity for good bibliographical tools.

IV. A BEACHHEAD FOR HISTORIANS OF WOMEN IN SCIENCE

Professional recognition and communication among scholars is also fundamental to advancing a field. Arguably, the subject of women in science did not exist at the time of the last critical issues conference, in 1957, and only a few sporadic publications marked its emergence over the following two decades. In 1981 in Bucharest, however, an International Commission on Women in Science, Technology, and Medicine was formed as part of the International Union for the History of Science (IUHS), and a commitment was made to incorporating the subject in all subsequent IUHS congresses. Before that there were only four papers on women listed in congress proceedings.[59] The commission's efforts led to conferences on women in science in Hungary in 1983 and again in 1991, hosted by Eva Vamos and the National

---

[58] Donna Haraway, *Primate Visions: Gender, Race, and Nature in the World of Modern Science* (New York: Routledge, 1989); see also Haraway, *Simians, Cyborgs, and Women: The Reinvention of Nature* (New York: Routledge, 1991), her collected essays. See also an essay review in philosophy of science by Helen E. Longino, "Feminist Standpoint Theory and the Problems of Knowledge," *Signs,* Fall 1993, *19*(1):201–212.

[59] Pam S. Codellas, "Ancient Greek Women Leaders in Science," *Fifth International Congress on the History of Sciences* (Lausanne, 1947); Elena Lotava, "L'influence des femmes dans le development de la science medicable turque," *Ninth International Congress on the History of Science* (Barcelona and Madrid, 1959); Joan Burstyn, "Women in American Science," *Tenth International Congress on the History of Science* (Ithaca, 1962); and Edythe Lutzker, "Edith Pechey-Phipson, M.D., Pioneer Woman of Victorian England and India," *Eleventh International Congress on the History of Science* (Warsaw, 1965). The United States delegation to the 1981 congress, chaired by Thomas Hughes and using a draft proposal written by Sally Gregory Kohlstedt (delegate) and Margaret Rossiter (alternate delegate), coordinated the effort to create the new commission.

Museum of Science and Technology in Budapest. In 1985 the commission sponsored a day-long symposium at the Berkeley congress, well attended by women from around the world, which resulted in the volume *Women in Science: Options and Access.*[60] At the 1993 International Congress in Zaragosa, Spain, more than forty papers were presented on the history of women in science, technology, and medicine, many dealing with Spain and other countries where the experiences of women have varied from those in North American and northern European settings. These individual papers underscore the need for more comparative work and more opportunities for discussion and publication.

In the United States courses on women and science are increasingly available in programs for the history, philosophy, or sociology of science and technology, although by no means universally available at even the best-recognized programs. A workshop on teaching the history of women in science at the History of Science Society's annual meeting in Seattle in 1990 had over a hundred participants, and subsequent workshops have been well attended. The Woman's Committee of the History of Science Society initiated a prize for published books and articles on women in science in 1986 and made the first award in 1987; awards have so far gone to four books and four articles.[61]

Optimism about the increase in literature must, however, be tempered with caution, because this enterprise has still had relatively little influence on the larger discipline of history of science in terms of research, publication, and teaching. It is true that the major journals relating to the history of science—particularly *Isis, History of Science,* the *British Journal for the History of Science,* and *Science, Technology and Human Values*—include articles, book reviews, and notes dealing with the history of women in science. Such materials, however, are often highlighted as "women's topics." Looking in greater detail at published materials, including other articles in those journals, reveals how little attention is being paid to the role of women even in books and articles dealing with well-researched periods and topics. It seems strange, for example, to find little mention of women in works on what Rossiter termed the "moderately feminized" area of biochemistry, which has received considerable attention recently, and only an oblique reference to women in Steven Shapin and Simon Schaffer's prize-winning account of the laboratory experiments in private homes in the seventeenth century.[62] Marcel LaFollette's study of scientific journalism, *Making Science Our Own,* systematically assesses women's participation and

---

[60] Commission on Women in Science, Technology, and Medicine, *Women in Science: Options and Access* (papers presented at the 17th International Congress on the History of Science, University of California at Berkeley, 31 July–8 August 1985) (Budapest: National Museum of Science and Technology, 1987).

[61] Leslie Burlingame organized the teaching session and Michele Aldrich coordinated the letter-writing campaign that raised the original endowment for the prize.

[62] Rossiter, *Women Scientists in America* (cit. n. 2), p. 289. Women get little attention in, e.g., Lily Kay, *The Molecular Vision of Life: Caltech, The Rockefeller Foundation, and the Rise of the New Biology* (New York: Oxford Univ. Press, 1993); F. L. Holmes, *Between Biology and Medicine: The Formation of Intermediary Metabolism* (Berkeley: Office for History of Science and Technology, Univ. California at Berkeley, 1992); Robert Kohler, *From Medical Chemistry to Biochemistry: The Making of a Biomedical Discipline* (Cambridge: Cambridge Univ. Press, 1982); and Mikuláš Teich with Dorothy M. Needham, *A Documentary History of Biochemistry, 1770–1940* (Rutherford, N.J.: Fairleigh Dickinson Univ. Press, 1992). See also Steve Shapin and Simon Schaffer, *Leviathan and the Air-Pump: Hobbes, Boyle, and the Experimental Method* (Princeton: Princeton Univ. Press, 1985).

problems in the field, but nearly all the discussions are segregated in a chapter that deals almost exclusively with women.[63] Problems of identification and of inclusion are apparently not easily addressed.

Turning to citation indexes, I found that while Rossiter's and Schiebinger's books, for example, were reviewed in the major journals for the history of science, they are not much cited there. Both authors are frequently referenced in articles published on social science and education; Rossiter is cited more frequently in journals of science. James MacLachlan's *Children of Prometheus: A History of Science and Technology* does not include women as a topic in the table of contents and lists just three individual women in the index. A recent bibliographical guide issued by the same press mentions only three books on women and science and concludes that their authors want to return science to a more integrated, more "holistic" view of nature; contributors to the guide ignore the substantial discussions in books like those of Margaret Rossiter, Evelyn Fox Keller, and Pnina Abir-Am.[64] There are few actual textbooks in the history of science, but even those recently produced for the Open University address the topics of women in childbirth and women in domestic technology, rather than including them as active participants in scientific and techno-logical research and application.[65] If the activities of women are to be integrated into the history of science, scholars must do more than tolerate discussion of "women and science" on the fringes of the discipline.

V. EQUITY AND FAIRNESS

Since studying the history of women in science is, for many of its practitioners, more than academic, perhaps some additional assessment is in order. Historical work has become a resource for social activists who believe that understanding the past can help direct current events and future policies. Since the 1970s, public and privately supported programs have encouraged women to go into science and engineering. Federal agencies have sponsored research to analyze the factors that inhibit women's involvement, while organizations like the Association of Women in Science (AWIS) and the Society of Women Engineers (SWE) gather and publish data on their membership and affiliated disciplines. Much of that literature is produced by social scientists as well as by active scientists who want to involve more women. For some the issue is one of equity; women should have an equal opportunity. Others hold the conviction that having more women will change the process and intellectual outcomes of scientific research.[66]

[63] LaFollette, *Making Science Our Own* (cit. n. 52).

[64] James MacLachlan, *Children of Prometheus: A History of Science and Technology* (Toronto: Wall & Thompson, 1988); and *Science in Society: An Annotated Guide to Resources* (Toronto: Wall & Thompson, 1989). Similarly, Helge Kragh, *An Introduction to the Historiography of Science* (Cambridge: Cambridge Univ. Press, 1987), raises important questions about objectivity, ideology, and various approaches to science with virtually no reference to women in science or feminist ap-proaches to the history of science.

[65] See, e.g., Gerrylynn K. Roberts, ed., *Sources for the Study of Science, Technology, and Everyday Life*, 2 vols. (London: Hodder & Stoughton for the Open University, 1988). These topics are im-portant, but not sufficient.

[66] Equity is the dominant theme in a special issue of *Science*, "Women in Science," 1992, *255*:1363–1388. The second and third such annual issues paid more attention to issues of gender: "Gender and the Culture of Science," 1993, *260*:383–432; and "Comparisons across Cultures," 1994, *263*:1467–1496. See also (on equity) Sandra Keith and Philip Keith, *Proceedings of the National Conference on Women in Mathematics and Science* (St. Cloud, Minn.: St. Cloud Univ. Dept. Mathe-

Women studied physiology as scientists and as medical practitioners, often engaging in laboratory work, as at the University of Minnesota's Physiology Laboratory, about 1940. Courtesy of the University of Minnesota Archives.

Those concerned with equity have engaged in political activity to effect institutional change. Although the United States failed to pass an Equal Rights Amendment in 1977, civil rights and affirmative action legislation pried open doors in federal agencies and universities, where the absolute numbers and percentages of women increased in the 1970s. The number of women and minorities increased significantly under the "stick" wielded by federal and state agencies and through the "carrots" provided by the National Science Foundation and several private foundations that made diversity a central theme in the 1970s and 1980s.

The number of women in engineering and physical science increased, but not nearly as rapidly as many anticipated.[67] Thus the question "Why so few?" remains. Psychologists and other social scientists continue to test the fundamental proposition that sexual differences have significant implications for the mental ability and manual dexterity of boys and girls. Sociologists point out the subtle (and not so subtle) ways in which group behavior excludes women or leads them to withdraw from scientific and technological work or study. Responses to these academic findings have led teachers and administrators to modify their own expectations, to investigate issues of discrimination and harassment by peers and faculty, to encourage women's

matics and Statistics, 1989), pp. 5–17; and (on gender) Sue V. Rosser, *Female-Friendly Science: Applying Women's Studies Methods and Theories to Attract Students* (New York: Pergamon, 1990).

[67] Recent data on women in science and engineering may be found in *National Survey of Women and Men Engineers: A Study of the Members of 22 Engineering Societies* (New York: Society of Women Engineers, 1993); and Betty M. Vetter, *Professional Women and Minorities: A Total Human Resource Data Compendium,* 11th ed. (Washington, D.C.: Commission on Professionals in Science and Technology, 1994).

networks, to build mentoring systems at every rung on the educational and employment ladder, and to address the personal issues of home and family that face women with particular intensity.

Many of the arguments made to increase support for women in science in the past decade focused on projected "shortages" of women needed in the labor force—but that argument has not created long-term opportunities for women, and in these tight economic times it invites a perception of women as a surplus and elastic labor force that can be pushed back out of the labor market, as after World War II.[68] Less likely to change with circumstances are arguments that point to the need for equal access to opportunities and to the equal capacities and enthusiasms for scientific study shown by girls and boys.

From the archivists who gather data to the bibliographers who make the data accessible, from the historians and researchers who investigate their meaning to the audiences that interpret historical events for reasons of scholarship, teaching, and activism, from the publishers who solicit and accept materials to the journals that review their results—all of us will shape what is learned in the future about the history of women in science, technology, and medicine.

VI. CONCLUSION

Virginia Woolf wrote near the end of her classic essay, *A Room of One's Own:*

> Women—but are you not sick to death of the word? I can assure you that I am. Let us agree, then, that a paper read by a woman to women should end with something particularly disagreeable.
> But how does it go? What can I think of? The truth is, I often like women. I like their unconventionality. I like their subtlety. I like their anonymity.

Like Virginia Woolf, those who study women and science for the most part do enjoy the company as they investigate past practice. The research itself holds the compelling drama of new discoveries, even as it provides important if complex evidence regarding the issue of gender and science. Moreover, these women's lives demonstrate their unconventionality, their subtlety, and their ingenuity, as well as the reasons for their anonymity.

The momentum for studying women and science is evident. The multiplicity of recent accounts of women in science warns scholars that the categories of sex and gender have been unstable over time, and they reveal often surprising contrasts in the lives and perceptions of women in different times and places; caution is thus warranted when generalizing. Nonetheless, current research provides glimpses into how such studies might transform the history of science so that it incorporates not only the subject but also the methods and sensibilities we have uncovered together in our search for the position of women in science and scientific thinking.

[68] There is an extensive literature on women workers during and after World War II, and Margaret Rossiter's sequel volume (cit. n. 2) will reveal the pattern for women scientists. For the current discussion see, e.g., Elizabeth Ivey, "Recruiting More Women into Engineering and Science," *Engineering Education,* 1985, *78*:46–58; and Sheila Widnall, "AAAS Presidential Lecture: Voices from the Pipeline," *Science,* 1988, *248*:1740–1745; and the three special issues of *Science* cited above (n. 66).

KNOWLEDGE CONSTRUCTED BY CHRONOLOGY AND GEOGRAPHY

Pierre Duhem, arguing that there was no "irreducible antagonism between the scientific spirit and the spirit of Christianity," set the research agenda for historians of medieval science in the second half of the twentieth century. See pages 62–63. Photograph courtesy of the AIP Niels Bohr Library, E. Scott Barr Collection.

Among those historians who supported Duhem's defense of the importance and originality of medieval scientific thought was Lynn Thorndike, in the first four volumes of his History of Magic and Experimental Science *(1923–1958). Photograph courtesy of the Chemical Heritage Foundation.*

Medieval Science and Its Religious Context

By David C. Lindberg*

I. HISTORIOGRAPHIC REFLECTIONS

To JUDGE FROM THEIR SCHOLARLY OUTPUT, many historians of medieval science in the twentieth century have perceived the critical problem confronting them to be the rescue of the object of their research (medieval science) from the ignominy into which it fell as a result of four or five hundred years as a convenient whipping boy, first for European and subsequently for American intellectuals endeavoring to establish and explain the superiority of their own cultures. Attitudes first developed by the humanists of the Renaissance, further articulated by the *philosophes* of the Enlightenment, and given canonical form by a variety of polemicists in the nineteenth century portrayed the Middle Ages as a period of catastrophic decline, a time of intellectual darkness and decadence, during which the human intellect succumbed to ignorance and superstition or yielded to the fetters of scholastic theology. I have addressed these historiographic developments elsewhere and will not repeat myself here.[1] But I would like to offer a reminder of the strength of nineteenth-century opinion and the overblown rhetoric and shrill tones in which denunciations of medieval scientific efforts were then frequently expressed, by quoting from a pair of influential nineteenth-century writers.

The first is the distinguished British historian and philosopher of science William Whewell (1794–1866), writing about mid century:

> In speaking of the character of the age of commentators [the medieval period], we noticed principally the ingenious servility which it displays;—the acuteness with which it finds ground for speculation in the expression of other men's thoughts;—the want of all vigor and fertility in acquiring any real and new truths. Such was the character of the reasoners of the stationary period from the first; but, at a later day, this character . . . was modified by new features. The servility which had yielded itself to the yoke, insisted upon forcing it on the necks of others; the subtlety which found all the truth it needed in certain accredited writings, resolved that no one should find there, or in any other region, any other truths; speculative men became tyrants without ceasing to be slaves; to their character of Commentators they added that of Dogmatists.[2]

*Department of History of Science, 4143 Helen White Hall, 600 North Park Street, University of Wisconsin, Madison, Wisconsin 53706.
[1] David C. Lindberg, "Conceptions of the Scientific Revolution from Bacon to Butterfield: A Preliminary Sketch," in *Reappraisals of the Scientific Revolution,* ed. Lindberg and Robert S. Westman (Cambridge: Cambridge Univ. Press, 1990), pp. 1–13.
[2] William Whewell, *History of the Inductive Sciences,* 3rd ed., 3 vols. (London: John W. Parker, 1857), Vol. I, p. 237.

The second is John Addington Symonds (1840–1893), an early follower of the great Renaissance historian Jacob Burckhardt and author of an extremely influential multivolume history of the Italian Renaissance. Symonds contrasted the spirit of the Middle Ages with that of the Renaissance in this way:

> During the Middle Ages man had lived enveloped in a cowl. He had not seen the beauty of the world, or had seen it only to cross himself, and turn aside and tell [count] his beads and pray. Like S. Bernard traveling along the shores of the Lake Leman, and noticing neither the azure of the waters, nor the luxuriance of the vines, nor the radiance of the mountains with their robe of sun and snow, but bending a thought-burdened forehead over the neck of his mule: even like this monk, humanity had passed, a careful pilgrim, intent on the terrors of sin, death, and judgment, along the highways of the world, and had scarcely known that they were sightworthy, or that life is a blessing. . . . The Renaissance shattered and destroyed [the fixed ideas of the ascetic medieval Church], rending the thick veil which they had drawn between the mind of man and the outer world and flashing the light of reality upon the darkened places of his own nature. For the mystic teaching of the Church was substituted culture in the classical humanities; a new ideal was established, whereby man strove to make himself the monarch of the globe. . . . The Renaissance was the liberation of the reason from a dungeon, the double discovery of the outer and the inner world.[3]

Such opinions were further popularized at the very end of the century by Andrew Dickson White, whose enormously influential *History of the Warfare of Science with Theology in Christendom* (1896) helped to give them a currency among the educated public that they retain to this day.

With this nineteenth-century opinion as background, we can begin to understand the fortress or siege mentality evident in the writings of many twentieth-century historians of medieval science. The opening volley in a counterattack was fired by the French physicist-philosopher-historian Pierre Duhem (1861–1916). In the course of searching for historical examples to buttress his philosophy of science, Duhem undertook an investigation of the origins of statics. This search led him first to the writings of Jordanus de Nemore, subsequently to the works of Albert of Saxony, Jean Buridan, Nicole Oresme, and others. Study of their works led Duhem to the conclusion that "the mechanics and physics of which the modern world is justifiably proud proceed, by an uninterrupted series of scarcely perceptible improvements, from doctrines professed in the heart of the medieval schools. The pretended intellectual revolutions were most often merely slow and long-prepared evolutions, the so-called renaissances merely unfair and sterile reactions."[4] The origins of modern science, if Duhem was right, are to be found not in the repudiation of medieval scholasticism but in the cultivation of natural philosophy within the medieval schools. As for the role of the medieval church, Duhem argued that the popular "teaching that pretends to have established the irreducible antagonism between the scientific spirit and the spirit of Christianity is the most colossal and most audacious lie that has ever attempted to dupe the people."[5] In his view, the church was not the

---

[3] John Addington Symonds, *Renaissance in Italy,* Part I: *The Age of Despots* (New York: Henry Holt, 1888), pp. 13–14. On Symonds's influence see Wallace K. Ferguson, *The Renaissance in Historical Thought: Five Centuries of Interpretation* (Boston: Houghton Mifflin, 1948), pp. 204–205.

[4] Pierre Duhem, *Les origines de la statique,* 2 vols. (Paris: Hermann, 1905–1906), Vol. I, p. iv.

[5] From a letter written in 1911. The French text is given in Hélène Pierre-Duhem, *Un savant français: Pierre Duhem* (Paris, 1936), and quoted in translation by Stanley Jaki, *Uneasy Genius: The Life and Work of Pierre Duhem* (The Hague: Nijhoff, 1984), p. 399 (with minor modifications).

villain of the piece, inhibiting the development of a fruitful scientific tradition, but a vital participant in the emergence of modern science.

Duhem fleshed out his thesis in a series of multivolume works: *Les origines de la statique* (2 vols., 1905–1906), *Etudes sur Léonard de Vinci* (3 vols., 1906–1913), and *Le système du monde* (10 vols., 1913–1959). Early support for Duhem's defense of the importance and originality of medieval scientific thought came from Charles Homer Haskins (1870–1937) in his *Studies in the History of Mediaeval Science* (1924) and *Renaissance of the Twelfth Century* (1927); and from Lynn Thorndike (1882–1965) in the first four volumes (1923–1934) of his *History of Magic and Experimental Science.*[6] Although Haskins and Thorndike offered a wealth of important and suggestive data and (in Thorndike's case) significant aids to research, which shaped and facilitated the work of other scholars, it was Duhem, more than they, who set the agenda for historians of medieval science in the second half of the twentieth century.

The decades immediately after World War II saw rapid development of the history of science as a profession and an explosion of research on the history of medieval science. Among the early leaders were Anneliese Maier, Ernest Moody, Marshall Clagett, Pearl Kibre, E. J. Dijksterhuis, and Alistair Crombie. Their work is far too rich and complicated to be cleanly dichotomized, but I think it may be useful nonetheless to identify two components: the interpretive and the textual.[7]

The most influential of the *interpreters* was undoubtedly Anneliese Maier (1905–1971), the most provocative Alistair Crombie (1915–). Both took their cues, in one way or another, from Duhem. Like Duhem, both focused on mechanics and the physical sciences, with little or no attention to the biological or medical realm; both devoted themselves to conceptual analysis of medieval texts (what some would refer to as an "internalist" approach); and both bestowed considerable attention on the question of continuity between medieval and early modern science, seeing it as one of their principal tasks to answer Duhem's question about the medieval contribution to the origins of modern science—as the titles of some of their major works make evident.[8] Maier's achievement was to bring a richer and more sophisticated understanding of philosophical context to the reading of the texts; this enabled her to offer a far more cautious, more nuanced, and less anachronistic analysis of the medieval scientific achievement and to deflate some of Duhem's more extreme claims on behalf of medieval originality. For example, she decisively refuted Duhem's strong claims regarding the relationship between the medieval theory of impetus and Galileo's concept of inertia, arguing persuasively that impetus theory did not anticipate inertial mechanics but merely functioned as a "connecting link" between Aristotelian and modern mechanics.[9]

---

[6] Thorndike's *History of Magic and Experimental Science* ran to eight volumes in all (1923–1958), covering European science through the seventeenth century.

[7] Of course, to read, edit, or translate a text is to interpret it, so that textual work is inevitably interpretive. But the distinction I have in mind is between the close, analytical interpretations demanded by work on the texts themselves and interpretations of a broader, more synthetic, and more adventurous sort.

[8] See, e.g., Maier's first "Studien zur Naturphilosophie der Spätscholastik," entitled *Die Vorläufer Galileis im 14. Jahrhundert* (Rome: Storia e Letteratura, 1949); and Alistair Crombie, *Robert Grosseteste and the Origins of Experimental Science, 1100–1700* (Oxford: Clarendon Press, 1953).

[9] See, e.g., Anneliese Maier, "Die naturphilosophische Bedeutung der scholastischen Impetustheorie," *Scholastik*, 1955, *30:*321–343; trans. in *On the Threshold of Exact Sciences: Selected*

If the Duhem thesis evoked cautious criticism from Maier, it elicited enthusiastic—his own term—support from Crombie. In the early 1950s Crombie issued a pair of manifestos concerning the relationship between medieval and early modern science, in the form of two books. In the first of these, a survey of medieval and early modern science published in 1952, Crombie argued that "it was the growth of . . . 13th- and 14th-century experimental and mathematical methods that brought about the movement which by the 17th century had become so striking as to be called the Scientific Revolution."[10] Crombie elaborated on this theme in the second book, published a year later, maintaining that the critical feature of early modern science was its possession of the proper methodology for the practice of science, the methodology of experimentation, and that this methodology was a creation of the later Middle Ages:

> The thesis of this book is that a systematic theory of experimental science was understood and practised by enough philosophers [of the thirteenth and fourteenth centuries] for their work to produce the methodological revolution to which modern science owes its origin. . . . What seems to be the first appearance of a clear understanding of the principles of modern experimental science is found in the writings of the [thirteenth-century] English logician, natural philosopher, and scholar, Robert Grosseteste.[11]

Crombie's books provoked a swift and hostile reaction from specialists on the history of early modern science, and thus the continuity debate in its modern form was born. I have offered my view of this debate in a recent book and will devote no further space to it here.[12]

If Maier and Crombie offered broad, occasionally daring, interpretations of medieval scientific achievements (often with an eye on the relationship between medieval and early modern science), a somewhat different road was taken by Marshall Clagett (1916–). Although no stranger to interpretation (particularly of a fairly narrow, analytical sort), Clagett commented frequently on the need for reliable texts and translations of the major medieval scientific treatises if interpretive efforts were to be properly grounded.[13] And he and his students (and *their* students and many of their students' students) therefore placed priority on textual scholarship. Clagett himself produced important editions and translations (always accompanied by summaries and analyses) of important mechanical treatises in his *Science of Mechanics in the*

Writings of Anneliese Maier on Late Medieval Natural Philosophy, ed. and trans. Steven D. Sargent (Philadelphia: Univ. Pennsylvania Press, 1982), Ch. 4.

[10] A. C. Crombie, *Augustine to Galileo: The History of Science A.D. 400–1650* (London: Falcon, 1952), p. 273, revised as *Medieval and Early Modern Science,* 2 vols. (Garden City: Doubleday Anchor, 1959). See the interesting retrospective review by Bruce S. Eastwood, "On the Continuity of Western Science from the Middle Ages: A. C. Crombie's *Augustine to Galileo,*" *Isis,* 1992, *83:* 84–99. For "enthusiastic" see Crombie, *Grosseteste,* 2d ed. (1962), p. v.

[11] Crombie, *Grosseteste,* (cit. n. 8), pp. 9–10.

[12] David C. Lindberg, *The Beginnings of Western Science: The European Scientific Tradition in Philosophical, Religious, and Institutional Context, 600 B.C. to A.D. 1450* (Chicago: Univ. Chicago Press, 1992), pp. 355–360. See also Eastwood, "On the Continuity of Western Science from the Middle Ages" (cit. n. 10). The most formidable of Crombie's opponents was Alexandre Koyré; see esp. his review of Crombie's *Grosseteste:* "The Origins of Modern Science: A New Interpretation," *Diogenes,* Winter 1956, *16:*1–22.

[13] See, e.g., Marshall Clagett, *The Science of Mechanics in the Middle Ages* (Madison: Univ. Wisconsin Press, 1959), introduction, p. xxii. See also Edward Grant and John E. Murdoch, eds., *Mathematics and its Applications to Science and Natural Philosophy in the Middle Ages: Essays in Honor of Marshall Clagett* (Cambridge: Cambridge Univ. Press, 1987), introduction, p. ix.

Middle Ages (1959) and the work he edited with Ernest Moody, *The Medieval Science of Weights* (1960); of Nicole Oresme's *Tractatus de configurationibus qualitatum et motuum* in his *Nicole Oresme and the Medieval Geometry of Qualities and Motions* (1968); and finally (and monumentally) of mathematical texts in his exhaustive, multivolume edition and translation of the medieval Archimedean corpus (1978–1984). The Duhemian influence on these textual efforts is evident in Clagett's choice of texts (most obviously in the mechanical texts selected for inclusion in his *Science of Mechanics*), in his continuing attention to the question of continuity between medieval and early modern science, and in the frequent references to Duhem scattered throughout his works.[14]

I have made no systematic study of the scholarly progeny of Maier, Crombie, Clagett, and others of their generation. It seems clear, however, that it was only Clagett who found himself in favorable institutional circumstances (perhaps created through his own efforts) that permitted him to pass on his vision of the history of medieval science to substantial numbers of graduate students, who in turn passed it on to *their* graduate students; and it was therefore the Clagett program of textual scholarship that predominated through the 1960s, 1970s, and 1980s. But I do not want to overstate the case: the lines of influence are complex; interpretive and textual efforts have never been quite as distinct as my analysis might suggest; and, of course, scholarly lineage is by no means the only factor that determines scholarly inclination. The point is simply that Clagett's stress on the text has remained visible among historians of medieval science to the present. And the result has been a multiplication of texts and translations, so that the textual resources at our disposal today are vastly superior to those available thirty or twenty or even ten years ago.

II. CRITICAL ISSUES IN THE HISTORY OF MEDIEVAL SCIENCE

With this brief historiographic sketch as background, can we identify the "critical problems" currently confronting historians of medieval science? I do not claim omniscience on this subject, but I think we can, with confidence, identify some of them.

First, there is plenty of textual work yet to be done. In the area that I know best, the history of ideas about light and vision, we now have usable versions of most of the important texts; but I would be surprised if there is another medieval subject for which that claim can be made, and there are many subjects the surface of which has barely been scratched. We are particularly needy when it comes to biomedical subjects, commentaries on the *libri naturales* of the Aristotelian corpus, alchemy, astrology, and other subjects now frequently marginalized under the rubrics "occult" or "pseudoscience." The need to expand our purview to include marginalized subjects or disciplines requires emphasis. There is no justification for historians of science excluding certain subjects simply because they have been excluded from the canon of modern scientific disciplines. We miss a critical opportunity if we do not explore the full range of medieval approaches to the natural world.[15]

[14] In *Science of Mechanics* (cit. n. 13), p. xxii, e.g., Clagett states explicitly that the purpose of the volume he edited with Ernest A. Moody, *The Medieval Science of Weights* (Madison: Univ. Wisconsin Press, 1960), was "to present the complete texts so that historian and scientist alike could judge the claims of Duhem for the importance of medieval statics."

[15] I am grateful to Joan Cadden for her forceful presentation of this point in her commentary on this paper: "Critical Problems in Medieval Science: Domestic and Imported," pp. 5–8 (unpublished).

Second, I believe it is time to reconsider the balance between textual and interpretive scholarship and strengthen our commitment to the interpretation of the texts already at our disposal. Indeed, I would like to stress the importance of interpretive efforts of a broad, synthetic sort—the sine qua non of any attempt to communicate the fruits of our research to a larger public. (If we have no interest in addressing that larger public, then I do not see that we are left with any adequate justification for our research, especially the portion of it that is publicly funded.)

Although the "Clagett program" stressed the importance of textual scholarship and analytical efforts closely connected to the texts, syntheses with varying degrees of breadth have begun to appear with increasing frequency in the past ten or fifteen years. One thinks (to sample widely from different genres) of Edward Grant's *Much Ado about Nothing* (1981) and his many other cosmological studies; of Nancy Siraisi's books on medieval medicine, especially her recent *Medieval and Early Renaissance Medicine* (1990); and of the intellectual biographies of Robert Grosseteste by James McEvoy (1982) and Sir Richard Southern (1986). But these represent no more than a beginning. We still need synthetic efforts of every sort and at every level. The need is pressing in fields like the science of motion, astronomy, astrology, mathematics, alchemy, and natural history, where we have neither a comprehensive history of the subject in general nor even book-length studies of the major pieces. And we have gone over forty years without a serious attempt at an overall synthesis of medieval science, since Crombie's *Augustine to Galileo*. My own synthetic effort has recently appeared, but surely there is room for more.[16]

Third, there is the question of continuity. I certainly do not believe that research on the history of medieval science should be dominated by it. Quite the contrary: it is now a cliché (albeit an important one) that we need to approach medieval science as an enterprise interesting in its own right, rather than scouring it for anticipations of early modern science. However, once we have devoted our attention to medieval science for its own sake, there is no reason at all why we should not return to the question of continuity and discontinuity between medieval and early modern science. We must not allow our revulsion at the horrors of whiggism to frighten us into ignoring an extremely important historical issue. No doubt the continuity-discontinuity question needs to be reformulated, so that we cease to squabble with our neighbors who specialize on the early modern period about the relative merit of scientific developments in our respective periods. We need rather to study the ways in which early modern science appropriated and transformed (if indeed it did) pieces of the medieval scientific achievement. Stated in such terms, there is nothing to fear from the question.

Fourth, historians of medieval science have fallen far behind their colleagues specializing in later periods in the attempt to understand the social and institutional context of the scientific enterprise. Science was no more autonomous and isolated, no more situated in a social and institutional vacuum, during the medieval period than in more recent eras; and we cannot pretend to have fully grasped the nature and significance, or even the content, of medieval science until we have thoroughly contextualized it.

One extremely important aspect of social context is the question of support or

---

[16] Crombie, *Augustine to Galileo* (cit. n. 10); and Lindberg, *Beginnings of Western Science* (cit. n. 12).

patronage and the related questions of motivation and market. Joan Cadden, in her commentary on the original version of this paper, urged that we begin to pay serious attention to the neglected area of vernacular scientific literature and the conditions of its production and consumption. This, she argued, will compel us to consider such trends as the laicization of science in the later Middle Ages—the movement by which science and natural philosophy were removed from the exclusive control of the universities and clerical culture. As an example of this trend, she pointed to the many vernacular translations commissioned by the French king Charles V in the second half of the fourteenth century, for the purpose (we are told at the end of one of the translations) "of animating, exciting, and moving the hearts of young men who have subtle and noble talents and the desire for knowledge."[17] The development of vernacular and lay scientific effort (frequently, but not necessarily, linked) surely constitutes one of the most formidable and most critical problems facing historians of medieval science today.

Without in any way detracting from this last claim, I would like to defend the proposition that the primary patron of scientific learning throughout the Middle Ages remained the church. And the involvement of the church in the promotion of scientific activity takes us to the heart of an old and powerful stereotype that continues to dominate popular conceptions of the nature of medieval science. The need to dispose of that stereotype and replace it with a cautious, defensible account of the relationship between the church and its theology, on the one hand, and the scientific enterprise, on the other, is essential not only to our mission of educating the public, but also to our own understanding of major portions of medieval scientific activity. To this task I would like to devote the remainder of the present essay—not because I can defend the claim (which I have occasionally made in reckless moments) that it is *the* most critical problem facing historians of medieval science, but because it is an important problem about which I think I have something to say.

III. SCIENCE AND THE MEDIEVAL CHURCH

Much of the energy expended by seventeenth-, eighteenth-, and nineteenth-century writers who reviled the medieval scientific achievement was focused on the medieval church. From Francis Bacon to Andrew Dickson White, the church and its theology were blamed in some degree for the dismal state of medieval science. Bacon mourned the devotion of men of "wit" to theology. Voltaire wrote of the deleterious effects of clerical power and authority on medieval thought. Condorcet claimed that "the triumph of Christianity was the signal for the complete decadence of philosophy and the sciences." And Andrew Dickson White maintained that "the establishment of Christianity, beginning a new evolution of theology, arrested the normal development of the physical sciences for over fifteen hundred years." As for the "Thomist synthesis," in White's view the "alliance between religious and scientific thought" crafted by Thomas Aquinas "became the main path for science during ages, and it led the world ever further and further from any fruitful fact or useful

---

[17] Cadden, "Critical Problems in Medieval Science" (cit. n. 15), pp. 8–13. For the quoted phrase see Nicole Oresme, *Le livre du ciel et du monde*, ed. A. D. Menut and A. J. Denomy (Madison: Univ. Wisconsin Press, 1968), p. 731 (quoted by Cadden, p. 12).

method."[18] By the early years of the twentieth century, the opinion that the church was the villain in the affair had become a widely held article of faith.

Pierre Duhem, of course, was not an adherent of that particular faith. He labeled the claim that Christianity and science were fundamentally incompatible an "audacious lie" and sought to identify a constructive role played by the church and Christian theology. He came eventually to view the condemnation of 219 propositions (many of them scientific) by the bishop of Paris, Etienne Tempier, in 1277 as an attack on entrenched Aristotelian natural philosophy and, consequently, as the birth certificate of modern science. In his *Etudes sur Léonard de Vinci* he argued that

> Christian orthodoxy therefore required, it seems, the renunciation of various principles of peripatetic physics, especially the impossibility of void, the immobility of the world, and the necessity that this world be unique. Affirmed by the condemnations, which had been supported by the doctors of the Sorbonne, these requirements were accepted not only at Paris but also at Oxford; they impressed on scholastic science, in France as well as in England, a new orientation that obliged it to deviate from the Aristotelian tradition at many points. . . . If we must assign a date to the birth of modern science, we would without doubt choose this year 1277, in which the bishop of Paris solemnly proclaimed that several worlds could exist and that the collection of celestial spheres could, without contradiction, be animated by a rectilinear movement.[19]

The church was thus implicated at the deepest level in the events that led to the abandonment of Aristotelian physics and cosmology in favor of new (and, in the long run, modern) physical and cosmological principles.

The *popular* twentieth-century response to Duhem's arguments and those of his adversaries has been approximately what we should have expected. There has been a fair amount of scuffling, at a quasi-scholarly level, between critics and defenders of the church. Thus Bertrand Russell, in his *Religion and Science* (1935), reasserted Andrew Dickson White's warfare thesis; and a variety of Christian apologists, most belligerently Stanley Jaki, have extended and defended Duhem's position.[20] Moreover, both opinions have been effectively disseminated in communities favorably disposed to their respective messages—though it appears to me that outside of conservative Christian circles the warfare thesis is overwhelmingly dominant.

But the response of *professional* historians of medieval science has been curiously muted. We all acknowledge (when pressed) that there *was* a Christian context, and we deal with it when we must, but in general the community of historians of medieval science has, since Duhem, steered a course away from religious issues. I am fully aware that this is a very broad, and therefore risky, generalization, to which there are many exceptions. I know that there is a very good and growing research

---

[18] Marquis de Condorcet, *Sketch for a Historical Picture of the Progress of the Human Mind,* ed. Stuart Hampshire, trans. June Barraclough (London: Weidenfeld & Nicolson, 1955), p. 72; and Andrew Dickson White, *A History of the Warfare of Science with Theology in Christendom,* 2 vols. (New York: Appleton, 1896), Vol. I, pp. 375, 381. On Bacon, Condorcet, and Voltaire see Lindberg, "Conceptions of the Scientific Revolution" (cit. n. 1), pp. 5–9.

[19] Pierre Duhem, *Etudes sur Léonard de Vinci,* 3 vols. (Paris: Hermann, 1906–1913), Vol. I, p. 412.

[20] Bertrand Russell, *Religion and Science* (Oxford: Oxford Univ. Press, 1961); Stanley Jaki, *The Road of Science and the Ways to God* (Chicago: Univ. Chicago Press, 1978); Jaki, *The Origin of Science and the Science of its Origin* (South Bend, Ind.: Regnery Gateway, 1978); and Jaki, *The Savior of Science* (Washington, D.C.: Regnery Gateway, 1988). For two critiques of Jaki's position see Ronald L. Numbers' review of *The Road of Science* in *Church History,* 1981, *50:*356–357; and David C. Lindberg, review of *The Savior of Science,* in *Isis,* 1990, *81:*538–539.

literature dealing with specific instances of the interaction between medieval science and medieval Christianity, and I can think of a few attempts at broader analysis. Amos Funkenstein's bold and provocative *Theology and the Scientific Imagination from the Middle Ages to the Seventeenth Century* (1986) is an example of the latter, though I judge its center of gravity to be in the seventeenth century rather than the Middle Ages and its conceptual focus to be fairly narrow. Nicholas Steneck's *Science and Creation in the Middle Ages: Henry of Langenstein on Genesis* (1976) is also a step in the right direction, though again there are important limitations.[21] If you go looking for a book that offers a full, responsible, and reasonably up-to-date analysis of the relationship between science and religion in the Middle Ages, I do not think you will find it. Indeed, forget the "up-to-date" criterion: go in search of any book written since Duhem that offered what would have been judged a full and responsible account as of its publication date, and I do not believe you will be able to find that either.

If this is true, it raises an obvious question: why have we tended to sidestep the role of religion? I cannot read minds, but I have hunches, which I am happy to report. In the first place, the tendency to give religion a low profile in our scholarship is, I would suggest, a product of the internalist program that has dominated the historiography of medieval science, combined with the perception of Christianity as an external force—an intrusive element not properly part of the story, a loud and occasionally disruptive spectator rather than a legitimate player in the game. Reinforcing this view of the matter is a positivist conception of science, no doubt rapidly disappearing among professional historians of science, but sufficiently strong over the past fifty years to have shaped much of the scholarship on which all of us must currently rely.

Second, I do not think there can be any question that throughout most of the twentieth century there has been a persecution complex or a siege mentality among historians of medieval science. This frame of mind can be explained historically as the product both of centuries of abuse, going back to the humanists of the Renaissance, and of the running battle, waged throughout the twentieth century with historians of early modern science, over the relative importance of medieval and early modern contributions to the origins of modern science. Perhaps also we have harbored the suspicion or the fear that Whewell, White, and the others who believed that medieval religion had a deleterious effect on medieval science were right. Whatever the exact sources of this siege mentality, the natural reaction to it has been to look for ways of rehabilitating medieval science; and one of the most obvious ways of doing that has been to shuck off the religious husk and concentrate on the kernels of real (= modern) scientific achievement. In short, if religion is the major source of embarrassment for medieval science, as traditionally portrayed, then the obvious solution is to reduce the visibility of religion by treating science as an autonomous intellectual pursuit.

But the obvious solution proves to be an inadequate solution. In the first place, those who wish to portray medieval science as an intellectually autonomous activity

---

[21] Amos Funkenstein, *Theology and the Scientific Imagination from the Middle Ages to the Seventeenth Century* (Princeton: Princeton Univ. Press, 1986); and Nicholas H. Steneck, *Science and Creation in the Middle Ages: Henry of Langenstein (d. 1397) on Genesis* (Notre Dame, Ind.: Univ. Notre Dame Press, 1976).

cannot merely assume it to have been so, but must demonstrate that it *was* so and show us how this extraordinary state of affairs came into being. Although I do not believe this can be accomplished, I would find the attempt interesting. Second, insofar as medieval science was not autonomous, but flourished by virtue of service to church, state, or other patron, we need to undertake a close examination of the power relationships, considering how they affected the practice and the content of science.

This is a plea, then, for a broader, more inclusive historiography of medieval science—specifically, a historiography that acknowledges the pervasive influence of Christianity within medieval Christendom and makes an attempt to come to terms with the impact of Christian thought, practice, and institutions on the content and practice of natural philosophy. It is not a repudiation of textual scholarship or studies of technical detail (both of which I value highly). It *is* the claim that if we stop there, if we try to write the history of medieval science without paying attention to the religious or theological presence, we will inevitably render a partial and distorted picture of the medieval scientific enterprise. And, of course, we need to avoid the extremes: the one extreme, of denying or apologizing for Christian influence; and the other extreme, of overstating the extent and power of that influence. Finally, a point that requires special stress because of the frequency with which it is denied or overlooked: the product of our labors will be seriously flawed if we fail to acknowledge Christianity as a legitimate player in the game.[22] This may not be true of the game of science as it is played in the twentieth century, but surely we must forgive medieval scholars for playing by the medieval rules and judge their achievement accordingly.

IV. TOWARD AN UNDERSTANDING OF THE ENCOUNTER BETWEEN CHRISTIANITY AND SCIENCE

If I wanted to play it safe, I would stop here; having identified some critical problems confronting historians of medieval science, I have perhaps fulfilled my mandate and could call it quits with a clear conscience. But this is not the first time I have called for serious scholarly effort on the interaction between science and religion in the Middle Ages. And I feel obliged, therefore, to go a step further and sketch the shape of the history that I think might result. I would certainly not be so bold as to suggest that I can solve this particular problem; that will occur only when we as a community of scholars have written a nuanced history of medieval science with religion as an integral part of the story. But perhaps I can point the way with a series of proposals about the ways in which I believe science and Christianity interacted during the medieval period. I present these proposals not as an exhaustive list or as final answers, but in the spirit of dialogue. And, owing to limitations of space and time, I

---

[22] By *legitimate* I mean simply that nothing in the world view or the social fabric of the Middle Ages would have led medieval scholars to deny the church and its theology an epistemological role or a substantial degree of epistemological authority with regard to questions about the visible world. Even as allegedly modern a thinker on these matters as Galileo did not dispute the church's epistemological authority, but only its exegetical principles. See the study of this question by Richard J. Blackwell, *Galileo, Bellarmine, and the Bible* (Notre Dame, Ind.: Univ. Notre Dame Press, 1991), esp. p. 77.

can present them only in skeletal form, without the historical flesh that would, I hope, lend them greater plausibility.[23]

Science and the Classical Tradition

It is important at the outset to define the expression *medieval science*. As a first approximation, I employ "medieval science" and its synonym "medieval natural philosophy" to refer to both a process and a product: the process by which Western scholars received, assimilated, criticized, modified, and extended the fruits of Greek and Muslim thought about nature; and the intellectual products of this process, expressed in lectures and texts.

There are no doubt many qualifications that would improve this definition, but two stand out. First, I am happy to acknowledge that indigenous European elements (folk medicine, for example) were from time to time incorporated into this largely classical natural philosophy. Second, in identifying reception and assimilation as central features of medieval science, I do not mean to suggest that the medieval scholar was incapable of exploring nature with his own eyes and intellect, or that he never did so, but to make clear that scholars in Western Christendom learned how to do science—that is, how to think about nature—largely by imitating their Greek and Muslim predecessors.[24] This linkage to the classical tradition was undoubtedly the fundamental, defining characteristic of the great majority of medieval science— a point that requires stress because of persistent attempts by Christian apologists such as Stanley Jaki to demonstrate that the decisive elements of the medieval scientific tradition were borrowed from the Christian, rather than the classical, tradition. My primary object in the remainder of this paper, therefore, is to understand the continuation, appropriation, and modification of the classical tradition. And the question is how that process was influenced by the Christian context within which it occurred.

Science as Handmaiden

If the most prominent feature of medieval scientific knowledge was its classical origin, surely its second most conspicuous characteristic was the "handmaiden" status by which it was frequently justified. St. Augustine played the critical role in defining this handmaiden status and helping it to become a central feature of the medieval world view. Augustine admonished his readers to set their affections on

---

[23] The argument that follows has had several oral presentations: at a colloquium sponsored by the Commission Internationale d'Histoire Ecclésiastique Comparée, Geneva, Switzerland, August 1989 (published as "Science and the Medieval Church: A Preliminary Appraisal," in *Les églises face aux sciences du moyen âge au XXᵉ siècle: Actes du colloque de la Commission Internationale d'Histoire Ecclésiastique Comparée tenu à Genève en août 1989,* ed. Olivier Fatio [Histoire des idées et critique littéraire, 300] [Geneva: Droz, 1991], pp. 11–27); as the Richard S. Westfall Lecture, Bloomington, Indiana, March 1990; as a fireside chat, Bellagio Study Center of the Rockefeller Foundation, April 1991; and as the keynote address at the 50th anniversary meeting of the American Scientific Affiliation, Wheaton, Illinois, July 1991. I am grateful for criticisms and suggestions received at these public lectures. Some of the prose here is borrowed, with revisions, from "Science and the Medieval Church."

[24] I do not use *imitating* pejoratively; imitation, after all, is as central to the practice of modern as of medieval science.

the celestial and eternal, rather than the earthly and temporal, and warned against attachment to temporal knowledge for its own sake. But Augustine also acknowledged that the temporal could serve the eternal and chastised those who have opened Christianity to ridicule by refusing to take knowledge about nature from those who possess it:

> Even a non-Christian knows something about the earth, the heavens, and the other elements of this world, about the motion and orbit of the stars and even their size and relative positions, about the predictable eclipses of the sun and moon, the cycles of the years and the seasons, about the kinds of animals, shrubs, stones, and so forth, and this knowledge he holds to as being certain from reason and experience. Now it is a disgraceful and dangerous thing for an infidel to hear a Christian, presumably giving the meaning of Holy Scripture, talking nonsense on these topics.

Accordingly, "if those who are called philosophers . . . have said things that are indeed true and well accommodated to our faith, they should not be feared; rather, what they have said should be taken from them as unjust possessors and converted to our use." The pagan sciences are neither to be loved nor to be repudiated; rather, they are to be appropriated, disciplined, and put to use. The sciences must accept subordinate status, as the handmaidens of religion and theology.[25]

Not everybody agreed with Augustine, of course, but his view came to be widely adopted throughout Christendom. It dominated scientific activity during the early Middle Ages and continued to be reasserted thereafter, whenever the pursuit of natural philosophy required justification. Even a combative eccentric like the thirteenth-century Franciscan Roger Bacon, notorious for his repudiation of authority and scholastic method, devoted a scholarly lifetime to reclaiming pagan learning for the faith—that is, to demonstrating that, purged of a few errors, Greek and Arabic science and mathematics could serve as the disciplined handmaidens of the church. "There is one perfect wisdom," Bacon argued in his *Opus maius,* "and this is contained in holy Scripture, in which all truth is rooted. I say, therefore, that one science is mistress of the others—namely theology, for which the others are integral necessities and which cannot achieve its ends without them. And it lays claim to their virtues and subordinates them to its nod and command." And just to be sure that nobody missed the point, Bacon concluded: "Every investigation of mankind that is *not* directed toward salvation is totally blind and leads finally to the darkness of hell."[26]

If Bacon's position seems simple and unambiguous, the handmaiden formula as applied broadly to the Middle Ages surely is not. There are many complications. One of them is that the formula covers a spectrum of attitudes, running from what

---

[25] Augustine, *The Literal Meaning of Genesis,* trans. John Hammond Taylor, S. J., 2 vols. (New York: Newman, 1982), Vol. I, pp. 42–43; and Augustine, *On Christian Doctrine,* trans. D. W. Robertson, Jr. (Indianapolis: Bobbs-Merrill, 1958), p. 75. It may be worth noting that Augustine's use of the feminine term *handmaiden* here had nothing to do with notions of feminine inferiority but probably derived simply from the grammatical gender of the Latin noun *philosophia;* the mistress, *theologia,* was also feminine. On Augustine's attitude toward the natural sciences see David C. Lindberg, "Science and the Early Church," in *God and Nature: Historical Essays on the Encounter between Christianity and Science,* ed. David C. Lindberg and Ronald L. Numbers (Berkeley/Los Angeles: Univ. California Press, 1986), pp. 26–28, 31, 35–38; and Lindberg, "Science as Handmaiden: Roger Bacon and the Patristic Tradition," *Isis,* 1987, 78:522–524.

[26] Roger Bacon, *The Opus Majus of Roger Bacon,* ed. John Henry Bridges, 3 vols. (London: Williams & Norgate, 1900), Vol. III, p. 36. For a full discussion of Bacon's campaign on behalf of the new science see Lindberg, "Science as Handmaiden," pp. 527–535.

is sometimes called the "Tertullian view," which was fearful of philosophical or scientific effort and tended to keep it on a very short leash, to the far more liberal position of Thomas Aquinas, who felt that the handmaiden named "philosophy" had amply demonstrated her usefulness and her reliability and therefore should be offered enlarged responsibilities, elevated status, and increased freedom.[27] We need to keep this diversity of opinion in mind if we are to do justice to the complexity of the historical situation.

The picture is also complicated by the need to distinguish two aspects of handmaiden status. The first is patronage: the support and encouragement offered by Christian theology and the church for scientific activity. The second is supervision: the exercise of control over the resulting activity and its intellectual products. And how are these two aspects of handmaiden status linked? Can there be patronage without supervision? No doubt, but such a state of affairs is surely exceptional. What is *not* exceptional is patronage with *varying degrees* of supervision. Aquinas, for example, wanted to reduce the level of supervision, because he believed the handmaiden would perform more effectively under those conditions.

And finally, when we think about the handmaiden formula, we need to distinguish between justification and motivation—the former comprising the reasons available to justify the existence, support, and practice of natural philosophy (most likely to be rolled out, for defensive purposes, when natural philosophy was under attack), the latter consisting of the factors that explain an individual's choice of natural philosophy as a vocation or an avocation (or, more narrowly, the individual's decision to embark on a specific scientific investigation). The handmaiden formula was invoked primarily on behalf of the former. As for the latter, motivation (because of its personal and affective character) is notoriously difficult to get a handle on, and all we can say with assurance is that most medieval natural philosophers must have devoted themselves to the study of nature because it seemed to possess some kind of general social or religious sanction, because it offered suitable rewards (income, status, personal satisfaction, and the like), and because they were good at it.

With these complications and qualifications in mind, then, the proposition that I am defending is simply this: that the overwhelming majority of medieval achievements in natural philosophy, including many of those that we now most admire, emerged from activities justified in the minds of both the participants and their patrons by the Augustinian formula of science (or natural philosophy) as handmaiden.[28] Whether handmaiden status was beneficial or detrimental is a question treated below.

The Implications of Handmaiden Status

What are we to make of this handmaiden status accorded to scientific knowledge? Specifically, does handmaiden status represent a blow against the scientific

[27] For the Tertullian view see Etienne Gilson, *Reason and Revelation in the Middle Ages* (New York: Scribners, 1938), pp. 5–11; for Aquinas's position see Lindberg, *Beginnings of Western Science* (cit. n. 12), pp. 231–234.

[28] I am convinced that considerably more thought needs to be given to the justification for scientific activity in the late medieval period (the fourteenth and fifteenth centuries), where the evidence is less plentiful and more ambiguous than for the early and high Middle Ages. Were there, for example, secularizing forces gaining strength in the late medieval period that need to be introduced into the picture?

enterprise? or modest, but welcome, support? If we compare the support for science offered by the medieval church (and its educational system) with that provided by the government of a modern Western democracy, the church will appear to have failed abysmally. But such a comparison is obviously unfair. If, instead, we compare the support given to the study of nature by the medieval church with the support available from any other contemporary social institution, it will become apparent that the church was *the* major patron of scientific learning. Its patronage was no doubt limited and selective, but limited and selective patronage is presumably better than no patronage at all. Remove the church from the picture, and the Middle Ages would not have been transformed into a secular, scientific utopia, fully equipped with governmental agencies for the support and encouragement of scientific research. Remove the church from the picture, and there is an enormous amount of serious intellectual activity that would not have occurred.

But a critic determined to view the medieval church as an obstacle to scientific progress might argue that the handmaiden status accorded the scientific enterprise is inconsistent with the existence of genuine science. True science, this critic would maintain, cannot be the handmaiden of anything, but must be totally autonomous; consequently, the subordinate, "disciplined" science that Augustine and Aquinas sought is no science at all.

The reply to this critic is really quite simple: a scientific enterprise that is *simultaneously* the recipient of social support *and* autonomous (a case of patronage without supervision) is an extremely rare creature—existing mainly in the minds of mythmakers. In the real world, to acquire social support is to give up some measure of autonomy. And most important developments in the history of science have emerged *not* from autonomous science (certainly not from scientific activity that was viewed by its *patrons* as autonomous), but from science in the service of some ideology, social program, or practical end. The question throughout most of the history of Western science has not been *whether* science will function as handmaiden, but *which* mistress it will serve.

What was the effect of handmaiden status on the content and methodology of medieval science? The handmaiden named "science" or "natural philosophy" was given considerable latitude in the performance of her duties, as I shall argue below, but there were also significant constraints; there can be no question that handmaiden status made a difference. These constraints took the form of fundamental theological assumptions and boundary conditions. Christian theology offered a broad framework within which medieval scholars endeavored to interpret the whole of human experience. When the universe was conceived, it was conceived in Christian terms, as the creation of a benevolent, omnipotent God, inscrutable in his ways. The universe was thought to be contingent on divine will and pervaded with purpose and providence. Truth was defined in theological terms, and certain pieces of it (including truths about nature) were held to have been given in Scripture.

But it is difficult to get much mileage out of a claim as general as this. If we want to understand more specifically what it meant for theology to impinge on the content of science, we need to distinguish between different sorts of scientific subject. In *technical* subjects such as mathematics, astronomy, optics, meteorology, medicine (in its more technical manifestations), and natural history, there were virtually no constraints and no limitations. When Greek and Arabic literature on these subjects became available in the course of the twelfth and thirteenth centuries, it contained no unpleasant theological or philosophical surprises; it was superior to anything Latin Christendom currently possessed; it filled an intellectual void; and it received an enthusiastic welcome. If theology impinged at all on such disciplines, it did so in ways that were either beneficial or inconsequential for the natural philosopher. For example, theological considerations may have called attention to the importance of astronomical or optical studies, but they did little or nothing to influence theoretical outcomes in either discipline. The medieval scholar could follow reason or inclination wherever it led and defend almost any position he wished.

Insofar as there was conflict, it appeared in the *broader disciplines* that impinged on world view—specifically, metaphysics, cosmology, psychology, and (to a modest degree) physics. Here a genuine clash of world views was possible; and it is not difficult to find instances of Christian theology adjudicating questions of natural philosophy. For example, Christian theology cast suspicion on a variety of Platonic and Aristotelian tendencies (such as Platonic or Neoplatonic pantheism and Aristotelian determinism); and it certainly placed a damper on any inclination to defend materialistic philosophies like that of the ancient Epicureans. Through the doctrine of creation, it exerted a controlling influence on cosmogonical speculations, including the question of the eternity of the world; it affected attitudes toward astrology (of which it generally disapproved), touched deep metaphysical issues (the relations between accidents and their subjects) through the doctrine of the Eucharist, placed serious constraints on what one thought about miracles, influenced theories of motion and space, and impinged powerfully on theories of natural law (where it gave rise to the idea of natural laws extrinsically imposed by an omnipotent Creator). In severe cases, such as the nature of the soul and the question of the eternity of the world, the scholar found himself struggling to reconcile principles of Christian theology with the philosophical presuppositions of Greek and Arabic natural philosophy. In short, Christian theology was an omnipresent reality, performing a steady selective and shaping function.[29]

The Church as a Coercive Agent

Does it follow that the church employed coercive measures? Church authorities were certainly capable of coercion when they judged vital theological doctrines to be at stake, as we see from a long series of theological condemnations. And scientific ideas were sometimes implicated as well; Tempier's notorious condemnation of 1277 included philosophical or scientific doctrines that appeared to him and his advisors to impinge dangerously on the doctrines of divine freedom and omnipotence or to cast doubt on the freedom of the human will. But the level and frequency

---

[29] There is no adequate treatment of these subjects, but for useful material see, e.g., John E. Murdoch, "The Development of a Critical Temper: New Approaches and Modes of Analysis in Fourteenth-Century Philosophy, Science, and Theology," *Medieval and Renaissance Studies,* 1978, 7:51–79; Funkenstein, *Theology and the Scientific Imagination* (cit. n. 21); Steneck, *Science and Creation* (cit. n. 21); James McEvoy, *The Philosophy of Robert Grosseteste* (Oxford: Clarendon Press, 1982); John E. Murdoch and Edith D. Sylla, eds., *The Cultural Context of Medieval Learning* (Boston Studies in the Philosophy of Science, 26) (Dordrecht: Reidel, 1975); Pierre Duhem, *Medieval Cosmology: Theories of Infinity, Place, Time, Void, and the Plurality of Worlds,* trans. Roger Ariew (Chicago: Univ. Chicago Press, 1985); James A. Weisheipl, O. P., *Nature and Motion in the Middle Ages,* ed. William E. Carroll (Washington, D.C.: Catholic University of America Press, 1985); Edward Grant, *Studies in Medieval Science and Natural Philosophy* (London: Variorum Reprints, 1981); and Lindberg, *Beginnings of Western Science* (cit. n. 12).

of interference in the scientific enterprise were far less than the stereotype of the Middle Ages would suggest. Some famous cases of alleged scientific coercion had nothing to do with science. Despite a widespread popular belief that Roger Bacon was imprisoned for his attack on authority and his urgent assertion of a novel scientific methodology, for example, Bacon in fact represented very old methodological traditions, and his imprisonment, if it occurred at all (which I doubt), probably resulted from his sympathies for the radical "poverty" wing of the Franciscan Order (a wholly theological matter), rather than from any scientific novelties that he may have proposed.[30]

In fact, medieval natural philosophers had remarkable freedom of thought and expression—a level of freedom that compares favorably with that available in European universities of the sixteenth and seventeenth centuries. Surely no *scientific* or *philosophical* orthodoxy was imposed on them by the church. All medieval scholars knew that there were theological limits, and knew approximately where they were, but these undoubtedly seemed broad and reasonable rather than narrow and restrictive. And as long as they stayed clear of a handful of sensitive theological doctrines, scholars could think and say almost anything they wished. To give but a single example, Aristotle's philosophy no sooner became available in the twelfth and thirteenth centuries than philosophers in the medieval universities embarked on an elaborate process of scrutiny and criticism, in which every claim Aristotle made was fair game. Indeed, the church encouraged and accelerated the process of criticism when it condemned aspects of Aristotelian philosophy in 1277.

But (I hear my reader murmur) how can we talk about freedom of expression and theological limits in the same breath? Can there be limits without coercion? This is a delicate question, and the answer no doubt depends on what we mean by "limits" and by "coercion"; but I think the answer is a qualified yes. By and large, Christian theology was not imposed on medieval scholars by force, but imparted by persuasion. They subscribed to it because they agreed with it; and they agreed with it because it provided a satisfying framework within which to interpret their experience of the world. That the religious authorities felt free (or obliged) to resort to coercion when persuasion failed in no way contradicts the claim that the overwhelming majority of medieval scholars were never tempted to overstep the bounds of orthodoxy and therefore never experienced the disapproval, let alone the coercive power, of the church.

Insofar as there *were* struggles, as there certainly were, we must not imagine the medieval scholar to have been a hapless victim, trapped in an unequal battle with the religious establishment; or caught in the crossfire between impersonal forces of Christian theology, on the one side, and classical philosophy and science, on the other. If there was any crossfire, it was between scholars; and if scholars were the victims of the conflict, they were also its perpetrators. Christian theology and

---

[30] On the condemnation of 1277 see Fernand van Steenberghen, *Aristotle in the West,* trans. Leonard Johnston (Louvain: Nauwelaerts, 1955); Steenberghen, *Thomas Aquinas and Radical Aristotelianism* (Washington, D.C.: Catholic Univ. of America Press, 1980); and Roland Hissette, *Enquête sur les 219 articles condamnés à Paris le 7 mars 1277* (Louvain: Publications Universitaires, 1977). On Bacon, see Lindberg, "Science as Handmaiden" (cit. n. 25). For those who find this characterization of Bacon surprising, I would like to explain that the Bacon myth refuses to die because for centuries Bacon has served as a symbol of the destructive power of religion on intellectual freedom, and those who hold this opinion—whether they hold it as serious scholars or as casual observers— are extremely reluctant to relinquish one of their most potent illustrations.

the classical scientific tradition existed as living traditions only insofar as there were scholars prepared to defend theological and scientific claims. These scholars entered the fray of their own volition, and defended their respective views with all of the ability they could muster and with no less sincerity than one finds in the average twentieth-century academic. The scholars who struggled to integrate Christian theology and the classical tradition of natural philosophy did not do so under compulsion from somebody in a position of authority; the initiative was theirs.

The Naturalization of Greek and Arabic Science

A funny thing happened on the way to modern science. In a turn of events that I do not believe anybody could have predicted, the church moved from wary and sometimes hostile observer of the classical tradition to become the patron of Greek and Arabic natural philosophy. It did so, in the first place, by presiding over a synthesis of Christian theology and classical natural philosophy. Although many details of this process remain obscure, the broad outlines are well known. Philosophers and theologians of the thirteenth and fourteenth centuries—including Robert Grosseteste, Roger Bacon, Albert the Great, Thomas Aquinas, Jean Buridan, and Nicole Oresme (to name a few of the more prominent figures)—tackled the problem of accommodating Greek and Arabic natural philosophy to Christian theology, *and* vice versa.[31] They were so successful in achieving their goal that by the sixteenth century Aristotle had taken on the appearance of a Christian saint. The remarkable fact, of which we must never lose sight, is that in the end the church did *not* reject the classical tradition, but appropriated it, creating what we must regard as a new world view and setting Christendom on a course that had profound implications for the subsequent history of Western science and, indeed, for the history of civilization.

The church became the patron of Greek and Arabic natural philosophy, in the second place, through its support of the schools and universities, which (in northern Europe) were under its authority and protection. In those universities natural science found a secure institutional home for the first time in history and became the common intellectual property of the educated classes—a position from which it would never be dislodged. Specifically, Aristotelian natural philosophy became a central feature of the arts curriculum, while Galenic medicine became entrenched in the faculty of medicine, and the mathematical sciences survived on the periphery.[32]

[31] The classics on this problem are by Etienne Gilson, including *History of Christian Philosophy in the Middle Ages* (New York: Random House, 1955); *Le Thomisme: Introduction à la philosophie de saint Thomas d'Aquin,* 5th ed. (Paris: Vrin, 1948); and *Reason and Revelation in the Middle Ages* (cit. n. 27). See also David Knowles, *The Evolution of Medieval Thought* (New York: Vintage, 1964).

[32] For introductory studies of the medieval universities see John W. Baldwin, *The Scholastic Culture of the Middle Ages, 1000–1300* (Lexington, Mass.: Heath, 1971); and William J. Courtenay, *Schools and Scholars in Fourteenth-Century England* (Princeton: Princeton Univ. Press, 1987). On the arts curriculum see also James A. Weisheipl, O.P., "Curriculum of the Faculty of Arts at Oxford in the Fourteenth Century," *Mediaeval Studies,* 1964, *26:*143–185; Weisheipl, "Developments in the Arts Curriculum at Oxford in the Early Fourteenth Century," *Mediaeval Studies,* 1966, *28:*151–175; Guy Beaujouan, "Motives and Opportunities for Science in the Medieval Universities," in *Scientific Change,* ed. A. C. Crombie (London: Heinemann, 1963), pp. 219–236; and the relevant articles in J. I. Catto, ed., *The Early Oxford Schools,* Vol. I of *The History of the University of Oxford,* ed. T. H. Aston (Oxford: Clarendon Press, 1984). On the curriculum of continental universities see Pearl Kibre, "The *Quadrivium* in the Thirteenth Century Universities (with Special Reference to Paris)," in *Arts libéraux et philosophie au moyen âge: Actes du quatrième congrès international de philosophie médiévale, Université de Montréal, 27 août-2 septembre 1967* (Montréal: Institut d'études médiévales, 1969), pp. 175–191; and Nancy G. Siraisi, *Arts and Sciences at Padua: The Studium of Padua*

Grading the Medieval Church

Finally, we come to the question of merit. Did Christianity ultimately make a positive or a negative contribution to the scientific movement? Was medieval science better off or worse off because of the Christian auspices under which it was pursued? I raise this question only because long experience has taught me that almost everybody, when confronted with an account of science and religion in the Middle Ages, demands an answer to it.

But they won't get an answer from me. This is one of those impossible, counterfactual, "what if" questions to which there is no acceptable answer except that we do not know. Under what auspices would the large quantity of science patronized by the church have been pursued if there had been no church? Who can say—unless we are prepared to conclude that it was either the auspices of the church or no auspices at all? What would the Middle Ages have been like without Christianity? We will never know; we can only be certain that they would not have been *the* Middle Ages that we have all come to know and love.

However, the question is not merely unanswerable; it is also pernicious. It is pernicious, in the first place, because it invites us to identify villains and victims or benefactors and beneficiaries; and that, in turn, is an invitation to grade the past on a scale of modern values, reflecting twentieth-century (or the historian's own) conceptions of the proper epistemological status and the appropriate power relations of science and religion.

It is pernicious, in the second place, because it buys into the conflict or warfare model developed by Renaissance humanists, elaborated in the course of the eighteenth and nineteenth centuries, and accepted as the appropriate framework within which to view the problem even by those who deny the humanist answer. Even Duhem and those who share his view of the constructive role played by the medieval church affirm the categories within which the traditional condemnation of Christianity was expressed: not conflict, but its negation or opposite, cooperation; not war but peace. But this conflict model seems to me a totally inadequate framework for analyzing the relationship between science and religion in the Middle Ages. I do not, of course, deny the existence of conflict or its converse, cooperation, on an occasional, episodic, specific, and individual basis. But I do not believe they were the most important forms of interaction. I would like to propose *accommodation* as an alternative model. And I would like to stress that the historical actors between whom this accommodation took place were not impersonal reifications of science and Christianity, but individual scholars and churchmen (many of whom were *both* scholars *and* churchmen), struggling honestly, and with all of the resources at their disposal, to solve a collection of exceedingly difficult problems. The goal of these medieval scholars was to remain true to theological principles in which they deeply believed, while heeding the explanatory power of inherited philosophical systems and the testimony of the senses. And their achievement was an intricate process of accommodation and adaptation, which deeply shaped both scientific and theological belief.

---

before 1350 (Toronto: Pontifical Institute of Mediaeval Studies, 1973). On the medical curriculum see Siraisi, *Taddeo Alderotti and His Pupils: Two Generations of Italian Medical Learning* (Princeton: Princeton Univ. Press, 1981); and Siraisi, *Medieval and Early Renaissance Medicine* (Chicago: Univ. Chicago Press, 1990), Ch. 3.

The scientific theories and practices that emerged in the early modern era, then, were not primarily the offspring of Christian theological principles, nor of the classical tradition (once Christianity "got off its back"), but of a complex amalgam to which Christian theology, the classical tradition, various medieval institutions, and many other cultural traditions and social forces all contributed. It is to this enormously complicated interaction that we must look if we wish to understand the essential character of the medieval scientific achievement, and the parentage of the world in which we live.

History of East Asian Science: Needs and Opportunities

By *Nakayama Shigeru**

OBJECTIVE AND VALUE-FREE SCHOLARSHIP is no more possible in the history of science than in any other field. This essay will describe and assess assumptions and values reflected in the historiography of East Asian science. My own viewpoint is that of a historian educated in both Japan and the United States and experienced in research on the history of science, ancient and modern, in China, Japan, and the West.

The vague if long-standing dichotomy between East and West has generated much shallow debate. In what follows, when I speak of Eastern and Western science, I refer to complementary, well-defined traditions. The historical center of Western science moved from Babylonia to classical Greece to Hellenistic Egypt, India, and the Middle East, and thence to medieval Europe, thus to the modern-day Western world. In the East the center remained in China until the period of European expansion; Korea, Japan, and Vietnam remained cultural satellites. Hence, unlike Indian and Arabic sciences, from which the main current of Western scientific thought developed, China and East Asia provide us with an independent counterculture of science.

I. A TYPOLOGY OF WESTERN APPROACHES TO EAST ASIAN SCIENCE

George Sarton and the Orientalists

When George Sarton, shortly after founding *Isis* in 1913, initiated the *Isis Critical Bibliography,* he devoted a section each to Chinese and Japanese science and invited librarians from China and Japan working in the United States to contribute. He had tried to learn East Asian languages without much success, but he remained curious about and respectful of Far Eastern scientific developments. He did not venture to make generalizations about them, believing that "although I do not know anything about East Asia, I am sure that something important was going on there, and want to know about it."[1] This attitude was shared by the first generation of Western specialists on premodern East Asian science, who began working in this field because they were fascinated by the unique and exotic features of ancient non-Western cul-

---

* STS Center, School of International Business Management, Kanagawa University, Hriatsuka, Japan.

Indexers please note: surnames of all East Asian scholars, including the author, precede given names in this article and are **boldfaced.**

[1] Personal communications in his final days.

tures. Of this generation of Orientalists, which included Léopold de Saussure, Willy Hartner, and to some extent Henri Maspero, Joseph Needham is the last living representative. The motivations of these scholars were of course highly individual: Saussure's interest was to a large extent antiquarian; Hartner was seeking common themes and issues in the great astronomical traditions, all of which he studied using original sources; Needham's scholarship has served a profoundly ecumenical vision of the unity of science worldwide. (Although such scholars were called "Orientalists," this term came to connote a low estimate of the culture studied and sympathy with imperialist goals. I therefore use the more neutral term *premodernist* below.)[2]

Professionalization and Relegation to the Periphery

After World War II the history of science rapidly became a professional academic field, particularly in the United States. As postgraduate education evolved, the generation of historians of science who came of age in the late 1950s and early 1960s replaced the amateurish charm of their predecessors with narrow specialization in a well-defined discipline. One result is that specialists in Western science without a strong foundation in East Asian languages avoid anything more than passing references to China and Japan in their own writings, leaving it mainly to East Asian specialists, in fear of stepping outside their own familiar field of expertise. They have been made aware that China has its own major technical traditions, thanks to Needham's monumental *Science and Civilisation in China,* the first volume of which appeared in 1954. Nonetheless, as work in the overall discipline has accumulated, few general texts provide serious coverage of Asian science, technology, and medicine, although at least some acknowledge their limitations by including the word *Western* in their titles.[3]

This peripheralization of East Asian science was also institutionalized in the *Isis Critical Bibliography.* When I. Bernard Cohen took over the editorship in 1953 and "modernized" the journal, he combined the coverage of China and Japan in a single section called "The Far East (to c. 1600)." This endpoint of circa 1600 reflected the assumption that since the rise of modern science in the West, technical activity in the rest of the world has conformed to its goals. One can scarcely agree, however, with the implication that the East Asian tradition was replaced immediately. Assimilation did not begin until the late nineteenth century. The intervening seventeenth to early nineteenth centuries pose distinct research problems as Asian and occidental cultures gradually confronted each other, in the arena of science as well as elsewhere.

---

[2] The appearance of Edward W. Said, *Orientalism* (New York: Pantheon, 1989), helped discredit the original term. On Needham see **Nakayama** Shigeru, "Joseph Needham, Organic Philosopher," in *Chinese Science: Explorations of an Ancient Tradition,* ed. Nakayama and Nathan Sivin (Cambridge, Mass.: MIT Press, 1973), pp. 23–43.

[3] E.g., John G. Burke, ed., *Science and Culture in the Western Tradition: Sources and Interpretations* (Scottsdale, Ariz.: Gorsuch Scarisbrick, 1988), a very popular introductory text, which does not devote a single page to Islamic, Indian, or East Asian science. More egregiously, a recent reference work that included a grand total of two pages on China and Japan, three on Islam, and less than two on "Hindu science" out of five hundred does not hesitate to call itself without qualification *Dictionary of the History of Science* (ed. W. F. Bynum, E. J. Browne, and Roy Porter [Princeton: Princeton Univ. Press, 1981]).

(Re)locating Center and Periphery

One issue relevant to the attempt to integrate the East Asian tradition with the Western view of history of science is the issue of parochialism, more particularly, of locating periphery and center. Parochialism can take various forms. Although I have published on the history of premodern and modern European, American, Chinese, and Japanese science, technology, and medicine, I am often treated by Western historians of science as a narrow specialist in Japanese or East Asian science. Admitting that as I reside in Japan, my work can be seen as geographically peripheral, I claim that scholars like myself usually have a wider perspective than those in Berkeley or Paris. Since scholars at the "periphery" know what is going on at the "center," and few of the latter are interested in what is going on at the periphery, it is more correct to see those at the center generally as narrow specialists in the history of Western science.[4] Furthermore, the main current of the Western scientific enterprise is now the common heritage of the history of science profession, capably summarized at the textbook level all over the world. Even in Asia the definition of the discipline draws on the Galilean-Newtonian tradition. Thus East Asian historians of science are involved in the same enterprise as Westerners.

The narrow approach of the "center" reveals itself in other ways. I am often invited to serve on the editorial board of international journals. Many are only nominally international, accepting foreign contributions only if written in acceptable English. *Isis,* for example, which still calls itself an "international review," ceased some years ago to accept articles in any language but English. I often find it difficult to think of their editors, some of whom are extremely narrow in intellectual outlook and unaware of non-European cultures, as "international." It would be better if the editors of international journals were scholars who were able to read a non-Western language as well as being capable editors of English. Perhaps Indian scholars are particularly qualified.

The Modernizers

Although the "Orientalists" were drawn to the special character of early Eastern science, an entirely different motivation for research soon appeared, first in the work of East Asian scholars[5] and then, in slightly altered form, among Western historians and social scientists. These scholars view acquisition of Western science and technology simply as a means of modernizing non-Western societies. Modernization implies conforming to European and American ways and changing any attitude or institution that interferes with doing so. Uncritical modernizers evaluate their subject matter according to how closely it approximates the scientific practices and institutions prevalent in the West. (In this essay I use *modernist* to refer to someone who studies the modern period, and *modernizer* for someone who, explicitly or not, takes this ideological stance toward history, described in more detail below.)[6]

---

[4] See **Nakayama** Shigeru, "The Shifting Centers of Science," *Interdisciplinary Science Reviews,* 1991, *16*(1):82–88.

[5] **Ogura** Kinnosuke's prewar writings on the modernization of Eastern mathematics, such as *Sūgakushi kenkyū* (Studies on the history of mathematics), two vols. (Tokyo: Iwanami, 1935–1948), are a good example.

[6] For an example of the modernizers' conception of science see William Beranek, Jr., and Gustave Ranis, *Science, Technology, and Economic Development: A Historical and Comparative Study* (New

Most Western historians of science unconsciously adopted this view. Until the late 1960s, hardly any questioned whether the criteria of the modernizers were valid for measuring the achievements of non-Western science. They even applied the same criteria to premodern East Asian science. In research of this kind the key question was whether Asian scientists achieved some item of modern knowledge earlier than their European counterparts. To be sure, Joseph Needham reversed the earlier tendency to use priorities as an argument for the inferiority of Asian cultures. He drew on a broad command of the Chinese literature to persuade Western readers that before modern times Eastern technologists were more innovative than Westerners. But the issue remained priorities. Needham's strategy of evaluating ancient Chinese science by modern European criteria paradoxically encouraged most of his followers around the world, including those in China, to accept the modernization perspective uncritically. It undermined the exemplary value of his own passion for comparative study.

Alternative Views of Science

The intellectual climate in the late 1960s and early 1970s, with its critiques of imperialism and the myth of value-free science, encouraged a new view of alternatives to conventional Western science. Structural anthropology provided a promising framework of analysis to account for the emerging Third World on its own terms. At the same time some observers argued that the history of science, by glorifying modern science as the quintessence of the modern European heritage, had become the last stronghold of Western chauvinism.[7]

In these new circumstances I was commissioned in 1977 to present a paper countering the assimilation thesis at the International Congress of the History of Science. Up until the 1960s, the papers I wrote in English for the Western audience were neither understood nor accepted unless they were framed to fit the assumption that the goal of East Asian science should be to assimilate modern Western science. The 1977 essay, entitled "Alternative Science of the East," argued instead that East Asian science should be appreciated in terms of its own unique paradigms, and its development measured by its own yardstick.[8] Although Needham, for example, always maintained optimistically that in the future Eastern and Western science will converge in an ecumenical synthesis of "modern science," as he defines it, I believe that unless external pressure had been applied, the two traditions would have developed along their own normal lines, diverging so much that easy synthesis would have been impossible.[9]

Once we accept this alternative stance, using modern Western science as a yardstick no longer appears attractive. As a leading member of the history of Western

---

York: Praeger, 1978). Modernists are not necessarily modernizers, although in China and Japan the two categories tend to coincide.

[7] It can be pointed out that the history of science is largely the domain of white researchers. Few black scholars attend international conferences in the history of science, perhaps because they, unlike Asians, may find no scientific tradition in their past.

[8] See **Nakayama** Shigeru, "Alternative Science of the East," in *Human Implications of Scientific Advance: Proceedings of the XVth International Congress of the History of Science,* ed. Eric G. Forbes (Edinburgh: Edinburgh Univ. Press, 1978), pp. 36–44. See also Nakayama, *Academic and Scientific Traditions in China, Japan, and the West* (Tokyo: Univ. Tokyo Press, 1984).

[9] Nakayama, "Joseph Needham" (cit. n. 2); and Nakayama, "Alternative Science" (cit. n. 8).

science, recently retired, put it, "Comparison with the West is instructive to Western readers but of less importance than the intrinsic fascination of the Chinese material."[10] This is the approach of the post-Needham generation of Western historians of East Asian science, such as Nathan Sivin. They address their message to the Western academic community of the history of science, by now firmly established, where the thesis of modernization, once taken for granted, is now being critically reexamined. In order to obtain a unique position in the community, they are now trying to understand more deeply than before the intrinsic values and structures of Eastern science. True, traditional native scholars also sought out those intrinsic values; their perspective, however, was not comparative. The "alternativist" generation should be able to make more informed comparisons. This may be the only way to integrate the Asian experience into the world history of science, as Needham originally suggested.[11]

We might well argue that the complementary Eastern and Western traditions defined in the opening paragraphs of this essay form a better-balanced unit for study than the Western tradition alone. This comparative and relativistic approach also has its utility in studying the evolution of Western science. The characteristics of the latter might be better defined and its contours carved more clearly when considered from the perspective of another distinct culture.

Science and Western Expansionism

Modernizers are aware that what convinced the East of Western superiority was not Newton but Newcomen. The East began to modernize not because of the Newtonian paradigm or, more generally, the outcome of the Scientific Revolution in the seventeenth century, but because of military technology and its underlying power, engineering—both products of the Industrial Revolution that made the Western military threat overwhelming.

Whether the Industrial Revolution could have eventuated without the preceding Scientific Revolution can scarcely be considered without studying the non-Western historical experience. Although mechanistic views of nature were embedded in both revolutions in the West, elsewhere, in one country after another, industrial change has swept over societies upon which modern scientific education has had practically no impact aside from the training of a small corps of technocrats. Third-world scholars have generally concluded that the most deeply felt influence of the West was the imperialistic impact of the Industrial Revolution.

As the integration of the European Community proceeds, its ideologists have tended recently to seek a historical basis for a common identity in "European expansion" in the age of the great navigators. In order to avoid an imperialistic tone, they have tried to invent a substitute for the old Eurocentric approach by creating a new

---

[10] A. Rupert Hall, "A Window on the East" (review of Joseph Needham *et al., Science and Civilisation in China*), *Notes and Records of the Royal Society of London,* 1990, *44*(1):108.

[11] Nathan Sivin argues, e.g., that the premodern classification of scientific knowledge is quite different from that of Europe: see Sivin, "Science and Medicine in Chinese History," in *Heritage of China: Contemporary Perspectives on Chinese Civilization,* ed. Paul S. Ropp (Berkeley: Univ. California Press, 1990), pp. 164–197.

dialogue between Europe and the non-Western world, those who did the expanding and those expanded upon.[12]

It might be interesting to create a history of European expansion as seen from Asia. The wealth of native language resources would allow construction of a relatively impartial account, free of imperialistic bias. For instance, the European influence on Japanese science has already been more revealingly studied by Japanese scholars than by Europeans, who have generally made scant use of materials in the Japanese language and have often taken their sources at face value rather than scrutinizing the interests that shaped them.[13]

Non-Western scholars, however, have worked only on the one-way, single channel from Europe to their native land, and rarely have an overview of worldwide influence. European scholars, located at the hub of many radiating channels of influence, are well placed to investigate the whole enterprise of imperial expansion. Some important new approaches are emerging. Lewis Pyenson, for example, examines the spread of modern science in the context of cultural imperialism, quite a contrast to the conventional modernizers' point of view.[14] I call this new approach "critical modernist."

It would be premature to forecast what these new approaches will accomplish, but there is some hope that the alternativists and the critical modernists will replace the conformist and outworn orthodoxies of their predecessors.

II. OLD VERSUS NEW: SCHOLARS OF THE PREMODERN AND MODERN PERIODS

Differences in Work Habits and Training

In a number of respects the differences between the historiography of premodern and modern science suggest two distinct cultures. This dichotomy applies to students of the West as well as to historians of Japan and other non-Western countries.

Premodernists have few enough sources that they can control them, but modernists are overwhelmed by data. The former habitually browse in libraries and bookshops, looking for additional sources and manuscripts, while the latter must sift through ever-increasing source materials.

Premodernists cultivate a bibliophilic attitude, rather like a hobbyist's, but modernists now tend to use computer databases systematically in the interest of efficiency. Of the foreigners who come to work in Japan, premodernists tend to work in Kyoto, and modernists in Tokyo. In the Western world the premodernist tradition survives in Europe, as evidenced in the activity of the Needham Research Institute

---

[12] In a little over a year I was invited to three such meetings: "Science and Empire," Paris, April 1990; "European Federalism," Lisbon, October 1990; and "History of European Expansion: Technology Transfer between East and West since Vasco Da Gama," Leiden and Amsterdam, June 1991.

[13] James R. Bartholomew, *The Formation of Science in Japan: Building a Research Tradition* (New Haven: Yale Univ. Press, 1989), is a rare example in Western scholarship. One of the earliest attempts to translate a Japanese source into English is **Watanabe** Masao, *The Japanese and Western Science*, trans. O. T. Benfey (Philadelphia: Univ. Pennsylvania Press, 1990). There are hundreds of others worth translating.

[14] See, e.g., Lewis Pyenson, *Cultural Imperialism and Exact Sciences: German Expansion Overseas* (New York: Lang, 1985); and Pyenson, *Empire of Reason: Exact Sciences in Indonesia, 1840–1940* (Leiden: Brill, 1989).

in Cambridge and P. E. Will's group in Paris (Groupement de Recherche 798 du CNRS), while Americans dominate the modernists.

Premodernists have an antiquated aura of cultivation and erudition, while modernists identify with the social and political sciences. The former need formidable linguistic skills, while the latter are preoccupied with scientific methodology. As a corollary, premodernists today are primarily interested in the internal aspect of traditional science, but modernists are preoccupied almost exclusively with socio-institutional aspects.

Finally, in the history of science many premodern subjects can be dealt with by those trained exclusively in the humanities, while studies of the modern period often require insight into the behavioral patterns of the scientific community. Such insight is often impossible without some training in science.

Differences in Problems

The historiography of non-Western science is further complicated by a slightly different dichotomy, interest in indigenous science versus concentration on interaction with the West. This dichotomy parallels that between premodern and modern studies, but the demarcation is even clearer. In Japan those who work on pre-Meiji history of science differ noticeably in approach as well as background from specialists on the post-Meiji era. In China one sees an analogous difference between those doing research on the periods before and after the Opium War.

China's is the dominant culture for those who study premodern science. The paradigms that inform premodern science derive from the Chinese classics, written in the last five centuries B.C. The primary sources of science were usually written in classical Chinese even by people in Korea, Vietnam, and Japan, who shared the Chinese paradigms.[15] One can speak of all these sources as Chinese science in the same sense that works written in Greek by Syrians, Egyptians, and others are sources for what is called Greek science.

The cultural dominance of China does not hold for the twentieth century. Modernists may suggest, for example, that the earlier Japanese experience of modernization could instead provide a model for China—and other non-Western countries as well.

Differences in Audience Concerns

We have so far dealt mainly with Western scholarship written in a Western language and addressed to the Western academic community. The issues may be further complicated when we turn to the works of Eastern scholars who write in their native language and address a native audience. I myself, when writing in Japanese, choose very different topics and write in a very different style than when writing in English. In the latter case I try to take the viewpoint of an objective outsider or even a propo-

---

[15] Although Thomas S. Kuhn has abandoned the term *paradigm,* and it was never widely used by historians of science, I find it particularly applicable to classical times, when printing did not exist. Printed works guarantee that scientists in a given school will share new information, but in the days when academic and scientific traditions were handed down by copying manuscripts, paradigms were more easily maintained through many generations. Later efforts tended to be attributed to a single ancestral, paradigmatic, common figure or treatise and shared by all members of the school, as in the schools of Aristotle, Pythagoras, or Confucius. See Nakayama, *Academic and Scientific Traditions* (cit. n. 8), Chs. 1–2.

nent of the value of Eastern science. When writing in Japanese and to a native audience, which I can address more effectively, I can write more critically on the same subject.

Native audiences are usually interested in what helps to resolve issues of current concern to them, while international audiences tend to focus on unique local characteristics. To take an example from Japan, Confucianism in the Edo period is an important topic for native premodernists. It is less interesting to Chinese and Western premodernists, who consider Japanese neo-Confucianism a mere derivative of the Chinese orthodoxy. Similarly, in the field of the history of science non-Japanese may be interested in unique aspects of Japanese culture such as traditional mathematics (*wasan*) and native Japanese flora and fauna. These premodern topics, however, attract a Japanese audience less, for they assign them to the realm of minor arts and consider them irrelevant to later scientific development in Japan. Japanese audiences are instead more interested in, for example, the great precursors who translated Western science during the Dutch period, not because of unique features that invite international attention, but because of the topic's bearing on the practical business of assimilating Western science.

The differences of style between premodernists and modernists are amplified in the case of native scholars, because of the impact of modern nationalism. In approaching premodern subjects, natives and foreigners can keep a nearly equal historical distance. Natives working on modern subjects, however, sometimes find it difficult to avoid parochial viewpoints because of their involvement in their own societies. For instance, when comparing the modernization of Chinese and Japanese mathematics and science, **Ogura** Kinnosuke expressed uneasiness that so many more Japanese than Chinese sources were available to him, and concern lest his assessment might be colored either by nationalist pride or, conversely, excessive self-criticism. The detached appraisals of outsiders would be helpful in striking the proper balance.

How to bridge the two cultures and find a synthesis is an intractable problem deeply embedded in the disciplinary structure. A hasty synthesis would have little value.[16] All I can suggest is to begin extending our research horizon by applying the methods and problems nurtured in one camp to the other.

III. LINGUISTIC AND COMMUNICATION PROBLEMS OF WESTERN SCHOLARS

In the study of East Asian science, the formidable requirement for research in primary sources sharply divides those proficient in Asian languages from those who are not. This counts far more than boundaries of nationality. In a day of professionalization and specialization, few scholars are prepared to study many foreign languages extensively, although the most successful graduate programs in the United States require students who want to do research on Chinese science to learn classical Chinese and modern Chinese and Japanese in addition to at least one European language. Western historians who merely aim to make informed comparisons—a reasonable expectation for any historian of science—ought at least to know the

---

[16] I have tried to bridge this gap for Japan by dealing with the transitional period in **Nakayama** Shigeru, *Bakumatsu no yogaku* (Western learning at the end of the Tokugawa period) (Kyoto: Minerva shobō, 1984).

European-language sources on the topic under scrutiny, and to have the overview of Asian science that might be provided by undergraduate survey courses.[17]

I have already considered the dichotomy between native and foreign audiences; in this section and the next I will discuss the options available to different kinds of researchers. I find a basic tripartite division among scholars: nonlinguists who can only read and write European languages, and are thus not normally considered researchers in the field; linguists who can read the native language and write European languages; and native scholars who speak and write only in the native language, although they may read English or another European language.

Turning "Nonlinguists" into Researchers: Two Approaches

The trend toward increasing specialization is irreversible in any modern discipline, including the history of science: the result is an atrophy of comparative studies. Many books that claim to compare cultures or deal with worldwide topics merely collect specialist articles on various societies, with no common ground.[18] The few recent bold attempts at comparative history have had to transcend a linguistic imbalance. My *Academic and Scientific Traditions in China, Japan, and the West* compared Chinese and Greek institutions without a command of the Greek language, and Geoffrey Lloyd did the same in his recent *Demystifying Mentalities* with a limited knowledge of Chinese.[19] Such experiments would be improved by actual collaborations. One such work by Lloyd and Sivin, tentatively titled "Tao and Logos," is now under way. Outright collaborations can be supplemented by interdisciplinary, multicultural conferences or symposia designed to overcome the deep gaps between disciplines and imbalances due to the inherently limited viewpoint of the culture one belongs to.

One systematic way to overcome the current state of compartmentalization is to create language networks in various subject areas. The members would each specialize in one language area, and the network would sponsor sessions devoted to comparative topics at international gatherings. How such a network might operate can be illustrated for the history of astronomy. The old premodernist approach was on the whole philological. Scholars such as Willy Hartner mastered many languages and compared technical terms and astronomical parameters over a large span of the Eastern and Western hemispheres. Even the most linguistically versatile historians today, such as David Pingree and **Yano** Michio, cannot approach that span. In fact not all the older historians did. Otto Neugebauer, despite his deep knowledge of the Western tradition, never evinced interest in Chinese astronomy.

If a network of premodern astronomy were formed, E. S. Kennedy and his pupils could be consulted for data on Islamic astronomy, **Yabuuti** Kiyosi and his pupils for Chinese astronomy, or I myself for Japanese astronomy. This approach is most suit-

---

[17] Nathan Sivin, "Over the Borders: Technical History, Philosophy, and the Social Sciences," *Chinese Science,* 1991, No. 10, pp. 69–80.

[18] A typical example is A. F. Aveni, ed., *World Archaeoastronomy: Selected Papers from the Second Oxford International Conference on Archeoastronomy* (Cambridge: Cambridge Univ. Press, 1986). The Comparative Studies of Health Systems and Medical Care monograph series published by University of California Press contains few volumes that address comparative issues.

[19] Nakayama, *Academic and Scientific Traditions* (cit. n. 8); and G. E. R. Lloyd, "A Test Case: China and Greece, Comparisons and Contrasts," *Demystifying Mentalities* (Cambridge: Cambridge Univ. Press, 1990).

able for mathematical astronomy, where a comparison of astronomical parameters can be used to trace relationships between one area and another. A network could search more widely than any scholar can do alone.

Another topic that invites the network approach is comparison of names for flora, fauna, and natural products in different cultures—a favorite project of an earlier generation of Orientalists, who knew many languages. Such small-scale projects aside, a broad comparison of paradigms and characteristic approaches in each culture would be feasible on such an occasion as an international congress of the history of science or, better still, in a small specialist seminar.

Another approach turns a weakness into a strength. It is nearly impossible for a Westerner to search as widely as a native Chinese for historical sources in the Chinese language. Conversely, although the historiography of early China has been transformed by archaeological discoveries, particularly since 1970, many traditional scholars still depend almost wholly on written sources. Outsiders may open new perspectives by going beyond the written word. This is true for the modern as well as the premodern era.

Among the relevant innovations for modern science are sociological methods, such as formal surveys and statistical studies of groups, and anthropological methods, such as interviews and participant observation. We already have two successful examples in which foreign scholars studied the Japanese physics community without knowing much of the language. Male physicists found it easy to relax with them and tell inside stories to researchers who, as women and foreigners, were clearly outsiders.[20]

Bibliographies: Disciplinizing the Field

Because of new methods used to teach non-Western languages, particularly in the United States, an increasing number of Westerners are proficient in Chinese and Japanese, among them professional historians of science. They may bridge the gap between nonlinguists and native historians and play other international roles. Yet the state of affairs in Western language scholarship on Asian science still reveals some lacks.

One is for more bibliographies. For premodern subjects, Western scholarship— or, more precisely, scholarship written in Western languages—has been disciplinized. That is to say, those who work in this area know what other works are available in Western languages and are aware of the need to cite them in their own publications. The historiography of modern Asian science, however, has not been yet disciplinized, even in native-language works. In Western languages the situation is still worse. For instance, I have been asked repeatedly to write a short general survey of the history of modern Japanese science in English. That I have already written several does not release me from the obligation, since the editor always assumes that no one has read or will bother to look up the surveys already published. This repetitious activity contributes little or nothing to the normal accumulation of new knowledge. If I do not do it, whoever commissions the paper is unlikely to look into the

[20] See Lillian Hoddeson, "Establishing KEK in Japan and Fermilabs in the United States: Internationalism, Nationalism, and High Energy Accelerators," *Social Studies of Science,* 1983, *13:*1–48; and Sharon Traweek, *Beamtimes and Lifetimes: The World of High Energy Physicists* (Cambridge, Mass.: Harvard Univ. Press, 1988).

state of the field, and may well ask someone who has not seriously studied Japanese science and will merely add to the store of misleading information. The low level of disciplinization is particularly severe in government publications and the like, where the author is often an Eastern bureaucrat who has no idea how to carry out and write up a scholarly investigation. Such literature is often more useful as a primary than as a secondary source.

We might begin disciplinizing the study of modern science by making a comprehensive collection of sources, classifying them into primary and secondary sources, and compiling a computerized bibliography that is easily kept up to date. Even the premodern period could use more such work. For premodern Chinese science Joseph Needham's *Science and Civilisation in China* gives the most comprehensive bibliography available for sources in both Western and Asian languages. Sources cited in the earlier volumes are, of course, out of date. Nathan Sivin has recently published bibliographical essays and annotated bibliographies of scholarship in Chinese, Japanese, and Western European languages.[21]

Nothing analogous to these reference sources, however, is yet available for Japanese or Korean science. A bibliography of English-language works on modern Japanese science (with emphasis on physics) by Morris Low is the best available at present.[22] To supply that need is the aim of the *Bibliography of East Asian Science, Technology, and Medicine* that I have compiled with the help of **Ho** Peng Yoke, Sivin, Low, **Jeon** Sangwoon, and Christine Daniels, now in press at Garland Publishing.

IV. EAST ASIAN HISTORIANS OF SCIENCE: SOME PROBLEM CHOICES

Scientists who can read only English tend to think that all important writings are available in the English language. This is largely true in the community of physical science, where internationalization is most advanced today. Works in humanities and social sciences are instead still written in many languages, and the highly culture-bound concepts employed in such fields render translation extremely difficult. This is true of the history of science.

Most historians of science in East Asia never intend to write in English. They communicate in their native language, forming a local citation group, often with local paradigms. They can of course read English or another major European language. They freely cite Western works, but their own publications are practically never read in the Western world. Some authors who are not native speakers of English but have a scientific background are inclined to follow the convention of the scientific community and write in English. Even they, however, become aware of the culture-bound character of authorship, which imposes a number of limitations.

Premodernists as Textual Editors

For the premodern period, varying language skills have led to a de facto division of labor. While Western premodernists concentrate their energy on a limited number

---

[21] Nathan Sivin, "Science and Medicine in Imperial China: The State of the Field," *Journal of Asian Studies*, 1988, *47*(1):41–90. See also Joseph Needham *et al., Science and Civilisation in China* (Cambridge: Cambridge Univ. Press, 1954–).

[22] Morris F. Low, "The Butterfly and the Frigate: Social Studies of Science in Japan," *Soc. Stud. Sci.*, 1989, *19*:313–342.

of representative classical works, native scholars tend to search widely for new manuscripts and editing collections of primary sources, often of second- and third-order importance. For Japanese and Korean sources, manuscript searches have nearly exhausted the supply. For Chinese sources, there are plenty of printed works available because of the larger audience, and scholars have little free time to look for unpublished sources. Yet the number of researchers is also large and can be mobilized to search for, collect, and edit rather recent manuscripts and local gazetteers. Since mainland China was isolated from the rest of the world for some time, some scholars still maintain the old-fashioned approach of giving priority to the first appearance of concepts and events without much analysis of their historical environment.[23] This work can be done mechanically with the advent of computerized databases.

Modernists as Modernizers

Much as with the Eastern assimilation of modern Western science, native historians of science working on modern subjects try to make their analyses conform to Western historiography, whether positivist or Marxist. This leads them to find similarities between East and West, while Western Orientalists are interested in finding the differences. The native historians are predisposed to the modernizers' approach. They tend to take the West in general or the world's most advanced level of research, that of Germany in the past and the United States at present, as the model; to compare the state of affairs of the local scientific community with the model; and to bewail local institutional and cultural defects and to attack what they see as the indifference of their own government toward science.[24] Most of these modernist writers are ex-scientists who largely share the value system and frustrations of the international scientific community, to which the Western Orientalist's approach is unacceptable. Although modernists trained as professional historians of science are a little more sophisticated, most of them also, whether prewar Japanese Marxists or postwar proponents of democracy, have shared the psychology, if not the ideology, of this preoccupation with catching up.[25]

The Japanese are somewhat more relaxed and freer of this psychology nowadays, but Korean, Taiwanese, and mainland Chinese modernists are still obsessed with catching up with the West and, increasingly, with Japan.[26]

Peripheral Citation Groups

I mentioned in the opening of Section IV that East Asian historians may form local citation groups. Many such independent citation groups exist not only in the

---

[23] See **Nakayama** Shigeru, "The History of Science as Practiced in China," in *China: Development and Challenge,* ed. **Lee** Ngok and **Leung** Ci-Keung (Hong Kong: Univ. Hong Kong Press, 1981), pp. 287–296.

[24] A typical pseudohistorical example is **Qian** Wen-yuan, *The Great Inertia: Scientific Stagnation in Traditional China* (London: Croon Helm, 1985).

[25] **Nakayama** Shigeru, "The History of Science: A Subject for the Frustrated," in *Science and Society in Modern Japan,* ed. Nakayama, David L. Swain, and **Yagi** Eri (Cambridge, Mass.: MIT Press, 1974), pp. 3–16.

[26] See the publications of the Science Policy Section of Academia Sinica.

humanities and social sciences but in the natural sciences.[27] These last offer a special opportunity for those seeking a broader base for comparative studies of science and are an ideal research subject for those speaking the citation group's language. **Imanishi** Kinji's non-Darwinist ecology group is a notable example. I have reported on the postwar "grassroots geology" research movement centered on the charismatic **Ijiri** Shoji.[28] Even in theoretical physics, which is almost completely international, **Yukawa** Hideki's elementary particle group formed such an independent citation group in the secluded circumstances of World War II. Some such groups boast that they often have richer resources than those who can read only English, since they can draw on native as well as Western research. Their concepts often can hardly be rendered into English.

I have cited these Japanese groups from my own knowledge; analogous Chinese and Korean groups, cited only in their native language, should certainly exist. But they are not visible to foreigners, and their existence is seldom recognized as part of modern Western science. They may constitute a nuisance for those who strive to make a universal model of science, but they provide an illuminating counterexample for sociological study.

V. A NEW DEVELOPMENT: HISTORY OF SCIENCE AND SCIENCE POLICY

Japan's experience with modern science and technology has made it a special focus of studies that combine history and science policy. First, its technical modernization predates that of most other non-Western societies, including China. Second, its postwar development, alluded to above, seems unique. These two apparent success stories have led to approaches that could be described as applied history of science.

The UNU "Japanese Experience" Project: A Failed Approach

During the 1970s the United Nations University (UNU), with headquarters in Tokyo, launched a big project entitled "Technology Transfer, Transformation, and Development: The Japanese Experience."[29] The motive was to derive lessons from the Japanese experience in the nineteenth century for contemporary developing countries. The underlying ideology was purely that of the modernizers.

From the outset I was critical of the validity of such an approach at a time when the more prudent science policy planners in third-world countries were concentrating on "appropriate technology" or "alternative technology." It also seemed to me self-evident that the nineteenth-century context of science and technology is so totally different from the contemporary world as to invalidate simple historical analogies.

[27] Examples in the social sciences are **Maruyama** Masao's school of political science and **Otsuka** Hisao's school of economic history. For **Yanagita** Kunio's folklore group in the humanities see Bernard Bernier, "Yanagita Kunio's 'About Our Ancestors': Is It a Model for an Indigenous Social Science?," in *International Perspectives on Yanagita Kunio and Japanese Folklore Studies,* ed. J. Victor Koschmann *et al.* (East Asian papers, 37) (Ithaca, N.Y.: China-Japan Program, Cornell Univ. 1985).

[28] **Nakayama** Shigeru, "Grass-roots Geology: Iriji Shoji and the Chidanken," in *Science and Society in Modern Japan,* ed. Nakayama, Swain, and Yagi (cit. n. 25), pp. 270–289.

[29] See the Human and Social Development Programme—Japanese Experience series of reports published by UNU in 1979–1982.

Most of the works produced for this project by Japanese historians were addressed to a Japanese academic audience. Many of these writings have scholarly merit, but they fail to share the concerns of the Third World. Only a handful of historians of science and technology, such as **Hoshino** Yoshiro and **Nakaoka** Tetsuro, who did not belong to a tight disciplinary group of historians and were accustomed to addressing wider intellectual audiences, were sensitive to third-world problems. They later visited nearby third-world countries to discuss their research findings with local scholars. Among other findings, they discovered the plain fact that third-world people are not particularly inspired by the nineteenth-century Japanese experience. These historians modified their project, working on postwar Japan with international and particularly third-world collaborators, but because of the financial drain UNU could not extend the project.

The Business School Approach

When rebuilding its industrial sector, postwar Japan followed a different path from the rest of the world. Technical development was led by the MITI-industrial complex (that is, by close cooperation between the Ministry of International Trade and Industry and various industries), but conducted mostly within the private sector. In this sense the complex differed from the publicly sponsored alliance of industry and the Department of Defense in the United States. While academic science is in an impoverished state in Japan, corporate science flourishes. The privatization or capitalization of science has now taken an extreme form, in which management experts, rather than historians or sociologists, are most interested and concerned.

Work that explores industrial science is of the highest importance, especially since most works in the sociology of science today are still largely confined to topics relating to traditional academic science. Assessment of postwar Japanese science and technology took on a new dimension in the 1980s, initiated by science policy groups. To promote the national as well as corporate interest in science and technology, a society was formed in 1985, entitled Kenkyu Gijutsu Keikaku Gakkai (the Society for Science, Technology, and Planning) with a membership that included corporate scientists, bureaucrats, and some scholars from university business schools.

The focus of their attention is not the academic science that historians and sociologists of science are accustomed to dealing with, but industrialized science or corporate science practiced in secrecy, the research findings of which are seldom published. Mainly because of this secretive nature, historical investigation is very difficult, even though corporate science now dominates many fields in Japan. An increased trend towards privatization is likely to occur in the post–Cold War period, with the Japanese model of a privately dominated structure of science replacing the government-sponsored military industrial complex model dominant in the immediately postwar world.

The society's aim is to focus on the efficiency and cost performance of research investment. Their language is aggressive, largely borrowed from military strategy, and intended to promote survival in the competitive marketplace rather than the pursuit of disinterested scientific concepts. Meetings of the society are taken up with the success stories of ex-engineers, which may interest us to some extent as case studies, but often no generalizations can be made, as the strategy of each corporation is different. The members do not use sociology to study laboratory workers.

Similar changes are occurring in the West. In the mid 1980s I made a trip to the United States to lecture on the postwar development of Japanese science and technology. Most of the audience consisted of business-school management specialists rather than the premodernist academics with whom I had previously been familiar.[30] I found myself unintentionally involved, not in civilized Orientalist conversation, but in an adversarial and controversial discussion of national and business interests. This made me uncomfortable, since I was not prepared to speak for "Japan, Incorporated."

But consciously or unconsciously, we are inevitably involved in such nation-state viewpoints, and this is likely to continue as long as data, statistics, and indicators are all related primarily to the nation state as a unit. It is thus essential to make explicit our independence of nation-state interests. Unfortunately, comparative histories of modern science in East Asia have conventionally tended to be success stories about Japanese cases. We should seek to establish perspectives removed from questions of national pride.

A "Four Sector" Approach

To counterbalance the business-school defense of national and corporate interests, I have introduced a "four-sector" approach to analyzing postwar Japanese scientific and technological activity. It incorporates not only academic, government, and private corporate viewpoints, but also that of the ordinary citizenry. A preview was published by Kegan Paul in 1991, under the title *Science, Technology, and Society in Postwar Japan*. A massive Japanese-language version of this project, the product of team research by more than fifty historians of Japanese science, will soon be published by Gakuyo Shobō. I hope that it will be translated into English by the turn of the century.

Historians are not in a position to defend the viewpoint of nation-states or of business firms. Most traditional historians of science have depended on the value system of the academic community, of which they themselves are members. For most of them science is an object of observation, testing, and criticism, not of promotion. The disinterested position of historians must be linked with the concerns of citizens, who by definition have no vested interest in promoting scientific activity. It is only from this vantage point that the critical examination of nationalistic or industrialized development of science and technology is possible.

This new dimension to the history of science is not confined to Japan. Newly industrialized economies in Asia, such as Korea and Taiwan, are rising quickly, and eventually continental China, with its population of a billion, will emerge into scientific and technological parity with the rest of the world. We need to ask such questions of these countries as well, taking care not to adopt positions limited to a nation-state perspective, or simplistic comparisons with the United States or Japan.

---

[30] Representative publications are Sheridan Tatsuno, *The Technopolis Strategy: Japan, High Technology, and the Control of the Twenty-first Century* (New York: Prentice Hall, 1986); and Gene Gregory, *Japanese Electronics Technology: Enterprise and Innovation* (New York: Wiley, 1986).

KNOWLEDGE CONSTRUCTED BY DISCIPLINE

The Experimental Life Sciences
in the Twentieth Century

By Daniel J. Kevles* and Gerald L. Geison**

T HE LIFE SCIENCES have entered a wholly new era during the twentieth century, in terms of scale, institutional visibility, claims on resources, and social
consequences. Above all, experimental biology has come to be seen as the most
powerful force in the modern reconception of the nature of life and in the radical
transformation of medical practice. This transformation had diverse sources, but
none was more telling than the attempt to subject issues in late-nineteenth century
evolutionary and developmental biology to experimental scrutiny. That general research program led to the emergence of new disciplines such as embryology, cytology, endocrinology, the reproductive sciences, and genetics, which rapidly took on
lives of their own, independent of evolutionary debates, and produced a wide range
of conceptual and utilitarian triumphs.[1]

Experimental biology has been widely hailed for its role in unpacking the riddles
of heredity, notably through the introduction of Mendelian and molecular genetics,
and for its contribution to the production of newly vigorous agricultural crops, newly
specific preventive measures in public health, and newly efficacious therapies in
medical practice. Its practitioners and advocates could point to such utilitarian results in agriculture as hybrid corn and the green revolution; or to such benefits in
medicine as antiseptics, vaccines, serum therapies, replacement therapies like vita-
mins, insulin, and other hormones, and above all antibiotics like penicillin and other
specific remedies for infectious diseases.

Since the discovery of the double-helix structure for DNA, in 1953, the most
spectacular achievements of modern experimental biology have derived from molecular biology. Examples include the use of genetic mapping with restriction fragment
length polymorphisms to identify diagnostic markers for genes that figure in disease
and to track down those genes for the purpose of sequencing and analyzing them;
the deployment of recombinant DNA techniques to construct transgenic animals to
study gene function by observing their effects when they are inserted into foreign
organisms; and the introduction of foreign genes into plants to improve features

* Humanities Division 228–77, California Institute of Technology, Pasadena, California 91125.
** Department of History, Dickinson Hall, Princeton University, Princeton, New Jersey 08544.
We are grateful to Ray Owen for detailed critique of an earlier draft of this article and to John
Lesch and other participants for comments at the session at which it was discussed.

[1] See, e.g., Adele E. Clarke, "Embryology and the Rise of American Reproductive Sciences, circa
1910–1940," in *The Expansion of American Biology,* ed. Keith Benson, Jane Maienschein, and Ronald Rainger (New Brunswick, N.J.: Rutgers Univ. Press, 1991), pp. 107–132; and Jane Maienschein,
"History of Biology," *Osiris,* 1985, *1:*147–162.

ranging from disease resistance to market qualities. Some of the utilitarian promise implicit in this research has begun to be realized—for example, in genetically engineered organisms that produce unprecedented yields of milk, proteins, insulin, or human growth hormone, and very recently in the first efforts to apply gene therapy to human victims of genetic diseases.[2]

Such striking results in both "pure" and "utilitarian" work in the experimental life sciences could hardly go unnoticed by historians. On the basic biological side, we now have valuable studies of several major developments in the history of twentieth-century physiology, embryology, biochemistry, classical genetics, and molecular genetics. On the utilitarian, institutional, and social side of the story, we also have a growing body of work in the history of agricultural research and a very large, if not always distinguished, body of literature on the history of modern medical institutions, problems, and practices. In fact, the history of diseases has lately become a sort of growth industry, with major studies of such afflictions as cholera, tuberculosis, yellow fever, polio, and even the new scourge AIDS, to speak only of the more obviously somatic diseases.

However, a vast terrain remains to be explored in the technical history of molecular biology and its disciplinary relatives, where thus far scientist-participants (often with the help of journalists) have set a largely whiggish tone and agenda, celebrating successes while ignoring twists, turns, and failures. Equally important, the historiography of the experimental life sciences is, like the sciences themselves, enormously diverse and disparate. Works in the history of the medically related sciences, for example, take little cognizance of those involved with agriculture, while the development of molecular biology has been treated as something of a force unto itself, disconnected (until recently) from the rest of modern biology. Indeed, some of the richest, most accessible, and most significant needs and opportunities for historians of modern life science lie in explorations of the interplay between basic experimental biology and agricultural or medical practices.

How is the history of such a rich and diverse domain of twentieth-century science to be written? To unify our study of the rise of the experimental life sciences and make it easier to understand, we have imposed a common analytic framework on the disparate fields that constitute them. Our framework is based on the following three clusters of categories: goals, patronage, and institutions; concepts and research programs; and methods, instruments, and materials devised within a discipline or imported from without. We will first discuss these categories schematically, then briefly illustrate how they can be used to structure and clarify the history of two major areas of research in the experimental life sciences. Most of our discussion concerns developments in the United States, but we believe that our framework is applicable to the history of the modern experimental life sciences elsewhere. Of course, our illustrative subjects and themes deserve much fuller historical and multinational analysis.

Our aim here is not to challenge prevailing historiographic interpretations. In fact, there are no overarching interpretive schools in the historiography of twentieth-

[2] See, e.g., Diana Long Hall, "Physiological Identity of American Sex Researchers between the Two World Wars," in *Physiology in the American Context, 1850–1940,* ed. G. L. Geison (Bethesda, Md.: American Physiological Society, 1987), pp. 263–268; and Daniel J. Kevles and Leroy Hood, eds., *The Code of Codes: Scientific and Social Issues in the Human Genome Project* (Cambridge, Mass.: Harvard Univ. Press, 1992).

century biology. Nor do we presume to be comprehensive: the corpus of historical studies in modern biology is as vast as it is disparate. We seek only to draw attention to the large number of subjects that warrant historical investigation within this domain and to suggest how our framework might help to reveal their commonalities.

I. THE ANALYTIC FRAMEWORK

The goal of understanding, preventing, and finding therapies or cures for the diseases that beset people, animals, and plants has generated an enormous amount of experimental biological research. So has the effort to improve nutrition, growth, and fitness or quality in all three types of organisms. These broad utilitarian goals have given rise to abundant patronage. There are the large philanthropic foundations, like the Rockefeller and Carnegie foundations and the Wellcome Trust; the numerous eleemosynary agencies concerned with general medicine, like the Markey Foundation; and the still more numerous single-disease philanthropies, like the former National Foundation for Infantile Paralysis (now the March of Dimes Birth Defects Foundation), the Cystic Fibrosis and Multiple Sclerosis foundations, and the Muscular Dystrophy Association. In the United States key government agencies include state departments of public health and, at the federal level, the Food and Drug Administration, the U.S. Department of Agriculture, and the National Institutes of Health; the Atomic Energy Commission (now the Department of Energy) and its national laboratories such as that at Oak Ridge, Tennessee; several military agencies, notably the Office of Naval Research; and the National Science Foundation. The Medical Research Council has been an essential patron in Britain, as has INSERM (Institut National de la Santé et de la Recherche Médicale) in France. Utilitarian goals have also played a major role in the proliferation of diverse and numerous institutions where experimental life science research has been conducted: private research centers such as the Rockefeller Institute for Medical Research or the Pasteur Institute, as well as a host of public and private medical schools, agricultural schools, veterinary schools, bacteriological laboratories, and university departments of biology, biochemistry, and molecular biology.[3]

But if purposes and patronage did much to shape the orientations and problem choices of the research carried out in these institutions, the work itself took place within a conceptual space occupied by a set of inherited and evolving research programs that sometimes competed with one another. Among the most obvious of the defining concepts were and are those associated with Darwinian evolution through natural selection, the germ theory of disease, Mendelian genetics, structural biochemistry, and the genetic code.

The specific research programs pursued within these conceptual frameworks posed inherent technical challenges. Their resolution often hinged on innovations in methods, the identification or construction of appropriate biological materials, and the invention of new instruments. In physiology, for example, investigators of human

---

[3] Jane S. Smith, *Patenting the Sun* (New York: William Morrow, 1990), makes clear the value of studying foundations devoted to disease research. See also George W. Corner, *A History of the Rockefeller Institute, 1901–1953: Origins and Growth* (New York: Rockefeller Institute Press, 1964); Robert E. Kohler, *Partners in Science: Foundations and Natural Scientists, 1900–1945* (Chicago: Univ. Chicago Press, 1991), esp. pp. 265–391; Michael Morange, ed., *L'Institut Pasteur: Contributions à son histoire* (Paris: Editions la Decouverte, 1991); and the works cited in note 16 below.

reproduction sometimes arranged with local physicians to gain access to discarded embryos, ova, and ovaries from miscarriages, tubal pregnancies, and hysterectomies;[4] studies of intermediary metabolism were transformed by the micromanometer in the hands of Hans Krebs (of Krebs cycle fame);[5] embryology, developmental biology, and immunology found powerful new resources in the techniques of tissue culture and transplantation;[6] and neurophysiology attained a new level of sophistication through the use of such biological material as the giant squid axon and new instruments such as the string galvanometer (forerunner of the electrocardiograph) for recording and amplifying biological signals.[7]

Virtually every branch of modern experimental biology came to rely on standardized biological materials and carefully constructed "laboratory animals," whether supplied by commercial firms that arose to meet the need or produced in on-site laboratory colonies of *Drosophila,* yeast, slime molds, rats, mice, or guinea pigs among other organisms.[8] The infiltration of experimental biology by physicists, chemists, and their techniques helped foster the development of important new instruments. The ultracentrifuge, chromatography, electrophoresis, X-ray diffraction, and electron microscopy collectively opened the door to isolating and analyzing biological substances and ultimately understanding their structure and function.

The inherent technical challenges were often common across institutions and research programs. The difficulty of separating biological substances, determining protein structure, or assessing chromosomal and genetic features, for example, was the same whether the substances or proteins or chromosomes or DNA derived from a bacterium, a bee, or a bull. Cytogenetics was transformed by the advent, in the 1950s, of methods that permitted the clear karyotyping of chromosomes and, in the 1960s, of still other methods that allowed the identification of chromosomes by the pattern of bands they displayed upon treatment with a fluorescent chemical.[9] And

---

[4] Adele Clarke, "Research Materials and Reproductive Science in the United States, 1910–1940," in *Physiology in the American Context*, ed. Geison (cit. n. 2), pp. 332–335.

[5] Frederic L. Holmes, "Manometers, Tissue Slices, and Intermediary Metabolism," in *The Right Tools for the Job: At Work in Twentieth-Century Life Sciences,* ed. Adele Clarke and Joan Fujimura (Princeton: Princeton Univ. Press, 1992), pp. 151–171.

[6] See, e.g., Frederick B. Bang, "History of Tissue Culture at Johns Hopkins," *Bulletin of the History of Medicine,* 1977, *51*:516–537.

[7] See, e.g., G. H. Bishop, "My Life among the Axons," in *The Excitement and Fascination of Science: Reflections by Eminent Scientists,* 2 vols (Palo Alto, Calif.: Annual Reviews, 1965), pp. 3–20; Louise Marshall, "The Fecundity of Aggregates: The Axonologists at Washington University, 1922–1942," in *Perspectives in Biology and Medicine,* 1983, *26*:613–636; Marshall, "Instruments, Techniques, and Social Units in American Neurophysiology, 1870–1950," in *Physiology in the American Context,* ed. Geison (cit. n. 2), pp. 351–369; Robert G. Frank, Jr., "The Joseph Erlanger Collection at Washington University School of Medicine, St. Louis," *Journal of the History of Biology,* 1979, *12*:193–201; and Frank and Judith H. Goetzel, "The Alexander Forbes Papers," *J. Hist. Biol.,* 1978, *11*:429–435.

[8] See Garland Allen, "The Introduction of *Drosophila* into the Study of Heredity and Evolution: 1900–1910," in *Science in America since 1920,* ed. Nathan Reingold (New York: Science History Publications, 1976), pp. 226–277; Robert E. Kohler, *Lords of the Fly:* Drosophila *Genetics and the Experimental Life* (Chicago: Univ. Chicago Press, 1994); Clarke, "Research Materials and Reproductive Science in the United States" (cit. n. 4), pp. 323–350; Clarke and Fujimura, eds., *The Right Tools for the Job* (cit. n. 5); H. L. Foster, "The History of Commercial Production of Laboratory Rodents," *Laboratory Animal Science,* 1980, *30*:793–798; and Michael Lynch, "Sacrifice and the Transformation of the Animal Body into a Scientific Object: Laboratory Culture and Ritual Practice in the Neurosciences," *Social Studies of Science,* 1988, *18*:265–289.

[9] See T. C. Hsu, *Human and Mammalian Cytogenetics: An Historical Perspective* (New York: Springer-Verlag, 1979).

material, methodological, and instrumental innovations developed in one branch of experimental biology were often transferred advantageously to other branches.

It is well known that microbiology and molecular biology profited from new concepts and methods that came their way through cross-disciplinary interactions with physics and chemistry. Less well recognized is the benefit that physiology and microbiology have derived from the experimental materials provided by other disciplines, including notably physics, whose particle accelerators produced radioactive isotopes in abundance beginning in the 1930s. The isotopes first served as markers for tracking the course of chemicals through the body.[10] In recent years they have become sine qua non in molecular biological research, serving as tags for fragments of DNA employed for purposes ranging from basic gene analysis to forensic genetic fingerprinting.

Although some of the topics mentioned above have been the subjects of historical study, many more await their historians. Like agricultural experiment stations or the Pasteur Institute, chromosomal banding, restriction enzymes, or the polymerase chain reaction, they merit historical treatment in and of themselves. And in our judgment, all of these topics would benefit from integrated consideration of our categories. We illustrate the point here with two examples—neurobiology and animal virology, focusing in both cases on research conducted in the United States after World War I.

II. NEUROBIOLOGY

An important recent book on the history of "neuroscientific concepts" does not even bother to enter the twentieth century, boldly claiming that "by 1850 the foundations of modern neuroscience had been laid."[11] That claim would surely be disputed by those who have participated in the development of twentieth-century neurobiology. At a very general conceptual level, to be sure, some or even most of the basic issues had been posed and vigorously pursued by the mid-nineteenth century, but no stable consensus had emerged about several central problems, and a huge amount was yet to be learned about the details of the structural and functional features of the nervous system.

Even by the turn of this century, two or three of the most fundamental concepts in modern neurobiology were still under dispute or not yet fully developed. At the anatomical level, more than a few physiologists still preferred the "reticular" theory of the nervous system as a continuous cytoplasmic network rather than the ultimately triumphant "neuron" theory, according to which the nervous system was a complex arrangement of discrete individual cells—the latter theory being associated above all with Santiago Ramon y Cajal, the first (and so far only) Spanish recipient of the Nobel Prize in physiology or medicine.[12] Even among those who accepted the neuron theory by about 1900, its functional implications and significance had just

---

[10] J. L. Heilbron and Robert W. Seidel, *Lawrence and His Laboratory: A History of the Lawrence Berkeley Laboratory*, Vol. I (Berkeley: Univ. California Press, 1989), pp. 156–157, 34–39, 219.

[11] Edwin Clarke and Louise S. Jacyna, *Nineteenth-Century Origins of Neuroscientific Concepts* (Berkeley: Univ. California Press, 1987), p. 1.

[12] See, e.g., Susan M. Billings, "Concepts of Nerve Fiber Development, 1839–1930," *J. Hist. Biol.*, 1971, *4*:275–306; and Santiago Ramon y Cajal, *Recollections of My Life* (Philadelphia: American Philosophical Society, 1937).

begun to be explored, notably by the English physiologist Charles Scott Sherrington, another future Nobel laureate. It was Sherrington who introduced the now universally accepted terms for the functional units of the cellular nervous system—axon, dendrite, and synapse—and who focused on the synapse, the junction between separate nerve cells, as the physiologically most significant unit. In 1907, after a decade of delicate animal experiments, notably on decerebrated cats, Sherrington advanced his famous, if highly complicated, theory of the "integrative action of the nervous system."[13]

During the next half century neurophysiologists pursued a rich variety of specific problems and developed several major new concepts with the aid of sophisticated new techniques, electronic instruments, and recording devices. Far from being a stagnant field whose foundations had already been laid by 1850, neurophysiology continued to attract highly talented scientists, including a disproportionate number of future Nobel laureates. Until the rise of structural biochemistry and molecular biology, no branch of the experimental life sciences enjoyed such favor with the Nobel committee.[14]

Like biochemistry and molecular biology, neurophysiology attained its privileged status partly by demonstrating the fertility of the mechanistic approach to biological problems—specifically, by showing the extent to which extremely complex events in the nervous system could be explained by or "reduced to" electrical-chemical and other basic physical concepts. Only after World War II did the once glorious success of "classic" neuromuscular physiology begin to fade. The analytic categories outlined above offer a way of beginning to understand both the prolonged success of twentieth-century neurophysiology as such and its eventual diversification into such new fields as endocrinology, central nervous system physiology, and information or cognitive science—into, in short, "neuroscience" writ large.

Goals, Patronage, and Institutions

Given the crucial role of the nervous system in the distinctive features of animal and human life, neurophysiology has always held a special place among the branches of physiology. It might seem that generous patronage would have come to a field that sought insight into the mechanisms of locomotion and reflexes, the five special senses, and sensation in general—perception, paralysis, passion, and pain. Who could deny the appeal of a subject with such profound implications for the grandest philosophical issues of all—the "seat" of the mind, the interplay between mind and body, and the very nature of thought or the soul itself?

Yet the very pertinence of neurophysiology to these and other central human concerns could be a burden as well as a boon. At least through the mid-nineteenth century, experimental research on the nervous system could and did expose its practitioners to charges of atheistic "materialism," and the results of such research were sometimes seen as dangerous to established beliefs, authorities, and institutions. During our more secular century neurophysiologists have had rather less to fear from

---

[13] C. S. Sherrington, *The Integrative Action of the Nervous System* (New Haven: Yale Univ. Press, 1907). See also Judith P. Swazey, *Reflexes and Motor Integration: Sherrington's Concept of Integrative Action* (Cambridge, Mass.: Harvard Univ. Press, 1969).

[14] See D. M. Fox, M. Meldrum, and I. Rezak, eds., *Nobel Laureates in Medicine or Physiology: A Biographical Dictionary* (New York: Garland, 1990).

such philosophical and religious objections (even though, like other experimental biologists, they have continued to face widely publicized charges of cruelty to animals). In the secularized and specialized twentieth-century Western world the goals of experimental neurobiology became more narrowly defined and much less controversial. But abundant patronage did not then flow automatically to the field. Its claims to attention and resources were now assessed according to a different set of criteria: like most branches of the biomedical sciences, neurobiology was obliged to articulate its goals and to seek patronage in terms that met the shifting needs or demands of medical education and medical practice.

For that reason the most important force in the development of experimental neurobiology during the past century has been the general ascendance of "scientific" medicine, based on the premise that experimental biology would yield benefits for medical education, clinical practice, and human welfare that were at least commensurate with its high costs. Leaving aside for now the question of how fully this ideology of scientific medicine was or is justified by the actual results of basic research in various branches of the biomedical sciences—a crucial issue woefully neglected by historians and other analysts—there can be no doubt that the ideology is widely accepted by the medical profession, private philanthropies, government agencies, and the public in general.[15]

In the United States, the first really large-scale patrons of scientific medicine were the Carnegie and Rockefeller foundations, especially the latter. The Rockefeller Foundation contributed not only directly through the Rockefeller Institute for Medical Research and fellowships for a host of individual research projects across the country and indeed around the world, but even more importantly through its crucial role in the radical transformation of American medical education after 1910. Taking its lead from the famous "Flexner Report" of that year on medical education in the United States and Canada, the General Education Board of the Rockefeller Foundation indicated its readiness to distribute tens of millions of dollars to medical schools throughout the country on the condition that they adopt "Flexnerian" programs of reform.[16] The Carnegie Foundation for the Advancement of Teaching, although its support for similar goals was less sustained and less extensive in scale, was in fact the official sponsor of the Flexner Report. And in the year the report was published, 1910, the Carnegie Foundation gave $2,000,000 to the Washington University Medical School in St. Louis so that it could be reorganized along the lines of Flexner's recommendations.[17]

The Flexner Report, citing the German university system and the Johns Hopkins University as its models, called for eliminating the worst of the numerous proprietary medical schools then common throughout the United States and transforming the rest into university-based institutions that emphasized the "preclinical" sciences, laboratory training, and the research ethos of the German universities. Medical

[15] See Gerald L. Geison, "'Divided We Stand': Physiologists and Clinicians in the American Context," in *The Therapeutic Revolution: Essays in the Social History of American Medicine*, ed. Morris Vogel and Charles E. Rosenberg (Philadelphia: Univ. Pennsylvania Press, 1979), pp. 67–90; and John Harley Warner, "Science in Medicine," *Osiris*, 1985, *1*:37–58.

[16] Abraham Flexner, *Medical Education in the United States and Canada* (New York: Carnegie Foundation, 1910); E. Richard Brown, *Rockefeller Medicine Men: Medicine and Capitalism in America* (Berkeley: Univ. California Press, 1979); and Kenneth M. Ludmerer, *Learning to Heal: The Development of American Medical Education* (New York: Basic Books, 1985).

[17] Marshall, "The Axonologists at Washington University" (cit. n. 7), p. 613.

schools enticed by the enormous funds dangled before their eyes by the Carnegie Foundation and especially the Rockefeller Foundation thus found themselves encouraged or obliged to recruit research-oriented experimental scientists, often Ph.D.s instead of clinically-oriented M.D.s, to teach the preclinical subjects. The upshot, already clear by the 1920s, was a sudden move toward a nationally standardized approach to medical education and research remarkably similar in structure to the one that still prevails today.[18]

Whatever the general virtues and defects of the Flexner model—and it has been the target of increasing criticism during the past two decades or so—it indisputably opened huge new opportunities for experimental research in the life sciences. Medical schools everywhere in the United States created positions for the newly ascendant practitioners of experimental biology and erected veritable laboratory Xanadus in which they could conduct their own research as well as teach experimental science to aspiring physicians. Happily for physiologists, the Flexner Report called physiology "the central discipline of the medical school,"[19] and physiologists, including not least neurophysiologists, were among the major early beneficiaries of the Flexnerian revolution in medical education.

In fact, a preliminary scan of the general history of American physiology suggests that the period between the Flexner Report and World War II may have been a golden age for American neurophysiology. During those four decades, American neurophysiologists continued to enjoy their traditional dominance within the discipline—a dominance institutionally ratified, so to speak, when "all five of the papers at the first annual meeting of the [American Physiological] Society, in 1888, were on neural topics." By the 1913 annual meeting of the society, the proportion of papers on neurophysiological topics had "declined" to 36 percent, while at the 1930 meeting fully 42 percent of the papers presented had something to do with the nervous system.[20] During the first half of this century only cardiovascular physiology—then considered a closely related specialty in any case—came close to challenging the hegemony of neurophysiology within American physiology and its official society.

The highwater mark, perhaps, for American neurophysiology was the decade of the 1930s. That decade began with the formation of a highly influential, if small and informal, group known as the "Axonologists," a sort of dining club for self-appointed disciplinary leaders that met at the same time as, though separately from, the American Physiological Society. This practice did not always endear them to outsiders from other branches of the discipline, one of whom later reported that, at annual meetings of the society during the 1930s, the "Axonologists were the important people, and almost strutted through the corridors, dominated the meetings, being very conscious that they alone were in the frontiers of physiological discovery."[21]

If the Axonologists or other American neurophysiologists really did prance about during the 1930s, it is not hard to understand why. Almost all were fairly young, in their thirties or forties, and they were flush with the acknowledged success of the

---

[18] Flexner, *Medical Education in the United States and Canada;* and Ludmerer, *Learning to Heal* (both cit. n. 16).

[19] Flexner, *Medical Education in the United States and Canada,* p. 63.

[20] Marshall, "Instruments, Techniques, and Social Units in American Neurophysiology" (cit. n. 7), pp. 354 (quotation), 358.

[21] *Ibid.,* pp. 358 (quotation), 359; and Marshall, "The Axonologists at Washington University" (cit. n. 7).

precise results that flowed from their new techniques for amplifying and recording electrical signals from biological materials. Neurophysiology was also then a special favorite of the Rockefeller Foundation. Thus during the mid 1930s a small but significant contingent of Axonologists at Washington University in St. Louis received generous Rockefeller funding for their expensive cathode-ray oscillographs. As early as 1923 one of them, the future Nobel laureate Herbert Gasser, had "without his seeking it" received a fellowship from Abraham Flexner and the Rockefeller Foundation for a two-year leave of absence to travel abroad.[22]

After World War II, as the Rockefeller Foundation reassessed its priorities and as its funding for medical research was vastly outstripped by the infusion of governmental support from the National Institutes of Health and other agencies, neurophysiology lost some of its prewar swagger. For a while neurophysiologists continued to dominate the councils and publications of the American Physiological Society, which now included the *Journal of Neurophysiology,* founded in 1937. As late as 1958, in a remarkable survey of the discipline commissioned by the American Physiological Society with support from the National Science Foundation, Ralph Gerard—himself a leading neurophysiologist who had convened the first meeting of the Axonologists—estimated that "two-thirds of all laboratory experiments [within physiology] are in neural and circulatory physiology."[23]

By then, however, Gerard and other neurophysiologists who had once dominated the annual meetings of the American Physiological Society had began to display a more subdued, almost wistful tone about the place of their specialty within the discipline of physiology and experimental biology more generally.[24] Historians of modern physiology have not yet fully explored the impact of World War II on the field, including how it may have moved neurophysiology away from its classical focus and reshaped its relations with other disciplines.

Concepts and Research Programs

To the general historian of scientific ideas or culture, surely the most familiar concept in twentieth-century neurophysiology is Ivan Pavlov's notion of the conditioned reflex, especially as deployed by behavioral psychologists like B. F. Skinner. Much less attention has been paid to the details of spinal reflex physiology, even as elaborated by C. S. Sherrington in his general theory of the "integrative action of the nervous system." Another central concept, at once related to and yet very different from prevailing ideas in neurophysiology, was Walter B. Cannon's notion of "homeostasis," as popularized in his 1932 book *The Wisdom of the Body.* Not surprisingly, given its seemingly clear links to such ideas as evolution, adaptation, and equilibrium—and thus, more broadly, to American social theory—Cannon's concept of homeostasis has already attracted a fair amount of historical attention.[25] For similar

[22] Marshall, "The Axonologists at Washington University," pp. 618–619.

[23] Ralph W. Gerard, *Mirror to Physiology: A Self-Survey of Physiological Science* (Washington, D.C.: American Physiological Society, 1958), p. 231.

[24] See, e.g., *ibid.*; Bishop, "My Life among the Axons" (cit. n. 7); and Ralph W. Gerard, "The Organization of Science" (1952) in *The Excitement and Fascination of Science* (cit. n. 7), pp. 149–160.

[25] On Cannon, homeostasis, and equilibrium or functionalist models in American social thought see Walter B. Cannon, "Organization for Physiological Homeostasis," *Physiological Reviews,* 1929, 9:399–431; Cannon, "The Autonomic Nervous System, and Interpretation," *The Lancet,* 1930, 1:1109–1115; Cannon, *The Wisdom of the Body* (New York: Norton, 1932); Saul Benison, A. Clifford

reasons there is a steadily increasing body of historical literature on the theory of hormones, a theory that encompassed both the effects of nervous action on hormonal secretions and the role of hormones (or "chemical messengers") in the transmission of nervous impulses at the synaptic junction between neurons. English and American physiologists contributed the important work on homeostasis and hormones, which Continental physiologists were relatively slow to accept.[26]

These wide-ranging ideas may seem more exciting than the "list of the major themes and concepts in twentieth-century physiology" that we owe to Louise Marshall, a neuroscientist-cum-historian: "(1) The central nervous system localization for control of hormonal secretion and body homeostasis, (2) the identification of control of movement at several levels of the higher brain centers, (3) the characterization of the elements of the compound action potential, (4) the forces influencing neuronal regeneration, and (5) the electrochemical theory of nervous transmission."[27] This otherwise valuable list also omits the "all-or-none law," according to which a given fiber from any sort of tissue either responds maximally or not at all, the strength of the response being independent of the strength of the stimulus. First advanced in the case of cardiac tissue in the late nineteenth century, the all-or-none law was extended to ordinary skeletal muscle by World War I, and then to peripheral nerves and finally the central nervous system by World War II.[28]

Marshall's list of "major themes and concepts" could also be challenged on the grounds that it minimizes theoretical issues and seems skewed toward central nervous control at the expense of peripheral, decentralized "autonomy" in the form of ganglia or circulating chemical substances (in a word, hormones).[29] Although Marshall acknowledges elsewhere the importance of the chemical theory of nervous transmission and its challenge to central nervous control, adding these and other controversial issues might have led to a list that better revealed how pertinent twentieth-century neurophysiology is to broader philosophical and ideological concerns. And it is not merely politically fashionable to suggest that historians of neurobiology might have paid closer attention to controversies over the site of control of particular neurophysiological functions. At least for the outsider, a discussion of such controversies would also help to clarify the *technical* issues at stake.

Barger, and Elin L. Wolfe, *Walter B. Cannon: The Life and Times of a Young Scientist* (Cambridge, Mass.: Belknap Press of Harvard Univ. Press, 1987); Stephen J. Cross and William Randall Albury, "Walter B. Cannon, L. J. Henderson, and the Organic Analogy," *Osiris*, 1987, *3*:165–192; Donald Fleming, "Walter B. Cannon and Homeostasis," *Social Research*, 1984, *51*:609–640; and C. E. Russett, *The Concept of Equilibrium in American Social Thought* (New Haven: Yale Univ. Press, 1968).

[26] See Gerald L. Geison, *Michael Foster and the Cambridge School of Physiology: The Scientific Enterprise in Victorian Society* (Princeton: Princeton Univ. Press, 1978), esp. the discussion of national styles on pp. 331–355. See also *ibid.*, pp. 311–321, 353–354, and the sources cited there. On the closely related topic of reproductive physiology and sex research see, e.g., Merriley Borell, "Organotherapy and the Emergence of Reproductive Endocrinology," *J. Hist. Biol.* 1985, *18*:1–30; Diana Long Hall, "Biology, Sex Hormones, and Sexism in the 1920s," *Philosophical Forum*, 1974, *5*:81–96; Hall, "Physiological Identity of American Sex Researchers" (cit. n. 2); and Clarke, "Research Materials and Reproductive Science in the United States" (cit. n. 4).

[27] Marshall, "Instruments, Techniques, and Social Units in American Neurophysiology" (cit. n. 7), p. 359.

[28] See Robert G. Frank, Jr., "Instruments, Nerve Action, and the All-or-None Principle," *Osiris*, 1994, *9*:208–235; and G. L. Geison, "Keith Lucas," in *Dictionary of Scientific Biography* (New York: Scribners, 1970–1980), Vol. III, pp. 532–535.

[29] Cf. Geison, *Michael Foster and the Cambridge School of Physiology* (cit. n. 26), passim.

Methods, Instruments, and Materials

Forty years ago the American physiologist Walter Fenn wrote that "the whole history of physiology could be written in terms of new tools for research." Fenn, like many experimental scientists, did not need historians, philosophers, or sociologists to teach him about the importance of techniques and experimental systems, or "the right tools for the job," in the production of the conceptual knowledge that has been the traditional concern of historians and philosophers of science.[30] His point about the centrality of "new tools for research" in the development of physiology, though surely exaggerated, is especially apt in the case of twentieth-century neurophysiology.

Every account of twentieth-century neurophysiology makes it clear that conceptual developments in the field were so closely bound up with advances in methods, instruments, and materials that it seems almost artificial to draw a distinction between its conceptual and its technical sides. This point emerges with special clarity when one recognizes the intimate link between particular instruments and specific research programs in the field. Even before World War I several leading neurophysiologists made their mark chiefly through their technical skill, one prominent example being the Cambridge physiologist Keith Lucas. Though once described as "essentially an engineer," Lucas laid much of the groundwork for the extension of the all-or-none law from cardiac muscle to other tissues, a research program that was further pursued by his student, the future Nobel Laureate Edgar Douglas Adrian. The next generation of neurophysiologists expressed admiration for Lucas's technical skills, tinged with regret that he had died—in 1916, in an airplane crash—before he could profit from the new physiological instruments that became available after, and indeed largely because of, World War I.[31]

The crucial common feature of postwar instruments was their capacity to amplify and record bioelectrical phenomena without the distortion produced by the recording levers in such traditional instruments as the kymograph (a revolving smoked cylinder that inscribed traces of muscular contraction). During World War I a few American physiologists, notably Alexander Forbes of Harvard, became aware of the potential utility for physiological research of amplified electronic waves, as exemplified by the radio compass. Increasingly refined versions of instruments based on this principle reached sophisticated expression in the cathode-ray oscillograph devised by Herbert Gasser and his associates at Washington University. Their device helped clear the way for a newly precise analysis of the effects of individual fiber size and other features of nervous tissue. In 1944 Gasser and his senior colleague Joseph Erlanger were awarded the Nobel Prize "for their work on 'the highly differentiated functions of single nerve fibers,' [but] the award implicitly recognized Erlanger and Gasser's seminal role in developing the single most important instrumental tool in modern neurophysiology, the amplifier *cum* cathode ray oscillograph."[32]

Much more could and should be said about the process by which the amplified

---

[30] Fenn as quoted in Gerard, *Mirror to Physiology* (cit. n. 23), p. 245. Cf. Clarke and Fujimura, *The Right Tools for the Job* (cit. n. 5).

[31] See Frank, "All-or-None Principle" (cit. n. 28).

[32] Frank, "The Joseph Erlanger Collection" (cit. n. 7), p. 195. See also Frank, "All-or-None Principle" (cit. n. 28); and Frank and Goetzel, "The Alexander Forbes Papers"; Marshall, "The Axonolo-

cathode-ray oscillograph was developed—not least because it both represents a striking example of the importance of "tinkering" and manual skills in science and draws our attention to the relations between experimental physiologists and industrial corporations such as Western Electric. Several other examples along the same lines could be drawn from the history of modern neurophysiology, and in fact Louise Marshall has already emphasized the extent to which research programs and groups in the field were associated with the exploitation of instruments, including the microelectrode in Gerard's laboratory at the University of Chicago.[33]

Not all neurophysiologists welcomed this full-blown "instrumentalization" of the field. By 1952 Gerard himself offered the following admonitory comment:

> What is important, and a change in kind, is that the users of instruments are increasingly not their masters. Once, any physiologist could tinker a kymograph into good behavior and even make or have one made in the shop in the basement. Few today dare open the crinkle-finish black boxes purchased from some "radio" firm, and, even of those who do, a small number indeed could carry on without the services of an expert electronics engineer. This may be unfortunate, but it is certainly inevitable. Not only do instrument societies flourish now, but a formal discipline of instrumentology is rapidly becoming established—indeed, becoming subdivided into new specialties—so that a self-respecting physiology laboratory can hardly limp along with only (besides technicians) glass blower, mechanic and electrical factotum.[34]

III. ANIMAL VIROLOGY

Like neurobiology, virology has become a broad, central subject in twentieth-century experimental life science. Fundamentally important in and of itself, it links a number of essential branches of biology—in the early decades of the century, botany, plant pathology, human and veterinary medicine, and bacteriology; in the later ones, genetics, protein chemistry, cytology, and molecular biology. The field comprises three main branches—bacterial, plant, and animal virology. A few popular and scholarly studies have attempted to deal with the overall history of the subject; the best scholarly study is that by the British virologist A. P. Waterson and the historian Lise Wilkinson.[35] However, these studies are of necessity introductory, not least because only one branch of the field—bacterial virology—has been well studied historically.

The mid-century history of bacterial virology has received abundant historiographic attention because of the key role it played in the development of molecular genetics. A good deal is known about the work at its principal centers, notably the Pasteur Institute in Paris, Cambridge University, and the amorphous American phage school that formed in the mid 1940s around the study of phage—the term for viruses that prey on bacteria—as a means of getting at the physical and chemical

---

gists at Washington University"; and Marshall, "Instruments, Techniques, and Social Units in American Neurophysiology" (all cit. n. 7).

[33] Marshall, "Instruments, Techniques, and Social Units in American Neurophysiology."

[34] Gerard, "The Organization of Science" (cit. n. 24), p. 153.

[35] A. P. Waterson and Lise Wilkinson, *An Introduction to the History of Virology* (Cambridge: Cambridge Univ. Press, 1978). See also Sally Smith Hughes, *The Virus: A History of the Concept* (New York: Science History Publications, 1977), most of which is concerned with the period before 1900, and Peter Radetsky, *The Invisible Invaders: The Story of the Emerging Age of Viruses* (New York: Little, Brown, 1991), a pioneering popular account.

basis of genetics. The phage group's founders and guiding spirits were Max Delbrück, Salvador Luria, and Alfred D. Hershey. The trio were at different institutions—Delbrück at the California Institute of Technology, Luria at Indiana University, Hershey at Washington University—but they conjoined during summers at the Cold Spring Harbor Laboratory on Long Island, to do research and teach about phage to new recruits. The phage group's scientific hallmarks included using simple, uniform biological systems—for example, bacteria and phage isolated and bred to have standard characteristics—and studying them with quantitative experimental techniques. The crucial role of phage research in the early development of molecular biology was signified by the award of the Nobel Prize in physiology or medicine in 1969 to Delbrück, Luria, and Hershey.[36]

The principal object of historiographic attention in plant virology has been the tobacco mosaic virus, again because of its connection with the development of molecular genetics. Identified in the late nineteenth century, tobacco mosaic virus emerged as a model plant virus in the 1920s and was much studied thereafter. It provided information that illuminated bacterial and animal virology later on, the most dramatic being, as Wendell Stanley demonstrated in 1935, that it could be crystallized and thus analyzed as a physicochemical substance.[37] However, historians have written little about plant virology apart from the tobacco mosaic virus, and they have devoted still less study to animal virology.

Thus plant and animal virology are rich with historiographic opportunity. So, we would claim, is bacterial virology, for historians to date have tended to emphasize conceptual developments—the interplay between ideas and concepts on the one side and experiment and technique on the other. With some exceptions, it is generally acontextual, inattentive to features of the research environment—local, national, and international—that gave rise to fundamental advances or allowed them to occur.[38] While Waterson and Wilkinson emphasize conceptual accomplishment, they also point out that virology raises a variety of issues, including how new disciplines arise and scientific institutions are rearranged, how scientific research has been related to medical practice, and how it can depend heavily on instruments and methods.[39]

In all, virology is a prime subject for the type of treatment outlined in the analytical framework we have advanced. To illustrate the value of the framework and the historiographic needs and opportunities in the field, we here focus on its least studied branch—animal virology—while attending to plant and bacterial virology as

[36] See Ernst Peter Fischer and Carol Lipson, *Thinking about Science: Max Delbrück and the Origins of Molecular Biology* (New York: Norton, 1988), pp. 148–166; Robert C. Olby, *The Path to the Double Helix* (Seattle: Univ. Washington Press, 1974), pp. 225–226, 238–240; and Horace F. Judson, *The Eighth Day of Creation: The Makers of the Revolution in Biology* (New York: Simon & Schuster, 1979), p. 70.

[37] Waterson and Wilkinson, *Introduction to the History of Virology* (cit. n. 35), p. 120.

[38] Prominent exceptions include Pnina G. Abir-Am, "The Discourse of Physical Power and Biological Knowledge in the 1930's: A Reappraisal of the Rockefeller Foundation's 'Policy' in Molecular Biology," *Soc. Stud. Sci.,* 1982, *12*:341–382; Abir-Am, "Essay Review: How Scientists View Their Heroes: Some Remarks on the Mechanism of Myth Construction," *J. Hist. Biol.,* 1982, *15*:281–315; Abir-Am, "The Biotheoretical Gatherings: Transdisciplinary Authority and the Incipient Legitimation of Molecular Biology in the 1930's: New Perspective on the Historical Sociology of Science," *History of Science,* 1987, *25*:1–70; Lily E. Kay, "Cooperative Individualism and the Growth of Molecular Biology at the California Institute of Technology, 1928–1953" (Ph.D. diss., Johns Hopkins University, 1987); and Kay, *The Molecular Vision of Life: Caltech, the Rockefeller Foundation, and the Rise of the New Biology* (New York: Oxford Univ. Press, 1991).

[39] Waterson and Wilkinson, *Introduction to the History of Virology* (cit. n. 35), pp. viii–xii.

necessary and appropriate. Although ranging through much of the twentieth century, our discussion is centered on the period from the late 1920s, when animal virology was a nascent field at best, extremely limited not only in knowledge but in numbers of practitioners and arsenal of basic methods, to the 1960s, when it emerged as one of the leading fields of microbiology.

Goals, Patronage, Institutions

Animal viruses were often investigated because they cause disease in animals and human beings, particularly infectious diseases such as rabies, equine encephalitis, foot and mouth disease, yellow fever, influenza, and polio. An eagerness to deal with viral diseases (in plants as well as animals) prompted the establishment of patronage and institutions for viral research—for example, the Potato Virus Research Station at Cambridge University, first funded privately by the biologist R. N. Salaman; the viral research institution that the German Ministry of Agriculture created in 1910, placing it on the island of Riems so as to isolate the work from mainland farm communities; and the Division for Plant Pathology that the Rockefeller Foundation funded at Princeton in 1931, which counted Wendell Stanley among its first staff recruits.[40] No doubt research in plant and animal virology went on in many other agricultural research institutions, public and private, whose development and research programs expressed concern with the particular vulnerabilities of local crops and animal breeds.

Work on viruses threatening to human beings was naturally pursued in the medical arena. A key institutional locus was the Rockefeller Institute for Medical Research in New York City, where important attention was given to common infectious diseases. There Peyton Rous suggested in 1911 that cancer might result from viral infection, demonstrating that a nonfilterable agent, as viruses were initially termed, would transmit sarcomas in chickens. Between the 1930s and the mid 1950s the principal sources of funds for research and training in animal virology were philanthropic agencies concerned with combating infectious diseases. The Rockefeller Foundation, for example enlarged its long-standing concern with international health and development to include investigations in viral diseases transmitted by insects, among them yellow fever, dengue, and encephalitis; and the American Cancer Society came to play a role in virology after it began supporting research in 1948.[41]

The program of the National Foundation for Infantile Paralysis (NFIP) exemplifies the significance of such philanthropic agencies in fostering advances in animal virology. Established in 1938 by Basil O'Connor, Franklin Delano Roosevelt's former law partner and ongoing confidant, the NFIP raised money each year through its March of Dimes campaign—enough to provide an annual operating budget of almost $3 million in 1940, close to $20 million in 1945, and more than $50 million in 1953. Committed to fighting and eventually eliminating the disease of poliomyelitis, the NFIP used its money to explore the nature of the disease and to develop defenses against it.[42]

[40] Ibid., pp. 122, 130, 143.
[41] See The President's Review from the Rockefeller Foundation Annual Report, 1956 (New York: Rockefeller Foundation, 1956), p. 29; and James T. Patterson, The Dread Disease: Cancer and Modern American Culture (Cambridge, Mass.: Harvard Univ. Press, 1987), pp. 171–172.
[42] Smith, Patenting the Sun (cit. n. 3), pp. 82, 161.

In formulating and developing its program, the NFIP consulted biological and medical experts. The experts were well aware that poliomyelitis was caused by an animal virus that attacked the cells of the nervous system, but that little was understood about the virus itself or how to proceed in dealing with the disease. They apparently advised NFIP officials to mount a two-pronged attack: award research grants to advance knowledge of the polio virus in particular and of animal viruses in general; and give postdoctoral fellowships to promising young scientists so as to increase the number of trained practitioners in the field. The magnitude of its activities is suggested by the fact that even in 1953, when the National Institutes of Health (NIH) made microbiology an explicit commitment of its external grants program, providing some support for work in polio, the NFIP spent more than twenty-five times as much on polio research as did NIH, which then devoted the largest share of its grant money to cancer research. Between 1938 and 1956 the NFIP awarded 322 postdoctoral fellowships in virology and other fields related to polio, including 97 in microbiology. An official at the foundation estimated in 1956 that no fewer than one third of the virologists under 45 in the United States had been trained under NFIP fellowships.[43]

In the twenty years after 1938 NFIP grants went for work of pathbreaking significance across a broad spectrum of microbiology. By 1956, 1,870 papers had been published that acknowledged its assistance: roughly 10 percent were in basic biochemistry, 14 percent in basic physiology, and 20 percent in viruses and viral diseases other than polio. The foundation's grants included sizable subventions to Linus Pauling at the California Institute of Technology for research into the structure of proteins, nucleic acids, and their components, and to Wendell Stanley, who had moved to Berkeley, for inquiries into the physical and chemical properties of plant, animal, and bacterial viruses. Its postdoctoral awards included a fellowship to James D. Watson that supported him during the year he puzzled out the structure of DNA with Francis Crick.[44]

Concepts and Research Programs

The fight against polio involved research into the epidemiology of the disease, the isolation and identification of its causative viral strains, and the development of a vaccine against it. How other viral diseases have been approached awaits systematic historical investigation. One wonders what constituted basic research in animal viruses in the premolecular era, what concepts were brought to it, and what advances such research yielded.

Peyton Rous's demonstration that cancer might be an infectious viral disease led others to investigate that possibility in animals other than chickens. The research

---

[43] In 1953 the NIH polio research budget was $72,000; the NFIP's, $2 million. See Smith, *Patenting the Sun* (cit. n. 3), p. 249; and T. E. Boyd, memo to Basil O'Connor, "Contributions to Science in the Field of Poliomyelitis," March of Dimes Birth Defects Foundation Archives, White Plains, New York, [1956], pp. 19–20.

[44] Boyd, memo to Basil O'Connor, pp. 23–24, 21–22, 31–32; and James D. Watson, *The Double Helix* (New York: Athenaeum, 1968), p. 132. The NFIP awarded grants for work on the encephalitides virus at Berkeley; on human viral diseases at Harvard, some of which monies were given to Enders; and on animal and plant viruses and biophysical properties of viruses at the University of Pittsburgh, where the program was stimulated by the arrival of Salk in 1947. Boyd, memo to Basil O'Connor, pp. 20–22.

program apparently followed Rous's: attempt to stimulate tumor growth in a healthy animal by injecting a nonfilterable extract obtained from a malignancy in a cancerous one. The program failed: for twenty years after Rous's initial experiment, neither Rous nor anyone else was able to transmit tumorous growths by inoculation in mammals. (In 1908 two Danish pathologists had isolated a nonfilterable agent that induced fatal leukemia in chickens. However, since leukemia at the time was not considered to be a form of cancer, their results were not thought relevant.[45]) Where and how these experimental attempts were conducted and why they failed requires investigation.

Whoever did them, the failures led to widespread rejection of the idea that cancer had much if anything to do with viruses; yet the concept and the research program that accompanied it remained alive at the Rockefeller Institute. In 1931 a member of the Institute staff named R. E. Shope examined a freshly shot rabbit with tumorlike growths and showed that the condition was transmissible in rabbits by a nonfilterable agent. In 1932 Shope investigated a papilloma found among the wild rabbit population in Iowa and Kansas, demonstrating that this too was caused by a nonfilterable agent. Indeed, upon injection with the wild rabbit agent, domestic rabbits developed papillomas that were at first benign but then became malignant.[46]

Shope's results by no means moved theories of oncogenesis in a viral direction. Scientists by and large looked elsewhere for the causes of cancer, entertaining a variety of theories and pursuing diverse research programs in consequence. Like the Rockefeller research program, these theories and research programs deserve historical scrutiny. Among the plausible theories was the idea that cancer had something to do with genes. Indications of a genetic basis for cancer came from several notable clusters of evidence: The disease often ran in families, which suggested some hereditary predisposition to it; particular cancers occurred with high frequency in certain lines of mice; malignant cells multiplied into more malignant cells; and mutagenesis (for example, by radiation) could lead to malignancies. To the end of exploring the genetic theory of cancer, biologists at the Jackson Laboratory in Bar Harbor, Maine, bred pure strains of mice differing from one another in their frequency of cancer, hoping to find a clue to oncogenesis through the classical Mendelian methods of crossing and backcrossing. In a recent book the Swedish biologist George Klein recounts that the program produced a startling result: "The hybrid offspring from a cross between a high-breast cancer strain and a low-cancer strain developed breast cancer at a relatively high frequency if the mother belonged to the high-incidence strain and the father to the low-incidence strain, but the offspring had a low incidence of cancer if the opposite was the case." In 1936 John Bittner at the Jackson Laboratory traced the phenomenon to the transmission from mouse mother to child of what he called a "milk factor," which later was termed the mouse mammary tumor virus (MMTV). At the time Bittner was actually convinced that the milk contained a virus that increased risk of breast cancer in the mouse but was not sufficient to give the disease. (While 90 percent of the maternal strain of mice contracted breast cancer, no more than 30 percent of the offspring did, which suggested that susceptibility to cancer, arising perhaps from hormones, might be of comparable importance to viruses in generating the disease.) According to Klein, Bittner used the term "milk

---

[45] Waterson and Wilkinson, *Introduction to the History of Virology* (cit. n. 35), p. 159.
[46] *Ibid.*, p. 160.

factor" instead of "virus" because he was reluctant to challenge the prevailing orthodoxy that cancer had nothing to do with viruses, explaining, "If I had called it a
virus, my grant applications would automatically have been put into the category of
'unrespectable proposals.' As long as I used the term 'factor,' it was respectable genetics."[47]

The viral role in oncogenesis nevertheless continued to attract at least some biologists in the 1930s (it would be useful to know which of them and with what research
consequences). One of them was Emory Ellis, a biologist at the California Institute
of Technology, and the research consequences of his innovation were considerable.
Ellis had trained as a physical chemist and began to work with viruses when he
received a fellowship for cancer research. He was aware that specific viruses caused
diseases in plants, lysis in some bacterial species, and some cancerous growths in
animals, and that the malignancies seemed to require both the presence of the right
virus and the susceptibility of the cell. Ellis expected that learning more about the
nature of viruses would help one understand such malignancies and perhaps those
of other origins.[48] How to acquire that knowledge—what model system to adopt—
was the question.

Ellis and his colleagues were reluctant to work with an animal virus like that
responsible for rabbit papilloma because of the care, time, and money required when
working with a large animal colony. The cost of investigating a plant virus such as
the tobacco mosaic virus would be lower but still significant. To Ellis, it seemed clear
that the most advantageous model system to use was bacteriophage, which required
virtually no care, occupied little laboratory space, would yield results in a matter of
hours, and would—lending themselves to a technique developed by the French biologist Félix d'Hérelle, to whom Ellis acknowledged a debt—make their activity known
by the production of readily observable plaques on a Petri-dish bacteria lawn.[49]

Ellis recalled that there also "appeared to exist some formal similarities in the
processes of bacteriophagy, fertilization of egg-cells by sperm and infection in virus
diseases." He added, "If these do indeed have common aspects, even though taking
place in substrates as different as man and bacteria, then study of the process in the
system lending itself to quantitative study seemed likely to be the most rewarding."
The similarities as well as obvious differences among the three processes motivated
his detailed study of bacteriophagy, Ellis remembered. "We hoped that once we
understood it, we would be in a better position to understand virus-induced malignancies. It was this argument which led us to start work on bacteriophage."[50]

The arrival of Max Delbrück at Caltech in 1937 soon broadened the work on
phage into what became the phage school, which reworked the original program
into one of bacterial genetics as such. However, historians ought to remember the
original argument that brought Ellis to adopt bacteriophage as his model system.

[47] George Klein, *The Atheist and the Holy City: Encounters and Reflections,* trans. Theodore and
Ingrid Friedman (Cambridge, Mass.: MIT Press, 1990), pp. 120–122, quoting from pp. 121, 122.
See also Waterson and Wilkinson, *Introduction to the History of Virology* (cit. n. 35), p. 161.

[48] Emory L. Ellis, "Bacteriophage: One-Step Growth," in *Phage and the Origins of Molecular
Biology,* ed. John Cairns, Gunther Stent, and James D. Watson (Cold Spring Harbor, N.Y.: Cold
Spring Harbor Laboratory of Quantitative Biology, 1966), pp. 53–54. See also Fischer and Lipson,
Thinking about Science (cit. n. 36), p. 114; and William C. Summers, "How and Why Bacteriophage
Came to Be Used by the Phage Group," unpublished MS in D. J. Kevles's possession, 1991, pp. 2–4.

[49] Ellis, "Bacteriophage," pp. 54–56.

[50] *Ibid.* See also Waterson and Wilkinson, *Introduction to the History of Virology* (cit. n. 35), p. 103.

The argument locates an important root of the phage school not only in the genius and philosophical commitments of Delbrück but in the ongoing tradition of inquiry into the causes of disease, particularly the school of viral oncogenesis that goes back to Peyton Rous.

Techniques, Instruments, Materials

What prompted Ellis to reject work with animal viruses—the need to use live animals—was a major problem for animal virology. It had long been recognized that viruses would not grow outside the living cell, which meant that the most convenient place for growing them was live animals. The best live animals for the purpose were those that, like mice or rabbits, were small and reproduced relatively quickly. In the early 1930s mice were indeed adapted for the study of the human influenza virus. But many animal viruses could be not be cultivated in mice. The polio virus, for example, could only be grown in monkeys, which were employed early in the century to demonstrate that polio was a viral disease of the central nervous system. In the 1930s the only effective means of cultivating polio virus was to inject it into monkeys, let it grow, then harvest it by killing the animals. Even when small animals could be used, the in vivo constraints made studies of animal viruses in the laboratory expensive, time consuming, and cumbersome, largely beyond the kind of controlled experiments that might permit analyses of how viral infection or oncogenesis worked, how viruses reproduced, even what they comprised. Animal cultures compelled the virologist to try to deduce from the animal's reaction to infection some information on the properties and the nature of viruses.

As early as World War I scientists tried to get around the difficulty by resorting to tissue culture—in vitro accumulations of living and reproducing cells—as a medium for growing animal viruses. Much is known about the early history of tissue culture. Between 1907 and 1911 the Yale biologist Ross G. Harrison pioneered a fundamental type of the technique—the so-called hanging-drop method—for studying the development of nerve fiber tissue. However, Harrison's method did not provide tissue cultures suitable for animal viruses, and for reasons that historical study might expose, developing such cultures was not a simple matter.[51]

In Manchester in 1928, for example, H. B. Maitland and his wife, Mary Cowan Maitland, introduced a technique that kept cells viable for a short time and, though growth was minimal, allowed them to express enough activity to multiply certain viruses for study. Max Theiler used the Maitlands' technique when developing a yellow fever vaccine. However, the technique could not be used for isolation of a virus from a test material. In the years bracketing World War I, Alexis Carrel at the Rockefeller Institute devised ingenious methods of tissue culture that could be adapted to the in vitro cultivation of animal viruses. In 1927 he and his collaborator Tom Rivers exclaimed that "one finely pulped chicken embryo might be capable of producing as much vaccine as a calf." Still, Carrel's methods were extremely complicated, particularly the intricate set of procedures required to keep the culture free from bacterial contamination. Years later a professor at the Royal Caroline Insti-

---

[51] In 1928 Alexis Carrel, in "Tissue Culture in the Study of Viruses," part of the classic text *Filterable Viruses,* edited by Tom Rivers (London: Ballière, Tindall, & Cox, 1928), ascribed the lack of progress to reliance on "the comparatively crude procedure . . . derived . . . from the experiments of Harrison." Waterson and Wilkinson, *Introduction to the History of Virology* (cit. n. 35), pp. 72–73.

tute in Sweden would note that Carrel's was "a complicated ritual. . . . Tissue culture developed almost into a tissue cult, a mystery the secret rites of which were revealed only to a narrow circle of inaugurates with Carrel as their high priest."[52]

In 1931 A. M. Woodruff and E. W. Goodpasture reported an advantageous method: growing animal viruses on the sheets of cells formed by the extraembryonic membranes of the chick embryo inside the fertilized egg. Their method was comparatively successful and widely used during the 1930s. One of the leading pioneers in animal virology, F. Macfarlane Burnet in Australia, succeeded in growing the influenza virus in the developing egg, for example. As Waterson and Wilkinson note, "The egg can be seen as a particularly cheap and convenient experimental animal; by a stretch of imagination (and definition) it can perhaps also be seen as a very sophisticated kind of tissue culture, carrying its own medium, by the same token that W. Roux's frog embryo experiments are often seen as the beginnings of tissue culture."[53]

For all their utility, chicken embryos were not a suitable host for all animal viruses of interest. In 1936 Albert Sabin and Peter K. Olitsky tried to grow polio virus in chicken embryos and failed. They also failed to grow it in Maitland cultures of mice and monkeys. They succeeded only with human embryonic brain tissue. The result fostered the idea, mistaken as it eventually turned out, that the polio virus was strictly neurotropic; it also discouraged follow-up of that particular culture technique because human embryonic tissue was an unsuitable medium for culturing viruses that might be used in vaccinations.[54] The National Foundation for Infantile Paralysis remained eager to find a culture that was suitable for the polio virus. In the late 1940s it awarded funds for research to John Enders, a medical research scientist at the Boston Children's Hospital, where he headed a small group at work on tissue-culturing infectious viruses.

Enders became interested in the viral culturing problem while on the staff of the Harvard Medical School during the 1930s. His research was interrupted by the war, but in 1947 he resumed exploration of tissue culture in collaboration with Thomas H. Weller, who as a Harvard medical student had assisted him just before the war, and Frederick C. Robbins, Weller's roommate at medical school. Enders, Weller, and Robbins soon succeeded in growing mumps virus in cultured chicken cells with their innovative technique of continuous culture, periodically replacing the nutritive medium while leaving the viral culture intact. The collaborators then sought to apply their technique to the cultivation of varicella (chicken pox) virus in cultures of its natural host, human embryonic skin and muscle tissues. In 1948, appropriating some of these cultures, they managed to cultivate the polio virus, an achievement that was to earn them the 1954 Nobel Prize in physiology or medicine.

They originally had no intention of experimenting with the polio virus, but were aware of the mounting evidence that it might not be a strict neurotrope. They and

---

[52] "Physiology or Medicine 1954: Presentation Speech by Professor S. Gard, member of the Staff of professors of the Royal Caroline Institute," *Nobel Lectures in Physiology or Medicine, 1942–1962* (Amsterdam: Elsevier, 1964), p. 444. See also *ibid.,* p. 445; and Waterson and Wilkinson, *Introduction to the History of Virology* (cit. n. 35), pp. 144 (Maitlands), 68–73 (Carrel).

[53] Waterson and Wilkinson, *Introduction to the History of Virology,* pp. 76, 138–139.

[54] Gard, "Physiology or Medicine 1954" (cit. n. 52), p. 445. Earlier, MacFarlane Burnet and A. V. Jackson in Australia had reported growing polio virus under similar conditions, but neither they nor anyone else had followed up the breakthrough: Boyd, memo to Basil O'Connor (cit. n. 43), p. 30.

others found it difficult to see, for example, how the nervous system alone could produce the abundant quantities of polio virus found in the feces of many patients. They also had in a laboratory freezer a sample of the Lansing strain of polio virus sent them some time earlier by the NFIP. As they reported in their Nobel address, "Thereupon it suddenly occurred to us that everything had been prepared almost without conscious effort on our part for a new attempt to cultivate the agent in extraneural tissue." According to a later account by a member of the NFIP, Weller had prepared too many tubes of culture medium for the experiment with the chicken pox virus, so Enders suggested that he seed the cultures with some polio virus from the laboratory freezer.[55]

The demonstration that polio virus could be grown in non-nerve cell tissue cultures was a stunning part but not the whole of the Enders group's achievement. With the mumps virus, their technique involved growing cells suspended in fluids; for polio, they developed methods for growing them in a solid layer. They also devised methods for keeping track of the multiplication of the virus and for using cell cultures containing the virus to test for poliomyelitis antibodies. Perhaps even more significant, they made it possible to recover usable polio virus from feces or spinal cord suspensions by suppressing the bacterial contamination of these sources with the newly available antibiotics, penicillin and streptomycin, then centrifuging the sample. They thus eliminated the need to obtain polio virus via the laborious and time-consuming procedure of intracerebral inoculation of monkeys.[56]

The feat of the Enders group not only transformed polio virus production, emancipating it from the expensive use of live monkeys and pointing the way to large-scale production of a polio vaccine; it also promised to revolutionize animal virology by liberating the field in general from the grip of Carrel's tissue cult. It provided methods for growing animal viruses reliably and efficiently in vitro and for acquiring them in abundance. As Enders, Weller, and Robbins noted in their Nobel address, the application of antibiotics had made it "possible to apply tissue culture to the routine isolation of viruses from materials heavily contaminated with microorganisms" and "to use them under conditions and in numbers which in the past would have been quite unthinkable."[57]

Tissue culture was thus revolutionized by the Enders group's work. In short order many new animal viruses were discovered, including, by the mid 1950s, at least eighteen different immunologic types of the human adenoviruses.[58] It would seem obvious that this revolution in tissue culture and its consequences warrants historical investigation. The role of the new antibiotics in the revolution also raises the historiographic question of the effect of World War II on the postwar development of the life sciences. Many medical researchers went off to war. During the war Enders was

[55] John F. Enders, Frederic Robbins, and Thomas H. Weller, "The Cultivation of the Poliomyelitis Viruses in Tissue Culture" (Nobel Lecture, 11 Dec. 1954), in *Nobel Lectures in Physiology or Medicine* (cit. n. 52), pp. 448–467, on. p. 451; and Boyd, memo to Basil O'Connor (cit. n. 43), p. 30.

[56] "John F. Enders," in *Nobel Prize Winners: An H. W. Wilson Biographical Dictionary*, ed. Tyler Wasson (New York: H. W. Wilson, 1987), pp. 300–302; and Enders, Robbins, and Weller, "Cultivation of the Poliomyelitis Viruses in Tissue Culture," p. 457.

[57] Smith, *Patenting the Sun* (cit. n. 3), pp. 135–137; Boyd, memo to Basil O'Connor, p. 30; and Enders, Robbins, and Weller, "Cultivation of the Poliomyelitis Viruses in Tissue Culture," pp. 458–459.

[58] Enders, Robbins, and Weller, "Cultivation of the Poliomyelitis Viruses in Tissue Culture," p. 465.

a consultant on epidemic diseases to the secretary of war, and Weller, a member of the Army Medical Corps, was stationed at the Antilles Medical Laboratory in Puerto Rico, where he headed the Departments of Bacteriology, Virology, and Parasitology. One wonders how the war changed the outlooks of biological practitioners and affected their research programs.

Certainly the war affected the materials and instruments available to animal virologists. Although radioactive tracers were produced by cyclotrons before the war, the nuclear piles of the Manhattan Project and then the Atomic Energy Commission yielded them in still greater variety and abundance. In the postwar era such tracers exercised "enormous impact across the whole spectrum of biological research," to cite the judgment of Waterson and Wilkinson. They were indispensable, for example, to Alfred Hershey and Martha Chase in their classic demonstration that the viral protein coat is adsorbed on the surface of the host cell by its tail, which then injects the DNA of the virus into the cell.[59]

One instrument that benefited the study of animal viruses, which are too small to be seen under the ordinary light microscope, was the electron microscope—invented before the war for use in physics. In 1940 RCA funded (at $3,000 a year) a National Research Council fellowship for the purpose of exploring the instrument's biological applications, and the young biologist appointed to the fellowship, Thomas F. Anderson, began using the microscope to photograph the tobacco mosaic virus and bacteriophage. During the second year of his fellowship Anderson began to work with Salvador Luria, who had visited RCA to explore the possibility of using the electron microscope to check the size of some bacteriophages which he and a collaborator had just estimated from X-ray cross sections. In 1942 Max Delbrück joined the electron picture-taking.[60]

Perhaps the most important visual evidence that the electron microscope provided during the war years was that phage particles multiply inside the cells, rather than at their surfaces; until lysis occurs, the number of particles visible at the surface remains constant. This constancy also means that very few, if any, of the particles enter the cell, an observation that seemed to Delbrück to be of the "greatest consequence" and led him to revise his thinking about how phage reproduced. According to Anderson's later reflections, the electron microscope brought to the fore "the deeper mysteries of how the particles are organized, what the function of each part might be, and why the particles appear to remain on the surface of the host instead of diving into it like a respectable parasite. The resolution of these mysteries has been shown to require the intelligent application of additional methods of research—the microscope can only suggest solutions, not confirm them."[61]

After the war, as its technology and resolution improved, the electron microscope became an increasingly valuable adjunct to virological research, widely used in all three branches of the field. It revealed viruses as concrete objects to think about, permitted them to be distinguished from one another morphologically, and provided visual tests of theories concerning viral properties and behavior that were

---

[59] Waterson and Wilkinson, *Introduction to the History of Virology* (cit. n. 35), p. 108.

[60] *Ibid.*, pp. 105–106.

[61] Thomas F. Anderson, "Electron Microscopy of Phages," in *Phage and the Origins of Molecular Biology,* ed. Cairns, Stent, and Watson (cit. n. 48), pp. 63–64, 67, 77 (long quotation). See also the obituary of Anderson, *New York Times,* 13 Aug. 1991, p. C19.

arrived at by other means.[62] This bare outline of accomplishments suggests that the precise role of the electron microscope in virology and other branches of experimental biology deserves systematic historical analysis. Certainly it awaits historical scrutiny.

IV. CONTEXT AND CONTINGENCY

In 1969, we are told, there was "a good deal of handwringing by some members of the American Physiological Society" when a new group of "Young Turks" established the independent and interdisciplinary Society for Neuroscience.[63] The anxiety was not merely "institutional," in the narrow sense that the old guard in the American Physiological Society feared a loss of members to the new and independent group. Conceptual issues and prospects for future funding were also at stake. Neuroscience was concerned mainly with aspects of brain function instead of classical neuromuscular topics. As such, it reached out toward such nascent fields as cybernetics and cognitive science instead of the traditional and clinically oriented specialties of neuroanatomy, neurology, neurosurgery, and psychiatry.

In this new context some of those who had flourished during the golden age of "classical" neurophysiology might have begun to doubt the wisdom of the Faustian bargain they had made with the utilitarian goals of medicine as conceived between the two world wars. Like cardiology, which relied on similar kinds of noninertial graphical recording instruments, neurophysiology was more than a little unsure about its immediate clinical utility.[64] Some of the less arcane results of neurophysiological research did seem to have implications for neurological *diagnoses,* but its direct *therapeutic* benefits were hard to see or even imagine—with the possible exception, it was sometimes supposed, of the guidance it gave to neurosurgeons performing lobotomies and related operations. In the face of such doubts about the direct clinical utility of their research, "classical" neurophysiologists could no longer rely so confidently on the "pure" intellectual excitement that their work had once aroused. Even in the "purely" intellectual arena, if not only there, enthusiasm was shifting toward the new interdisciplinary field called neuroscience.

The case was quite different in animal virology. The field received increasing attention during the 1950s, partly because the electron microscope revealed the presence of viruses in animal tumor cells, partly because during the decade a number of viruses were demonstrated to stimulate malignancies. One such virus, found to generate several types of tumors in mice, rats, and hamsters, was named the polyoma virus in recognition of its multiple potencies. (Why biologists found so much viral

---

[62] Anderson, "Electron Microscopy of Phages," p. 77; and Waterson and Wilkinson, *Introduction to the History of Virology* (cit. n. 35), p. 155. See also Boyd, memo to Basil O'Connor (cit. n. 43), pp. 2–3.

[63] Marshall, "Instruments, Techniques, and Social Units in American Neurophysiology" (cit. n. 7), p. 359.

[64] On cardiology see the superb studies by Robert G. Frank, Jr., "The Telltale Heart: Physiological Instruments, Graphic Methods, and Clinical Hopes, 1854–1914," in *The Investigative Enterprise: Experimental Physiology in Nineteenth-Century Medicine,* ed. W. Coleman and F. L. Holmes (Berkeley: Univ. California Press, 1988), pp. 211–290; and Joel Howell, "Cardiac Physiology and Clinical Medicine? Two Case Studies," in *Physiology in the American Context,* ed. Geison (cit. n. 2), pp. 279–292. See also Geison, "Physiologists and Clinicians in the American Context" (cit. n. 15).

oncogensis, as the phenomenon had come to be called, in the 1950s when they could not find it in the 1920s and 1930s is another puzzle for historians to explain.)[65]

Animal virological research was also accelerated in the 1950s by the merger of innovations in tissue culture with the quantitative, plaque-counting methods developed in bacterial genetics. A principal locus of the merger was the California Institute of Technology, where animal virology came to occupy several biologists in a group headed by Renato Dulbecco (and partially supported by the National Foundation for Infantile Paryalysis). Dulbecco, who had learned phage-group methods in the laboratory of Salvador Luria, devised ingenious methods for culturing animal viruses in monolayers of human or animal tissue spread out on a flat dish. The methods made cellular degeneration arising from viral infection visible as a plaque. Applying the techniques of phage analysis to such cultures, Dulbecco and his collaborator Marguerite Vogt were able to pursue the type of genetic analysis of animal viruses, including polio viruses, that had been brilliantly accomplished with bacteriophage.[66]

The research of Dulbecco's group—which included Howard Temin and Harry Rubin—helped to establish animal viral genetics as an exciting field in its own right. It also suggested that the distinction between viral and genetic theories of oncogenic action was fuzzy, not least because Dulbecco and Vogt observed that the polyoma virus transformed—that is, caused to divide without restraint—hamster cells cultured in a laboratory dish. They also found that the virus quit reproducing in the transformed cells, which suggested, by analogy with the behavior of temperate phage, that its DNA had been incorporated into the genome of the cell itself, thus accounting for the transformation.[67]

By the 1960s, not only could viral genetics be pursued quantitatively in cell culture, but so also could animal-tumor virology—with the result, as James Watson later said, that "for the first time, thinking at the molecular level could begin." Tumor virology was additionally boosted by reports from a number of laboratories that the Rous sarcoma virus would induce tumorous growths not only in fowl but also in mammals, including mice, rats, hamsters, rabbits, and monkeys. Research on animal tumor viruses flourished, enlarging the texts published about them, forming a major branch of basic medical and biological science. In a sense the field had come full circle, moving from the seemingly dubious work of Peyton Rous into bacteriophage, then turning back to animal viruses via Dulbecco, among others. In 1966 the completion of the circle and the vitality of the field were recognized when Rous, at age eighty-five, shared the Nobel Prize in physiology or medicine.[68]

[65] Waterson and Wilkinson, *Introduction to the History of Virology* (cit. n. 35), p. 162; Ludwik Gross, *Oncogenic Viruses,* 2nd ed. (Oxford: Pergamon Press, 1970), pp. 10–13, 106, 264; W. Ray Bryan, "Peyton Rous," *Science,* 1966, *154:*364–365; "Peyton Rous," in *Nobel Prize Winners Biographical Dictionary* (cit. n. 56), pp. 889–890; J. Michael Bishop, "Oncogenes," *Scientific American,* 1982, *246:*81; and Natalie Angier, *Natural Obsessions: The Search for the Oncogene* (Boston: Houghton Mifflin, 1988), p. 5.

[66] Boyd, memo to Basil O'Connor (cit. n. 43), p. 31.

[67] Harry Rubin, "Quantitative Tumor Virology," in *Phage and the Origins of Molecular Biology,* ed. Cairns, Stent, and Watson (cit. n. 48), pp. 294–295; and Marguerite Vogt and Renato Dulbecco, "Virus-Cell Interaction with a Tumor-Producing Virus," *Proceedings of the National Academy of Sciences,* 1960, *46:*365–370.

[68] James D. Watson, "Foreword," in *Viral Oncogenes* (Cold Spring Harbor Symposia on Quantitative Biology, 44) (Cold Spring Harbor, N.Y.: Cold Spring Harbor Laboratory, 1980), pt. 1, p. xvii;

The scientific prospects of animal tumor virology helped generate a degree of boosterism for the field—a crash research program might find cures for cancer—and proclamations of that kind figured importantly in the creation of the so-called war on cancer in 1971, during the administration of President Richard M. Nixon. That war led to neither immediate therapies nor cure, but the huge investment of funds (several billion dollars) in the field accelerated the development of molecular biology and DNA technology in ways that are understood in outline but beg for systematic historical analysis.[69] Unlike the case with classical neurophysiology and neuroscience, what the cancer war did not realize in one way, it yielded in others, notably clinical payoffs such as DNA diagnostics and the immense stimulus that the molecular biological advances of the 1970s provided to the biotechnology industry. Then, too, animal virology as such has continued to flourish because of the role that viruses play in infectious disease and because practitioners in the field can point to unalloyed successes such as the polio vaccine and to dark challenges, notably the AIDS epidemic.

V. CONCLUDING REMARKS

We hope that our flexible analytic framework will be useful in accounting for the post-1960 transmutation of neuroscience and the further development of animal virology, as well as for other fields in the experimental life sciences. We wish to emphasize the importance of one category of that framework, the role of methods, instruments, materials. Until recently, it had been the topic most neglected by historians of the modern life sciences, perhaps because it has increasingly involved technological imports from other disciplines.[70] Here the historians were once in good company with those biologists whose resistance to recognizing the importance of materials and instrumentation was proportional to the sophistication of the instruments and materials on which they relied. Thus Professor Sven Gard, of the Royal Caroline Institute in Sweden, when, in 1954, he presented Enders, Robbins, and Weller for their Nobel Prize:

Gross, *Oncogenic Viruses* (cit. n. 65), pp. 136–139; "Rous, Francis Peyton," *McGraw-Hill Modern Scientists and Engineers*, 3 vols. (New York: McGraw-Hill, 1980), Vol. III, pp. 48–49; and Robert J. Huebner and George J. Todaro, "Oncogenes of RNA Tumor Viruses as Determinants of Cancer," *Proceedings of the National Academy of Sciences*, 1969, *64*:1087–1088. Rous shared the Nobel prize with Charles B. Huggins, who had demonstrated the role of hormones in the treatment of cancer, especially cancer of the prostate.
[69] Klein, *The Atheist and the Holy City* (cit. n. 47), p. 128.
[70] This long-standing neglect is now rapidly being rectified. See, e.g., Bruno Latour and Steve Woolgar, *Laboratory Life: The Construction of Scientific Facts* (Princeton: Princeton Univ. Press, 1987); Merriley Borell, "Extending the Senses: The Graphic Method," *Medical Heritage*, 1986, *2*:114–121; Borell, "Instruments and an Independent Physiology: The Harvard Physiological Laboratory, 1871–1906," in *Physiology in the American Context*, ed. Geison (cit. n. 2), pp. 293–321; Borell, "Instrumentation and the Rise of Modern Physiology," *Science & Technology Studies*, 1987, *5*:53–62; Timothy Lenoir, "Models and Instruments in the Development of Electrophysiology, 1845–1912," *Historical Studies in the Physical and Biological Sciences*, 1986, *17*:1–54; Lenoir, "Helmholtz and the Materialities of Communication," *Osiris*, 1994, *9*:185–207; Frank, "All-or-None Principle" (cit. n. 28); Alberto Cambrosio and Peter Keating, "'Going Monoclonal': Art, Science and Magic in the Day-to-Day Use of Hybridoma Technology," *Social Problems*, 1988, *35*:244–260; Lynch, "Sacrifice and the Transformation of the Animal Body into a Scientific Object" (cit. n. 8); R. J. Greenspan, "The Emergence of Neurogenetics," *The Neurosciences*, 1990, *2*:145–157; and esp. Clarke and Fujimura, eds., *The Right Tools for the Job* (cit. n. 5), including their extensive bibliography.

The electronics, radioactive isotopes, and complicated biochemistry of our age has threatened to turn medical science into something dangerously resembling technology. Now and again we need to be reminded of its fundamental biological elements. Against this background we express our admiration of the biological common sense, characterizing your approach to important medical problems, and of the wonderful simplicity of the solutions you have presented.[71]

Applied to our two case studies, our framework also calls attention to two important general points concerning the ascent and descent of disciplines. First, the rise and relative decline of "classical" neurophysiology indicates that the interplay between basic experimental biology and agriculture or medicine is not always marked by steady progress or uniformly effective results. Once-favored disciplines or specialities in the biomedical sciences can slip from their lofty perch if their clinical utility comes into doubt, and perhaps even more readily if they become intellectually less exciting than other specialties always ready to take their place. Second, the case of animal virology suggests that it is a mistake to think of medical or agricultural practices as "applied" experimental biology; in fact, the interplay has gone both ways, and medical or agricultural interests have often been essential to shaping developments in so-called basic research. Further, substitutes for a lack of immediate clinical payoff can be found in the richness of new intellectual programs and in the reward of unexpected utilitarian dividends. History is not only contextual; it is also contingent.

[71] Gard, "Physiology or Medicine 1954" (cit. n. 52), p. 447.

The History of Mathematics and *L'esprit humain*: A Critical Reappraisal

*By Joan L. Richards**

I N HIS CLASSIC WORK *The Structure of Scientific Revolutions* (1962), Thomas Kuhn gave historians of science a model of scientific change with a number of ordering concepts like "scientific revolution," "normal science," and the ever-elusive "paradigm." Investigating, sharpening, and testing these and related concepts has provided narrative structure for many subsequent historical investigations; refining, rejecting, and replacing these models has provided the focus of much philosophical work. Looking at Kuhn's work from the distance of almost thirty years, though, the importance of these specifics pales in the face of the message in the introduction, entitled "A Role for History." Here Kuhn pointed out the value of historical investigation as a way of coming to know the nature of science itself. Rejecting as inadequate to its history the view of science gleaned from a twentieth-century scientific education, he proposed that a new view be constructed from that history itself.[1]

From a historian's point of view, Kuhn's message can be seen as the history of science variant on the call Herbert Butterfield sent out eleven years earlier, to abandon "whig" history and cease to read the political past as a linear and progressive path to the present. Within the history of science its implications over the past three decades have been radical. When it was sounded, most people viewed the history of science as little more than a curious furbelow to enliven presentations of scientific knowledge and method. It was clear that historical researches in no way affected that knowledge and that method: Their validity and importance were firmly established elsewhere, not only in the marvelous successes of twentieth-century science but in clearly laid out philosophies that explained and justified those successes extrahistorically.

Subsequent historical research, pursued in the non-whig spirit Kuhn's work pioneered, has not changed the validity of Isaac Newton's theories, nor Albert Einstein's, nor Niels Bohr's. It has, however, played an active role in changing many people's perception of these theories and of their place in our understanding of the world and ourselves within that world. The willingness to entertain other views of scientific

---

* Department of History, Brown University, Providence, Rhode Island 02912.

This version of my article was much influenced by the comments of an anonymous reviewer and by conversation with Thomas Hankins and Kim Plofker.

[1] Thomas S. Kuhn, *The Structure of Scientific Revolutions,* 2d ed. (Chicago: Univ. Chicago Press, 1970).

knowledge in historical work was first battled out in debates about whether the historical development of science could be adequately understood through an "internalist" focus on modern scientific ideas alone, or whether the "externalist" view that included the larger conceptual picture in which those ideas were embedded had also to be considered. The broad-mindedness externalists expressed as a willingness to include and interpret past errors and distortions in descriptions of scientific development has slowly been expanded to a recognition that the boundaries of internal and external, of scientific truth and error, do not themselves stand fixed and immutable outside of the larger historical context. By expanding upon the political, social, and cultural implications of the phrase "matters of fact" in a particular time and place, for example, Steven Shapin's and Simon Schaffer's *Leviathan and the Air-Pump* (1985) raised questions about the phrase's self-evidence and timeless centrality. Theirs is but a particularly visible contribution to a growing literature exploring different historical manifestations of concepts as central to science as "experiment," "objectivity," or "scientific method." [2]

At present, the multiplicity of such studies is balancing if not overwhelming the search for a single model of scientific change that can embrace them all. Specific considerations of such models is often the motivation behind this or that investigation, but, at least for the nonce, no particular one seems adequate to the wealth of new materials and perspectives currently being explored. Sociology, anthropology, philosophy, historiography—all of these voices can be heard here or there, but what ties the effort together is none of these. It is much more the conviction that the insights gained by closely examining our past can provide the most enlightening view of our present and our science.

When he wrote *Structure,* Kuhn exempted mathematics from the model of dramatic, discontinuous change he constructed for the sciences. Even as he boldly elaborated upon the historically bound nature of the most imperviously empirical physical, chemical, and biological theories, he found no illustrations in mathematics. His non-whig approach both opened the sciences to the possibility that new understandings of them might be generated from their history and foreshadowed the current significant richness in the history of science, but the history of mathematics has not fared so well.

The difference can be seen in terms of the so-called externalist-internalist debate. At the 1991 "Critical Problems" conference represented by this volume, Shapin read what was essentially an obituary for this distinction in the history of science, remarking that "within a generation . . . [it] seems to have passed from the commonplace to the gauche." [3] The historians of mathematics at the same gathering, however, were actively engaged in drawing an ever sharper line to divide themselves into these two groups. The anonymous referee of this article (in the version given at the meeting) pointed to the problem and commented: "One of the critical problems in the history of mathematics is the reconciliation, once and for all, of the internalist-externalist debate."

I would differ only to insist that the division between these two camps is not only

[2] Steven Shapin and Simon Schaffer, *Leviathan and the Air-Pump: Hobbes, Boyle, and the Experimental Life* (Princeton: Princeton Univ. Press, 1985).

[3] Steven Shapin, "Discipline and Bounding: The History and Sociology of Science As Seen through the Externalism-Internalism Debate," *History of Science,* 1992, *30:*333–369.

a but *the* critical problem in the history of mathematics. At the very least, by insisting on the central importance of an issue that the historian of science Charles Gillispie adjudged "a passing schizophrenia,"[4] historians of mathematics are opening a yawning gulf between themselves and increasingly bored historians of science. More tragically and more centrally, the sheer nastiness of the discussion threatens to rip the field to shreds from within. Distinctions have been honed to razor sharpness and routinely used to slash viciously at those who disagree; as the same referee put it: "The fragmentation of and the lack of communication within the community of those interested professionally in the history of mathematics have militated against the dialogue necessary to bring about some sort of reconciliation on the point." Historians of mathematics need to find new ways to understand their differences if the field is to remain viable.

The obvious place to begin such an effort is by identifying the issues that divide us. In his recent book *We Have Never Been Modern,* Bruno Latour has treated this issue for the sciences and argued that there we need to reconsider radically the categories of the natural and the social that have served as the poles for the internalism-externalism debates. In actuality, he insists, scientific products—his examples include the ozone layer, electromagnetic fields, and Robert Boyle's air pump—are neither natural nor social, but rather hybrid mixtures that we insist on taking apart in our treatments of them. The internalists are those who insist on the primacy of the natural, the externalists on that of the social, but Latour argues that neither view is ultimately helpful, because history is played out in between.[5]

The nature pole does not have a neat mathematical analogue so it is not surprising that, like Kuhn before him, Latour leaves mathematics out of his analysis. Nonetheless, his attempt to reconsider the categories of the polarity that splits us can be helpful for historians of mathematics as well. The issues that divide historians of mathematics have not necessarily divided the people we study; we would do well to try to learn from them the origins and implications of our current situation. To do this Latour would say we need to become anthropologists of the past; I have phrased it somewhat differently by saying we need to become Kuhnian historians. Whatever the terminology, the crucial point is that we move beyond the present blindnesses, alliances, and rivalries that lock us in intellectual battle, to encounter the past on terms of its choosing.

For Latour the terms we want to transcend are *nature* and *society;* to follow his lead historians of mathematics would need to identify and understand what these poles might be for mathematics. A possible mathematical analogue to the nature pole is suggested by the title of Philip Davis and Reuben Hersh's bestseller *The Mathematical Experience.* Their emphasis on this personal category opens them to "the inexhaustible variety presented by the mathematical experience." With no intention of presenting "a systematic, self-contained discussion of a specific corpus of mathematical material," the book jumps blithely from Euclid to Bertrand Russell, from Plato to Imre Lakatos, reveling in the diversity of experience and interpretation that mathematics includes.[6]

---

[4] *Ibid.,* p. 345, quoting Charles C. Gillispie, "Scholarship Epitomized" (essay review), *Isis,* 1991, 82:94–98, on p. 97.

[5] Bruno Latour, *We Have Never Been Modern,* trans. Catharine Porter (Cambridge, Mass.: Harvard Univ. Press, 1993).

[6] Philip J. Davis and Reuben Hersh, *The Mathematical Experience* (Boston: Birkhäuser, 1980).

Davis and Hersh's focus on the essentially human category of experience suggests that the subject may possess a historically interesting past. The approach is potentially problematic, though, because experience may be too personal to have a recorded history. Whatever may be happening in the external world, our experience of it is our own. One may be lonely in a crowded room; find a lavish meal tasteless; be entranced by a performance that leaves others cold. By the same token one may be caught up by the beauty, richness, and transcendent power of mathematics in a way others are not. If this were all there was to it, mathematics would no more have a coherent history than does taste or joy.

But this is not all there is to it. Mathematics may be experienced in this kind of intensely personal way, but that does not mean that it is private; the conviction that mathematical ideas are shared can create a strong bond among mathematicians, both present and past. This tie may be as independent of external circumstances as the experience itself. The implications for history are illustrated in André Weil's discussion of Napier's logarithms.

> Consider the following assertion: logarithms establish an isomorphism between the multiplicative semigroup of numbers between 0 and 1, and the additive semigroup of positive real numbers. This could have made no sense until comparatively recently. If, however, we leave the words aside and look at the facts behind that statement, there is no doubt that they were well understood by [John Napier] when he invented logarithms, except that his concept of real numbers was not as clear as ours; this is why he had to appeal to kinematic concepts in order to clarify his meaning, just as Archimedes had done, for rather similar reasons, in his definition of the spiral.[7]

The historian may object to this cavalier dismissal of what divides Napier from his predecessors. For Weil, though, the historian's historical, linguistic, and cultural issues are irrelevant to the recognition of transcendent mathematical truth that binds him to his forebears, the experience that he describes as understanding "the facts behind the statement."

Interpreting their experience as universal creates a special relationship between mathematicians and their history. James Gleick has remarked on it as something dividing mathematicians from physicists.

> A branch of physics, once it becomes obsolete or unproductive, tends to be forever part of the past. It may be a historical curiosity, perhaps the source of some inspiration to a modern scientist, but dead physics is usually dead for good reason. Mathematics, by contrast, is full of channels and byways that seem to lead nowhere in one era and become major areas of study in another.[8]

Because old mathematics never dies, mathematical history is not whiggish in the strict sense of that term. Whigs are progressives, and the whig history decried in Kuhn's *Structure* and elsewhere is one that interprets the past as a march to the present. Kuhn's revolutions complicate the notion of progress, but they too serve to separate the present from the past. Weil's stance, on the other hand, joins past and present; it is marked by timelessness, not progress. Michael Crowe made this point

---

[7] André Weil, "History of Mathematics: Why and How," *Collected Papers,* Vol. 3 (New York: Springer-Verlag, 1978), pp. 434–442, on p. 438.
[8] James Gleick, *Chaos* (New York: Penguin Books, 1987), p. 113.

with an architectural analogy when he wrote: "Scattered over the landscape of the past of mathematics are numerous citadels, once proudly erected, but which, although never attacked, are now left unoccupied by active mathematicians."[9] Crowe's open landscape allows significantly more play than a unidimensional progressive path. It also usefully describes mathematicians' tendency to rediscover their past because it does not preclude reentering any given citadel from a new direction.

Even as it supports an internalist history that is not whiggish, recognizing the mathematical experience as central also effectively holds the anthropologist or Kuhnian historian at bay. Empirical history can have little to say about such a textually transcendent experience. People and their historical or cultural contexts may be placed in such histories for inspiration, or for the human interest their situations may afford: a classic example of the genre is E. T. Bell's *Men of Mathematics*.[10] However, neither people nor contexts have significant impact on the basic mathematical story line.

The mathematical experience can be an equally effective block against philosophers who try to encapsulate mathematics within a particular system. Here the dynamic can be illustrated by the fate of foundational debates in our century. These attempts to define mathematics through its logical foundations originated within the mathematical community and absorbed considerable mathematical attention for several decades around the turn of the century. As the century progressed, the questions raised continued to be of considerable interest to philosophers, but the mathematical community has moved on without them. Jeremy Gray recently noted about debates in the foundations of mathematics: "Either they are technical and accessible only to logicians, or they are epistemological and draw their examples from the [elementary] concepts we meet in school. The result in each case is a debate that does not interest, and does not seek to affect, working mathematicians." This example suggests that the attempt to communicate ineffable experience is by definition a doomed process. It produces words, arguments, and new ideas, but none of these are essential to the experience itself. Productive mathematicians are such regardless of how they conceive of the subject. As Reuben Hersh so neatly put it, the working mathematician is "a Platonist on weekdays and a formalist on Sundays." To philosophers this may be an impossibility, but mathematicians continue to do mathematics regardless.[11]

Despite the independence of the working mathematician, the dominant philosophical perception of mathematics grew out of the debates about foundations, an approach that, following Thomas Tymoczko, I label foundationalism. This approach, pioneered by Gottlob Frege, evolved from the myriad nineteenth-century challenges to the absolute validity of mathematical ideas like number, magnitude, or space; it attempted to derive all of mathematics from a system of mathematical logic. Frege's particular attempt foundered on the rock of Russell's paradox, but in the twentieth century logicism, set theory, and the metamathematics of formal systems were offered as alternative ways to fulfill his foundationalist vision. Although none of these

---

[9] Michael J. Crowe, "Ten Misconceptions about Mathematics and Its History," in *History and Philosophy of Modern Mathematics,* ed. William Aspray and Philip Kitcher (Minnesota Studies in the Philosophy of Science, 11) (Minneapolis, Univ. Minnesota Press, 1988), pp. 260–277.

[10] Erick Temple Bell, *Men of Mathematics* (New York: Simon and Schuster, 1937).

[11] Jeremy Gray, "The Nineteenth-Century Revolution in Mathematical Ontology," in *Revolutions in Mathematics,* ed. Donald Gillies (Oxford: Clarendon Press, 1992); and Davis and Hersh, *Mathematical Experience* (cit. n. 6), p. 321.

has succeeded, their common perception that the nature of mathematics is to be understood through some kind of logical construction of foundations continues to dominate twentieth-century perceptions of the subject.

For the Kuhnian historian, looking for the essence of mathematics in its history, this foundationalist vision is daunting. Not only does it completely remove mathematics from any human realm, but it locates it instead in areas like symbolic logic or set theory, which are inaccessible without intense and specialized training. The messiness of human history is completely irrelevant to understanding the nature of a subject so extrahumanly established. Even when considering mathematical issues far from those of foundations per se, the extrahuman attitudes and orientation of this definition carry significant implications for historical work.

The historical ramifications of regarding mathematics as extrahuman can be seen by turning to the relatively well researched area of eighteenth-century mathematics. In enlightenment France mathematics was seen to encompass not only "pure" subjects like arithmetic and geometry, but also "mixed" ones. This category—which, following Jean d'Alembert's classification in the *Preliminary Discourse,* included subjects like mechanics, geometric astronomy, optics, and the art of conjecture or analysis of games of chance—covers much that would today be classified as applied mathematics. However, whereas the term *applied* connotes bringing something from one area and imposing it on another, the term *mixed* blurs the boundaries between them. This distinction may seem so small as to be insignificant, but close historical studies have revealed that regarding fields like mechanics as mixed rather than applied often played a significant role in shaping mathematical work.

For example, in the middle of the eighteenth century virtually all of Europe's major mathematicians were embroiled in a controversy about the curve described by a vibrating string. The discussion began as a dispute between d'Alembert and Leonhard Euler about how much one should restrict the physical problem to fit existing mathematical notions and techniques. D'Alembert confined his discussion to traditional mathematical curves, an approach that did not accommodate the range of physical possibilities of such strings, but did accord well with mathematical ideas of legitimate curves. Euler, on the other hand, was willing to include any curve that could be drawn; this gave him a physically more adequate but mathematically more problematic approach. The controversy was essentially resolved along the lines suggested by yet another discussant, Joseph-Louis Lagrange, who suggested a technique based on a model of the string as a set of point masses taken to the limit.[12]

The issues fueling this controversy were multifarious. In addition to generating mathematical results, d'Alembert's wave equation was the first differential equation to be examined in detail, and the points he argued with Euler were important parts of what later became the theory of functions. It encompassed not only decisions about the criteria that bounded adequate definitions of curves but an entire spectrum of personal and institutional relations among the participants. Ignoring these issues as irrelevant because extramathematical makes the controversy itself vanish. This is the picture that Martin Kline, for example, draws in *Mathematical Thought from*

[12] Thomas Hankins, *Jean d'Alembert: Science and the Enlightenment* (Oxford: Clarendon Press, 1970), pp. 47–48; and J. R. Ravetz, "Vibrating Strings and Arbitrary Functions" in *The Logic of Personal Knowledge: Essays Presented to Michael Polanyi on his Seventieth Birthday, 11 March 1961* (Glencoe, Ill.: Free Press, 1961).

Ancient to Modern Times, where the mathematical material forged in the heat of the argument is neatly subsumed under a linearly progressive picture of mathematics.[13] This kind of historical editing removes the possible power and significance of historical study for an understanding of mathematics based on practice. Whether mathematics is approached as attempts to communicate ineffable experience or as the development of a formal system makes little difference to the historian; both approaches effectively marginalize the human element of the story.

The accounts written in this way may be of value to the practicing mathematicians and logicians, but to accept them as adequate to the history of mathematics would be to underestimate the purview of the subject. For the practitioner the essence of mathematics may lie in some kind of an experience of truth,[14] but the products of that experience transcend the personal processes of their creation or recognition. Whatever it may be for mathematicians, their subject plays important roles in our science, our culture, and the ways we think about ourselves. Scientific theories are constructed on mathematical models. The subject is taught in our schools to a cohort far greater than those who may eventually be creative within it. It has long played a central role in our attempts to understand ourselves, how we learn and know. True mathematicians may be working in an inaccessible personal space, but this space is embedded in a public one that both supports and interprets mathematical activity.

Since the publication of Kuhn's *Structure* there have been occasional attempts to explore whether mathematical history might be understood in ways that display the more human characteristics of change Kuhn identified in the sciences. Changing this approach to the history of mathematics entails confronting not only firmly held and impressively buttressed beliefs about the subject itself but also well-established practice, which both rests on and validates those beliefs. Both the technical power of foundationalist philosophy and the sociocultural power of a productive mathematical practice have been ranged against this kind of historical innovation.[15]

Recently, however, cracks are beginning to appear in this united front. A number of "maverick" philosophers[16] have mounted challenges against the foundationalist view of mathematics. Each particular attempt to secure mathematical foundations has been inadequate, these critics point out: Russell's paradox lies festering at the very heart of set theory, and Gödel's theorem undercuts any attempts to construct a wholly internally consistent system. Whereas for decades people have tried to move past these problems, convinced that the foundationalist ends for which they were constructed are sound, many now are not. This issue is also being raised from the perspective of mathematics as it is taught and practiced at present. The failed attempt at introducing "new mathematics" in the 1960s suggests how very far set theory really is from the experience of mathematics of both teachers and students. The same is also apparently true of mathematical practitioners.

One solution to this disparity lies in creating a new philosophy of mathematics,

---

[13] Morris Kline, *Mathematical Thought from Ancient to Modern Times* (New York: Oxford Univ. Press, 1972), pp. 478–484.

[14] Cogent statements to this effect by Jean Dieudonné and J. P. Cohen are quoted in Davis and Hersh, *Mathematical Experience* (cit. n. 6), pp. 321–322.

[15] For an informal overview of the literature see the section "History of Mathematics: A Brief and Biased History," in the "Opinionated Introduction" to *Modern Mathematics,* ed. Aspray and Kitcher (cit. n. 9), pp. 20–31.

[16] For the term see *ibid.*

one that will correspond to mathematical practice rather than to some abstract notion of logical consistency or system. This solution has appealed not only to dissatisfied working mathematicians but to the maverick philosophers as well. They are converging with the frustrated practitioner on the idea that mathematics should be defined by what the subject is for those who practice, learn, and teach it, rather than by an intrinsically flawed foundationalist construction that bears little relation to that experience. Because of their interest in creating philosophies of mathematics that reflect mathematical experience, Thomas Tymoczko labeled these mavericks "quasi-empiricists."[17]

In many ways these new philosophical perspectives promise relief to historians focusing on mathematics simply because they reopen the door closed earlier in the century to the importance of knowing how people have viewed and developed the subject. Mathematics has been practiced for many centuries, and those who wish to develop a philosophy of the subject based on this practice need to know a great deal that has been systematically ignored or cut out of most twentieth-century histories. Moving from Weil's characterization of Napier's logarithms to one that incorporates what Napier perceived requires a significant amount of research into material only peripherally mentioned in Weil's account. At the very least, then, the new philosophical approaches promise work for mathematical historians.

This is just the least, however. The maverick philosophers have opened the door to a richer history of mathematics, but not necessarily to the full range of non-whig possibilities. There is a danger that the new philosophers' models and programs could become as constraining to historians as were those of foundationalists before them. This can be seen, for example, in the work of Lakatos, whose *Proofs and Refutations* is as powerful a statement supporting a developmental, changing approach to mathematics as Kuhn's *Structure* is for the sciences. Although Lakatos relied heavily on historical sources in writing this work, he did not embrace Kuhn's new role for history. Instead his goal was "rational reconstruction." To this end he carefully picked and chose his examples to fit his philosophical goals, rather than opening himself to the multifarious practices his historical sources might have revealed. This was a conscious decision on his part, and Lakatos is not alone among philosophers in feeling that it may be best to handle historical argument "with the care necessary in dealing with any explosives."[18]

Explosives can be very dangerous when lit in confined spaces. Fireworks exploded in the open, however, send up showers of light in which the myriad fragmented particles form coherent patterns against the sky. The open-ended empiricism of history may threaten any fixed set of generalizations about mathematics, be they internal or external or some hybrid form. But the explosiveness that is dangerous to philosophy may be constructive for historians who are willing to consider the larger pattern even as they follow a particular trajectory with their own work.

---

[17] For an elaboration of some of these philosophical approaches see *ibid.;* the essays in *New Directions in the Philosophy of Mathematics,* ed. Thomas Tymoczko (Boston: Birkhäuser; 1985); or Philip Kitcher, *The Nature of Mathematical Knowledge* (New York: Oxford Univ. Press, 1983).

[18] Imre Lakatos, "A Renaissance of Empiricism in the Recent Philosophy of Mathematics?" in *New Directions,* ed. Tymoczko (cit. n. 17), pp. 29–48, on p. 155. For a statement of one historian's problems with this point of view see Joseph Dauben, "Abraham Robinson and Nonstandard Analysis: History, Philosophy, and Foundations of Mathematics," in *Modern Mathematics,* ed. Aspray and Kitcher (cit. n. 9), pp. 177–200, esp. pp. 179–183.

Thus it is here that the historian may part company with the philosopher or the sociologist or anyone else who tries to confine the possible forms mathematics may have taken in the past. Just as no single construction seems adequate for capturing the mathematicians' experience, it may be that no single philosophy will suffice to cover all that mathematics has been in the past. Lakatos's insights and those of other critics of foundationalism nonetheless open vast fields of largely uncharted territory to historians. The same rich array of approaches and methodologies that is being applied to the sciences waits to be explored with respect to mathematics. It is too early to know what such works would turn up. But for historians of mathematics, too long bound by a single, internally constrained view of their subject, the opportunity to try many lenses in viewing the past is wonderfully exciting.

One major area that has attracted historical attention on Kuhnian lines involves the claim that mathematics is cumulative. From the perspective of someone like Weil this might be a circular statement, since one can define real mathematics as that which has persisted without change through time. If one moves into the public space within which mathematics is discussed, however, the claim gains meaning and significance. The cumulative nature of the subject has long been hailed as one of its defining characteristics. However, careful attention to mathematical texts of the past soon dispels the illusion that the present subject simply encompasses all past ideas and techniques. Historical studies written during the past twenty years are contributing to new understanding of this aspect of mathematical change.

An earlier study, D. T. Whiteside's magisterial "Patterns of Mathematical Thought in the Seventeenth Century," should have dispelled any lingering illusions about the possibility of easy commerce, even on the most technical questions, between mathematicians of the past and the present. He clearly shows that seventeenth-century ways of proceeding in calculus were very different from those to be met in modern calculus classes.[19]

Historiographically, Whiteside's problem definition and approach are relatively traditional; he is essentially following the convoluted roots of the calculus. A broader history results from recognizing subjects not easily interpreted as roots—subjects once avidly pursued but now largely ignored. The whole area of geometrical construction, whether with compass and straightedge or by some other carefully prescribed method, has in our day been reduced to an arcane corner of geometry. For the past decade, however, H. J. M. Bos has been writing articles exploring changing approaches to problems of construction in the seventeenth century. His study "The Structure of Descartes' *Géométrie*," for example, shows how closely linked Cartesian algebra is to classical geometrical existence arguments based on construction. At the same time, Bos argues, the power of Descartes's algebraic methods made the whole constructive enterprise so easy as to be trivial. The very success of Descartes's *Géométrie* rendered its goals obsolete, and his mathematical successors developed them in new algebraic rather than classical geometrical directions.[20]

Bos's analysis points to the important role of challenging problems in keeping interest in a mathematical field alive. In our studies of projective geometry Lorraine

[19] D. T. Whiteside, "Patterns of Mathematical Thought in the Seventeenth Century." *Archive for History of Exact Sciences*, 1961, *1*:179–388.

[20] Henk J. M. Bos, "The Structure of Descartes' *Géométrie*," in *Descartes, Il Metodo e i Saggi: Atti del Convegno per il 350 anniversario . . .*, ed. G. Belgioso *et al.* (Rome: Istituto della Enciclopedia Italiana, 1990), pp. 349–369.

Daston and I emphasized a different set of factors involved in supporting a mathematical field of interest. Projective geometry, we argued, served very important ideological and philosophical functions in the cultures wherein it was pursued in the nineteenth century. What kept it alive was not so much the attraction and intricacy of its problems per se, but rather the particular value it was assigned in the larger philosophical and institutional matrix of mathematical practice. Changes in this larger complex had reduced interest in the subject by the end of the century even though its problems and technical capabilities remained as strong as ever.[21]

The development of Michael Crowe's ideas over the past two decades is cogent testimony to the power of this kind of historical work for views of mathematical development. A strikingly graceful recognition of how much his ideas have changed, Crowe's article "Ten Misconceptions about Mathematics and Its History" reexamines critically ten characteristics of that history he originally propounded in 1975. One is the claim that mathematics develops cumulatively. Though he still contends that real mathematics can never be invalidated, Crowe here emphasizes that it can be ignored, to lie unnoticed in past literature. The constructivist geometry of the seventeenth century, the projective geometry of the nineteenth, or Crowe's example of Hamilton's quaternions could all serve as examples of these deserted citadels. Crowe's image is not tied to a coherent philosophy of mathematics or its history. It is nonetheless a useful description of mathematics as experienced and practiced.[22]

Crowe's analogy is in many ways an attractive one that considerably widens the purview of the history of mathematics. However, the residual cumulativeness of the picture it paints may still be too narrow for the varieties of historical change mathematical historians are uncovering. Joseph Dauben and Judith Grabiner argue for revolutionary models of mathematical change that challenge the easy commerce mathematicians have claimed with their past.

In Dauben the challenge is muted because he argues that mathematical development can be at once cumulative and revolutionary. Examining Georg Cantor's development of transfinite numbers, for example, he explains: "This was conceptually impossible within the bounds of traditional mathematics, yet in no way did it contradict or compromise finite mathematics. Cantor's work did not displace, but it *did* augment the capacity of previous theory in a way that was revolutionary."[23]

In "Is Mathematical Truth Time-Dependent?" Grabiner explores what she sees as a truly revolutionary break in mathematical development between the eighteenth and nineteenth centuries in France. She sharply contrasts eighteenth-century perceptions of the nature of mathematics, characterized by a deep-seated faith in symbols and a mixed-mathematical emphasis on results, with the nineteenth-century focus on abstract concepts and proof. This change had significant impact on French mathematical practice. As formulated by Augustin Cauchy, for example, the nineteenth-century approach to calculus entirely banished the eighteenth-century study of divergent series from the realm of legitimate mathematics. Grabiner's study of changes fundamental enough to effect the complete destruction of previous mathematical

[21] Lorraine J. Daston, "The Physicalist Tradition in Early Nineteenth-Century French Geometry," *Studies in the History and Philosophy of Science,* 1986, *17*:269–295; and Joan L. Richards, "Projective Geometry and Mathematical Progress in Mid-Victorian Britain," *ibid.,* pp. 297–325.
[22] Michael Crowe, "Ten Misconceptions" (cit. n. 9).
[23] Joseph Dauben, "Conceptual Revolutions and the History of Mathematics: Two Studies in the Growth of Knowledge," in *Revolutions in Mathematics,* ed. Gillies (cit. n. 11), pp. 49–71, on p. 62.

practices, leads her to conclude that "mathematics is *not* the unique science without revolutions. Rather, mathematics is that area of human activity which has at once the least destructive and most fundamental revolutions."[24]

In developing her radically noncumulative view Grabiner, like Crowe, uses an architectural analogy. For her, outmoded citadels are not abandoned, however, they are destroyed and the material recycled into new buildings. Those thirsty for the assurance of cumulative knowledge may be comforted to know that "most of the old bricks will find places somewhere in the new structure."[25] But the building within which these bricks are embedded and the mortar holding them together can be radically different in different times and places.

Both Dauben's and Grabiner's pictures of radical historical breaks recognize as essential to mathematics interpretations and structures of meaning. If one sees merely the bricks—equations like $2 + 2 = 4$ or $a^2 - b^2 = (a + b)(a - b)$, or theorems like that of Pythagoras—then the picture may be cumulative. But a pile of bricks does not display the beauty, order, or functionality to be found in the edifices that can be constructed from them. It may always be true that $2 + 2 = 4$, but to some this would have meaning for experiments with fruit, for others it might be embedded in some abstract notion of a group, and for yet others it might be important for figuring distances on a map. These different interpretations may be seen as the mortar that holds the buildings together: they materially affect the form of mathematical development.

The interplay of mathematical forms with meaning, while vastly important, has only recently been treated by historians of mathematics. The variety of work done by those taking up the challenge is great enough to suggest that the relation between form and meaning is not a clearly defined historical constant. A quick movement through some studies spread from the seventeenth through the nineteenth centuries will illustrate this point.

Bos's study of constructive practice in the seventeenth century does not stop with technical developments that undercut interest in geometrical development. He recognizes how tightly and inextricably Descartes's geometry is interwoven with the larger arena of his thought. Specifically, Bos suggests that the "long chains of reasoning" that characterized the vision of science Descartes presented in the *Discourse on Method,* as well as the thought of his eighteenth-century posterity, did not refer to abstract notions of logic. Instead it rested on a specific mechanical interpretation of reasoning that Descartes defended in his geometrical work. Overall, Bos's work indicates the importance of balancing technical with contextual forces in assessing the development of mathematics. As he puts it, "processes [of mathematical change] operate both on the level of technique and on the level of motivation, meaning and sense of the mathematical enterprise."[26]

Michael Mahoney's detailed studies of Pierre de Fermat and algebraic development in the seventeenth century have led him to a similar recognition of the interactions of mathematical ideas with "canons of intelligibility." In a recently published

---

[24] Judith V. Grabiner, "Is Mathematical Truth Time-Dependent?" in *New Directions,* ed. Tymoczko (cit. n. 17) pp. 201–214.

[25] *Ibid.,* p. 212.

[26] H. J. M. Bos, "The Concept of Construction and the Representation of Curves in Seventeenth-Century Mathematics," in *Proceedings of the International Congress of Mathematicians at Berkeley,* ed. Andrew M. Gleason (Providence, R.I.: American Mathematical Society, 1987).

article he shows how shifting interpretations of the reasonable and comprehensible shaped mathematical development by affecting decisions about what was and was not mathematically acceptable.[27]

A powerful study illustrating the riches of the interaction is Daston's work on eighteenth-century probability theory. Daston focused on the final item in d'Alembert's list of mixed mathematical subjects—"the art of conjecture or games of chance"—and found that the new mathematical theory of probability was nurtured and shaped by ever-evolving concepts about the nature of human society and reason. Social issues are inextricably intertwined with mathematical ones in the picture she paints of probability theory. To focus on just one example, the game described in the famous St. Petersburg problem was constructed in such a way that the mathematical expectation of a positive outcome (winning more than one invested) far exceeded what any prudent person would actually be willing to bet. Thus there was a difference between the computed expectation and the practical reality that numerical value was thought to embody. Discussion of the St. Petersburg problem ranged over a huge area of economic, social, and psychological issues. Daston's book clearly shows that it is impossible to separate the eighteenth-century development of probability theory from this complex of shifting concerns.[28]

Grabiner's article maintaining the revolutionary nature of mathematical change is drawn from the collapse of the mixed mathematical tradition of the vibrating string and the St. Petersburg problem. In the postrevolutionary period, the approach that saw mathematics as mixed with mechanics and other fields was replaced by a recognizably modern one that strictly separated the applied from the pure. Yet the sheer power and comfortable familiarity of the polarized postrevolutionary vision ought not to blind us to the assortment of mathematical practices that continued to flourish in the nineteenth century.

My studies of mathematical development in Victorian England demonstrate that there is nothing absolutely compelling about the direction of mathematical change. Even as the French were moving towards the nineteenth-century approach Grabiner describes, the English were rejecting it in favor of one closer to the one she ascribes to the eighteenth. Despite a great deal of interest in and contact with French developments, English culture supported a very different mathematical tradition until well past the middle of the century.[29] German mathematical development presents yet another picture, as does the United States.

[27] Michael Mahoney, "Infinitesimals and Transcendent Relations: The Mathematics of Motion in the Late Seventeenth Century," in *Reappraisals of the Scientific Revolution,* ed. David Lindberg and Robert S. Westman (Cambridge: Cambridge Univ. Press, 1988), pp. 461–491.

[28] The St. Petersburg problem was first posed by Nicholas Bernoulli in a letter to Pierre Montmort published in 1713. "Two players, A and B, play a coin-toss game. If the coin turns up heads on the first toss, B gives A two ducats; if heads does not turn up until the second toss, B pays A 4 ducats and so on, such that if heads does not occur until the nth toss, A wins 2^{n-1} ducats. Figured according to the standard definitions of expectation, A's expectation is infinite, for there is a finite, though vanishingly small probability that even a fair coin will produce an unbroken string of tails, and the payoff always grows apace. Therefore, A must pay B an infinite amount . . . to play the game." Lorraine J. Daston, *Classical Probability in the Enlightenment* (Princeton: Princeton Univ. Press, 1988), p. 69. See also Daston, "D'Alembert's Critique of Probability Theory," *Historia Mathematica,* 1979, 6:259–279; and Daston, "Probabilistic Expectation and Rationality in Classical Probability Theory," *ibid.,* 1980, 7:234–260.

[29] Joan L. Richards, "Rigor and Clarity: Foundations of Mathematics in France and England, 1800–1840," *Science in Context,* 1991, 4:297–319.

Taking national traditions seriously is one way to demonstrate the variety of meanings available even within contemporaneous mathematical traditions; another way of
approaching and analyzing such meanings is through biography. Thomas Hankins's
books on d'Alembert and on William Rowan Hamilton illustrate ways of integrating
mathematical work with an understanding of personal concerns and commitments.
The recent biography of William Thomson by Crosbie Smith and Norton Wise expanded this approach into a whole new dimension. The extraordinary breadth of
detail they included in their attempt to create the world of meaning in which Thomson lived swelled their biography to more than 800 pages.[30]

These studies model the potential breadth and richness available to the questing
historian who includes changing meanings as well as form in histories of mathematics. The enterprise has further implications as well. Attempts to incorporate meaning
into our understanding of mathematical development introduces considerable historiographical complications. How does one legitimately move from a mathematical
text to the world around it? What is the significance of a change in terminology
like that between mixed and applied mathematics? What constitutes evidence of
connection between a technical term, like *expectation* or *work,* and the ambient cultural use of the term? When is one justified in drawing connections along and across
such lines?

These issues are neither new nor peripheral. Weil's attribution of understanding
to Napier entails making very powerful interpretative decisions, and Kline's treatment of the vibrating string discussion entails a highly selective reading of texts.
The shape of any history is determined by decisions about this kind of issue. No one
transcends them, though some people are more explicit than others about what their
positions are.

It is the explosive power of these philosophical-historiographical issues, boxed
into artificially restrictive and polarized categories of internalist and externalist, that
threatens to shatter the history of mathematics. It is no solution to follow Lakatos
into some form of rational reconstruction, to insist upon a new form of internalism
based on another fixed view of the nature of mathematics. To do so would be to
abandon the creative power of historical studies envisaged by Kuhn or Latour and
to cut ourselves off from the enlightening surprises that historical texts can reveal
when encountered as openly as is humanly possible.

Exploring various approaches to mathematical history will of course lead us to
draw multiple images of mathematics from them. This diversity can create many
unsettling divisions, especially when simplistically lumped into polar camps of internalists and externalists. But it need not be this way; the cacophony is less a threat
than a testimony to the richness of the mathematical tradition. Just as much mathematical work takes place regardless of the ultimately irrelevant distinction between
Platonism and formalism, the creativity of the history of mathematics lies on the
fractal boundary between internalism and externalism. We will learn little as long
as we treat this boundary as a neatly drawn battle line; using history as a way to
analyze its construction and implications is a much more fruitful endeavor.

In 1758, when J. E. Montucla wrote his *Histoire des mathématiques,* he opined

---

[30] Hankins, *D'Alembert* (cit. n. 12); Thomas Hankins, *Sir William Rowan Hamilton* (Baltimore:
Johns Hopkins Univ. Press, 1980); and Crosbie Smith and Norton Wise, *Energy and Empire: A
Biographical Study of Lord Kelvin* (Cambridge: Cambridge Univ. Press, 1989).

that "such a work well done, could be looked upon as the history of the human mind [*l'esprit humain*], since it is in this science more than in all others that man makes known the excellence of this gift of intelligence that God has given him to raise him above all other creatures." It is not hard to see why this quotation is a perennial favorite among historians of mathematics from Dirk Struik to Noel Swerdlow.[31] Behind its reassuring affirmation of the importance of our work lies both a challenge and a danger. The challenge is to move towards a history of mathematics as centrally human as that Montucla envisaged and to an admirable extent modeled. The danger is that we would stop there, with a monolithic eighteenth-century definition of the human and the rational guiding our understanding. In the late twentieth century people in all fields are struggling to recognize the amazing variety that has marked the human mind and spirit over the ages, a diversity much larger than any Montucla would have acknowledged. If the history of mathematics is to live up to his vision in this time, we must recognize and focus on the huge range of experience that we include under the heading of mathematics, and leave behind the destructive generalizations that fuel the simplistically polarized debate over internalism and externalism.

[31] Jean Etienne Montucla, *Histoire des mathematiques*, Vol. I (Paris, 1858), p. viii, quoting from a letter of Montmort to Bernoulli; Dirk J. Struik, "The Historiography of Mathematics from Proklos to Cantor," *NTM: Zeitschrift für Geschichte der Naturwissenschaft, Technik un Medizin*, 1980, *17*:1–22; and Noel Swerdlow, "The History of the Exact Sciences," precirculated MSS for Conference on Critical Problems and Research Frontiers (Madison, 1991), pp. 448–470.

KNOWLEDGE IN RELATED FIELDS

How far can "antiwhiggism" go? Admitting that one cannot be a pure antiwhig historian in Herbert Butterfield's sense does not mean that one can turn back the clock to George Sarton's vision of the field of history and philosophy of science. See pages 151–155 and 217–219. Photographs courtesy of the Chemical Heritage Foundation (Butterfield) and the Smithsonian Institution (Sarton).

Philosophy of Science and History of Science

By Thomas Nickles*

S INCE THE ORIGINAL CRITICAL PROBLEMS CONGRESS on the history of science, organized in 1957 by Marshall Clagett, there has been a near reversal in the relations of history of science and philosophy of science. Then methodologists confidently advanced normative theses that were supposed to be virtually immune to criticism by both scientists and historians. Historical work, too, merely furnished illustrative material for rational reconstruction by philosophers. In his essay on the two approaches in the Clagett congress volume, Father Joseph Clark's aim was to show how philosophy of science could benefit history of science.[1] According to Clark, the good historian should begin from twentieth-century positivist conclusions about what constitutes a scientific theory, law, explanation, confirmation, and so on. We now know, he said, that the correct method of science is the hypothetico-deductive method. Moreover, the up-to-date historian should not hesitate to impose these results on the historical interpretation of all periods, back to the Presocratics.

Only five years after the 1957 congress, and in the same year that the congress volume appeared, Thomas Kuhn famously asserted the priority of history to logic.[2] Subsequently, the new, *historical* methodologists of science sought empirical evidence for their claims, and they looked mostly to historical studies to provide it. (Whether and how historical information furnishes evidence for "philosophical" claims about science remains an issue today: see Sections II and VII.) Moreover, historical philosophers of science were the first to assert the disciplinary insufficiency of philosophy of science, claiming that the discipline lacks the resources to solve its own problems completely, not to mention those problems raised *for* it by neighboring disciplines.

Many philosophers of science now accept this last point, and nearly all agree that the positivist account of science is too narrow. However, it would be much too strong

* Department of Philosophy, University of Nevada, Reno, Nevada 89557.

This contribution includes, in revised form, about half the content of my paper of the same title printed in the Madison conference volume, "Critical Problems and Research Frontiers." I thank my Madison commentator John Beatty and also Gaye McCollum, Marcello Pera, Nancy Nersessian, Rachel Laudan, Elliot Sober, Marga Vicedo, Maurice Finocchiaro, Steve Fuller, Ernan McMullin, and Gary Cage for suggestions.

[1] Joseph Clark, S. J., "The Philosophy of Science and the History of Science," in *Critical Problems in the History of Science,* ed. Marshall Clagett (Madison: Univ. Wisconsin Press, 1962), pp. 103–140.

[2] Thomas Kuhn, *The Structure of Scientific Revolutions* (Chicago: Univ. Chicago Press, 1962). *History* can designate either sequences of real-world events or scholarly writing about them. I shall usually rely on context to determine which I mean.

to suggest that most philosophers are historical methodologists or that the relations between history of and philosophy of science are simple and unproblematic. Quite the contrary. The early enthusiasm for an integrated history *and* philosophy of science (HPS) was not excessive, and it has since waned. In the 1960s new journals, departments, and programs were founded. By the 1980s many of these, whether marriages of convenience or more intimate relationships, had gone out of existence or had been transformed, usually by substituting a sociological component for philosophy.[3] Part of the explanation is that HPS programs were formed by philosophy joining an *internalist* history of science that was about to go out of fashion. And thanks to disciplinary changes, both historians and philosophers of science now feel somewhat more comfortable in their home departments than they did in 1965. Gradually, historians and philosophers realized that studying a common object—"science"—did not automatically create disciplinary compatibility. Some began to question whether the two disciplines *were* studying the very same object. The kind of work done by philosophers remained, indeed remains, markedly different from that done by historians. For example, historical work is extensive and philosophical work intensive. Historians spread out and attempt to achieve full coverage, to "fill gaps in the literature"; and they tend not to infringe on someone else's territory until such time as a major reevaluation seems necessary. By contrast, philosophical work within any generation centers on a few big problems, and philosophers love to lock horns over the same issues. Whereas in philosophy there is a premium on disagreement (as reflected in the structure of professional meetings and the purpose of commentators), in history there is more of a premium on agreement.

In this respect, sociology of science is closer to philosophy. Indeed, sociology of knowledge, whether or not it is considered a replacement for the P in HPS, has become considerably more philosophical than is professional history of science. To be sure, over the past two decades, the social studies of science have grown into a big, loosely organized, international, multidisciplinary enterprise that largely envelops history of as well as historical philosophy of science. However, as the largest and most established contributor to science studies, history of science has retained more disciplinary autonomy than the other components have.

Historians may also have tired of the company of philosophers because they find philosophers who dabble in history more dangerous to the historical enterprise than ardent positivists. However interesting and influential his work may have been, Imre Lakatos helped give historical philosophy a bad name, as did the "conjectural histories" of other followers of Karl Popper. The latter inspired L. Pearce Williams to declare that philosophers should not be allowed to do history.[4]

A deeper historicist complaint is that the philosophers' whole conception of scien-

---

[3] Still, the philosophy of the 1970s and 1980s was more historically sensitive than that of the early 1960s, when several HPS programs were founded. On the marriage see Ronald Giere, "History and Philosophy of Science: Intimate Relationship or Marriage of Convenience?" *British Journal for the Philosophy of Science,* 1973, *24:*282–297; and Richard Burian, "More than a Marriage of Convenience," *Philosophy of Science,* 1977, *44:*1–42.

[4] See I. B. Cohen, "History and the Philosophy of Science" (plus comments by Peter Achinstein and others), in *The Structure of Scientific Theories,* ed. Frederick Suppe (Urbana: Univ. Illinois Press, 1974), pp. 308–373; Imre Lakatos, "History of Science and its Rational Reconstructions," *Philosophical Papers,* Vol. I: *The Methodology of Scientific Research Programmes,* ed. John Worrall and Gregory Currie (Cambridge: Cambridge Univ. Press, 1978), pp. 102–138; and L. Pearce Williams, "Should Philosophers Be Allowed to Write History?" *Brit. J. Phil. Science,* 1975, *26:*241–253.

tific practice is wrongheaded, indeed, well-nigh theological, otherworldly, and transcendent, as their constant references to reason, rationality, and truth suggest. Such
critiques of philosophy are hardly new. In the *Economic and Philosophical Manuscripts* (1844) Karl Marx wrote: "Feuerbach's greatest achievement is to have shown
that philosophy is nothing more than religion brought into thought and developed
by thought, and that it is equally to be condemned as another form and mode of
existence of human alienation."[5] Today's postmodern, deconstructivist movements
sweep away Marx's own program along with every other progressive, liberal, humanist "Enlightenment" project (including modern science and positivist philosophy)
that promises to provide a coherent view of the universe and to confer meaning and
a better future upon human life. Postmodernists view all of these projects as totalizing grand schemes, secularized replacements for traditional religion; and when
Friedrich Nietzsche announced the death of God more than a century ago, he meant
that *all* gods are dead and that all attempts to disclose a master narrative as the story
of the universe, or even the story of "Man" (as a being with a fixed, crystalline
essence), are moribund. Many cultural critics and science studies experts dismiss
as transcendent all philosophy, and certainly any that regards science as rational,
progressive, or truth attaining.

But the fault for the failure of HPS programs does not lie entirely with the failure
of philosophers to become suitably historicized. For professional history of science
has become a big business by academic standards, and professional historians can
be an exclusive crowd. Furthermore, philosophers are now less sure of the value of
history to their projects. First, individual case studies no longer have the significance
they once enjoyed. Mary Hesse and others have criticized the standard "case study
method" of the 1960s and 1970s.[6] Too often philosophers yanked specially selected
cases out of time and historical context to support a favorite thesis. Too often they
drew conclusions about science-in-general from one or two cases from centuries
ago, before the scientific field in its modern form had yet emerged. And indeed,
historical case studies can be too much like the Bible in the respect that if one looks
long and hard enough, one can find an isolated instance that confirms or disconfirms
almost any claim. Second, some philosophers reject *historical* case studies as having
little evidential value and look instead to *contemporary* case studies or to social
scientific research. Third, historical case studies are now less attractive to some philosophers, as historians have largely abandoned the internalistic accounts of discovery and justification that philosophers found so congenial. Fourth, the emergence of
philosophy of biology and philosophy of cognitive psychology as lively specialty
areas has attracted away many scholars. While some do take a historical approach
to philosophy of biology, the historical dimension is largely lacking in cognitive
studies, where philosophers either do "foundations" work or actively engage in cognitive research. Cognitive and biological studies reflect the recent "naturalistic turn"
in epistemology, which has been largely a turn toward "nature" rather than toward

[5] Karl Marx, "Economic and Philosophical Manuscripts," trans. Tom Bottomore, in *Marx's Concept of Man*, ed. Erich Fromm (New York: Frederick Ungar, 1966), pp. 87–196, on p. 171.
[6] Mary Hesse, *Revolutions and Reconstructions in the Philosophy of Science* (Bloomington: Indiana Univ. Press, 1980). Yet some of the same social studies experts who ridicule philosophers' use
of historical cases are happy to generalize to present science and society from case studies such as
the fascinating seventeenth-century study of Steven Shapin and Simon Schaffer, *Leviathan and the
Air-Pump: Hobbes, Boyle, and the Experimental Life* (Princeton: Princeton Univ. Press, 1985).

history. Indeed, strong naturalism (based, say, on accepted biological and psychological claims) is difficult to reconcile with strong historicism.[7]

Fifth, many philosophers fear that strong historicism implies an untenable relativism. The new historical sensitivity generates serious problems about scientific change and incommensurability and, in the 1960s and early 1970s, led to the manifestos of Kuhn and Paul Feyerabend. Later, the Edinburgh strong program only confirmed the fears of relativism. One philosophical reaction to these fears was a strong reassertion of rationalism and realism.[8] In the ensuing debates philosophers, their collaborators, and critics produced hundreds of articles for or against realism, rationality, and relativism of various types. Although there are important exceptions, these debates generally focused the attention even of erstwhile historical philosophers of science away from the nitty-gritty details of how scientific research "works" and toward abstract, "philosophical" issues.

I. SOME HISTORICAL BACKGROUND

During the two millennia from the Greek and Roman writers up to the "modern" period, it would be difficult to find two disciplines further apart than history and philosophy. Insofar as we may speak of disciplines or forms of inquiry here, philosophy and history possessed distinct subject matters, goals, methods, and audiences and (at least officially) expressed themselves in different genres (logical argument vs. narrative and rhetoric). Philosophy amounted to thinking deeply and abstractly about the permanent and eternal, and thinking itself was the grasping of universals and the bringing of particulars under universals. Meanwhile, history encompassed any inquiry into particular matters of fact, whether natural or social—facts of geology, botany, civil events, and especially the actions of princes.[9] Philosophy aimed at certainty, at *episteme* or *scientia,* at demonstrative knowledge of general truths. History, like rhetoric, was concerned with *doxa* or probable opinion. And while historians described and explained activities of the actual world, philosophers sought the *summum bonum,* the best of all *possible* worlds. The very subject matters of philosophy and history were worlds apart!

In asserting the claims of philosophy to intellectual preeminence, Socrates and Plato attacked rhetoric and poetry in terms broad enough to damage history. At the beginning of the so-called modern period, René Descartes and Isaac Newton again

---

[7] See Thomas Nickles, "Integrating the Science Studies Disciplines," in *The Cognitive Turn: Sociological and Psychological Perspectives on Science,* ed. Steve Fuller, Marc de Mey, Terry Shinn, and Steve Woolgar (Dordrecht: Kluwer, 1989), pp. 225–256.

[8] These fears were aroused also by *philosophers* who strongly attacked realism by means of general methodological arguments and contemporary case studies. See, e.g., Bas van Fraassen, *The Scientific Image* (Oxford: Oxford Univ. Press, 1980); Nancy Cartwright, *How the Laws of Physics Lie* (Oxford: Oxford Univ. Press, 1983); and Arthur Fine, *The Shaky Game: Einstein, Realism, and the Quantum Theory* (Chicago: Univ. Chicago Press, 1986). For the realism debate see Jarrett Leplin, ed., *Scientific Realism* (Berkeley: Univ. California Press, 1984). For the rationality debate see Bryan Wilson, ed., *Rationality* (New York: Harper & Row, 1970); Martin Hollis and Steven Lukes, eds., *Rationality and Relativism* (Cambridge, Mass.: MIT Press, 1982); and John Robert Brown, ed., *Scientific Rationality: The Sociological Turn* (Dordrecht: Reidel, 1984). Two principal philosophical positions of the 1970s were Lakatos's defense of a historically constant methodology of science and Larry Laudan's attempt to redefine rationality in terms of progress, in *Progress and Its Problems: Toward a Theory of Scientific Growth* (Berkeley: Univ. California Press, 1977).

[9] See Barbara Shapiro, *Probability and Certainty in Seventeenth-Century English Thought* (Princeton: Princeton Univ. Press, 1983).

attacked history, rhetoric, and poetry and reaffirmed the attainability of *scientia* through use of right method. But their victory—based on dramatic progress in solving specific technical problems and on new, mechanistic world pictures—was temporary. For the limbo between genuine knowledge and ignorance (mere opinion) became ever more important. Historical information rapidly increased in both quantity and perceived value and forced businessmen, lawyers, churchmen, and officials to develop new ways of talking about knowledge. It became more and more acceptable to speak of fallible, *probable knowledge* (previously a contradiction in terms), and such knowledge eventually became more important than *scientia.* Some philosophers were deeply involved in the redefinition of knowledge. Writing near the end of the century, John Locke noted that even scientific knowledge consisted almost entirely of merely probable knowledge, claims that were at best practically certain, not metaphysically certain.

During the nineteenth century, history became an explicit part of methodology, in some quarters, and poetry and rhetoric sometimes were regarded more seriously. One major conjunction of philosophy and history was G. W. F. Hegel's historicization of Immanuel Kant. Subsequently, the left Hegelians, including Marx, put history in place of God and nature as the carrier of value, as the teleological medium that gave meaning to life and direction to human existence. One expression of this view is the idea of history (either the events themselves or their study) as judge: history will issue the last judgment. Another conjunction was William Whewell's attempt to derive scientific methodology (now more modest in its aspirations than in Newton's day) from the previous history of science, plus some philosophical presuppositions. By this time Newtonian optics as well as Newtonian chemistry had failed, and several "inductive sciences" (in addition to rational mechanics) had come to theoretical maturity. So a Whewellian historical survey of the forever-changed methodological landscape could yield positive historical guidance to methodology, where Francis Bacon's and Descartes's historical retrospectives had turned up nothing. There was now a history of discovery worth studying. Two or three decades later, Charles Darwin's theory began to naturalize philosophical thinking, so that by the early twentieth century American pragmatists such as Charles Peirce, William James, and John Dewey could depict the growth of knowledge in both naturalistic and sociohistorical terms. Meanwhile, French writers such as Gaston Bachelard and his student Georges Canguilhem were developing interesting views that anticipated Kuhn.

Then it happened for a third time! Philosophy returned to its ancient roots, reasserted its dominion over the intellectual world and, in the form of logical positivism and related forms of analysis, once again banished history, rhetoric, poetry, and, of course, what remained of metaphysics from the realm of the cognitively significant and methodologically relevant. The intellectual occasion was the linguistic turn and the revolution in modern, symbolic logic, which would lead to the computer age and a kind of neo-Enlightenment mentality. It was out of the unstable intellectual and social milieu of the interwar years that academic philosophy of science crystallized in the form of logical positivism.

Since 1957 historians and then sociologists and anthropologists of science have once again historicized science, now more thoroughly than ever. The irony is that science, the motor of the Enlightenment and the neo-Enlightenment, turns out to be a rather "romantic" affair. Thus the recent, neoromantic attacks on orthodox, computational, computer science and on the Western philosophical tradition that

extends back to ancient Greece. And thus the claim by neoromantic science studies
that a life in science is life as a member of a cultural community in which one learns
from a master how to understand and apply rules—or how to model one's own work
on concrete examples *rather than* to apply methodological rules. Enlightenment science presented itself as a debunker of myth in favor of a literal, true description of
the world. Ironically, this self-characterization itself amounted to the creation of a
new myth—according to present-day science studies, which sets itself the task of
demythologizing *science,* a task that the positivists believed they had already
achieved with their attack on metaphysics.

II. WHAT HAS HISTORY DONE RECENTLY FOR METHODOLOGY OF SCIENCE?

Kuhn's *Structure of Scientific Revolutions* did most to transform philosophers' concern with the logical structure of science into a concern for its historical structure—
or at least it sparked the great debate over how the historical and logical structures
could possibly be related.[10] There were two main types of answers to this problem
of conceptual change. The conservative wing retained the dominance of logical
structure and interpreted historical developments as a working out of the inner logic
of the position or program. In the beginning there was Hegel; in our time there
was Lakatos. The more radical wing made historical structure dominant—or rather
denied that there is any definite patterning of historical development either. History
undermines logic, but history also undermines (every pattern we read into) previous
history. Feyerabend took this line in *Against Method.*[11] Somewhere in between, and
containing elements of both Kuhn and Feyerabend, is the view that history is a suc-
cession of different, major *Denkstile,* within which different logico-mathematical
systems develop.[12]

In *Structure* Kuhn was already less conservative than Lakatos, but he still thought
that there is a definite pattern or rule of historical development for physical science—periods of normal science punctuated by crisis and, eventually, revolution.
According to radical historicists, there is no pattern that we can confidently extract
from history as an expectation for the future. There is no definitive hermeneutic
decoding of the meaning of history (as the sequence of events and world developments). History in this sense can, and probably will, undermine every such interpretation. (In statements of this sort, of course, *history* is an abstract term standing for
all manner of sociopolitical, intellectual, and natural-causal processes. We must be
careful not to reify history as an autonomous agency. Again, history itself is a dynamic thing that tends to undermine previous conceptions of itself.) All a priori
schemes will fail, but so will all schemes anchored empirically in the historical past.

---

[10] Kuhn, *Structure* (cit. n. 2). For the present point see Steve Fuller, "The Philosophy of Science
since Kuhn: Readings on the Revolution That Has Yet to Come," *Choice,* 1989, 27:595–601, on
p. 595. See also Fuller, "Is History and Philosophy of Science Withering on the Vine?" (plus peer
commentary), *Philosophy of the Social Sciences,* 1991, 21:149–244.
[11] Paul Feyerabend, *Against Method: Outline of an Anarchistic Theory of Knowledge* (London:
New Left Books, 1975). For a critique of Lakatos see Thomas Nickles, "Lakatosian Heuristics and
Epistemic Support," *Brit. J. Phil. Science,* 1987, 38:181–205.
[12] See Ludwik Fleck, *Genesis and Development of a Scientific Fact* (1935), ed. Thaddeus J. Trenn
and Robert K. Merton (Chicago: Univ. Chicago Press, 1979); and Ian Hacking, "Language, Truth,
and Reason," in *Rationality and Relativism,* ed. Hollis and Lukes (cit. n. 8), pp. 48–66.

For Bacon and Descartes, the poverty of history showed the *necessity* of method and methodology. For postmodernists, the richness of history and the diversity of human cultures and interests reveals their *impossibility,* which realization invites us to deconstruct reliance on uniform method.

This brand of historicism implies the nonexistence of any viable, a priori, foundational methodology. It suggests that if and when useful methods exist at all, they are not content neutral (as logic aspires to be) but content specific, that is, bound up with claims about how the world is—claims that imply that some modes of inquiry are more likely to succeed than others. Analytic philosophers will recall old debates over the proper parametric values to plug into Rudolf Carnap's continuum of inductive methods.[13] In a totally cracked universe, we cannot learn anything. Now our historical experience (indeed, our very existence) suggests enough patterning to say that our universe is not *totally* cracked, but it *is* cracked enough to undermine traditional approaches to knowledge. Despite this, the problem remains one for investigation.

A radical historicism implies the impossibility of any methodology at all, even useful heuristic guides for future research. A more moderate and hopeful historicism invites methodologists to exploit as an opportunity the epistemic slack left by the failure of rigid foundationalism. It makes a virtue of necessity by using the presumed underdetermination of future history to create intellectual space for methodological play. Feyerabend is the most visible example here. In fact, in his different moods, he may exemplify both the radical and semiradical positions just sketched. For him, nothing is fixed, not even the most basic goals or standards of the scientific enterprise. Everything is historically contingent. In effect, he says, we can hypothetically play with our deepest standards and goals and bet on the historical return—or even try to turn history in our direction, as Galileo did, by clever use of rhetoric. A considered violation of a methodological rule or constraint may lead to fruitful results, likely to be missed otherwise; indeed, this success might be parlayed into a *"new form of rationality* that will provide a rational justification for the whole procedure."[14] Forms of rationality (and there are more than one) are themselves products of historical development.

In the 1960s and 1970s, the historicization of methodology was usually pretty tame. Philosophers debated such (important) issues as whether it was necessary to introduce a "time variable" into confirmation theory (e.g., to avoid ad hoc theorizing and to safeguard predictive novelty) and whether theories properly so-called are logical entities or historically developing things. This was rather like asking today whether we need to add a "social" variable to methodological rules. Only gradually did a deeper sense of history find expression. According to Stephen Toulmin:

> By insisting on the radical character of scientific change, Kuhn completed the *historicization* of human thought that had begun in the eighteenth century, and so finally

---

[13] Rudolf Carnap, *Continuum of Inductive Methods* (Chicago: Univ. Chicago Press, 1952).
[14] Paul Feyerabend, "Changing Patterns of Reconstruction," *Brit. J. Phil. Science,* 1977, *28*:351–382, at p. 368 (his emphasis); cf. Ernan McMullin's commentary, "The Ambiguity of Historicism," in *Current Research in Philosophy of Science,* ed. Peter Asquith and Henry Kyburg (East Lansing: Philosophy of Science Assn., 1979), pp. 55–83, on p. 74.

undercut older views about the "immutable" order of nature and human knowledge. . . . Despite their thorough historicization of human affairs, Hegel and Marx too left this basic fabric of physical nature untouched. . . . [O]nce this lesson is taken to heart, the problems of natural science, too, become *hermeneutic* problems.[15]

Similarly, Alasdair MacIntyre defends the Vichian thesis that "scientific reason turns out to be subordinate to, and intelligible only in terms of, historical reason."[16] Logical argument is subordinate to narrative. Of course, truly radical historicists may feel uncomfortable with any talk of historical *reason*.

Rather than appeal to the distinctive features of historical interpretation and narrative, over against natural science, many historical methodologists have attempted to make straightforward appeals to historically described cases as scientific (or metascientific) data. On this view, methodology itself is an empirical science, and the relation (or one relation) of philosophy to history of science is that of theory to evidence.[17] Since philosophy of science is methodology of science, history serves to confront philosophy with "real" science as practiced but in such a way as to include developments over time rather than merely "time slices." The most explicit of these appeals to history of science as evidence was Larry Laudan's project at Virginia Polytechnic Institute to treat methodological claims about scientific change as hypotheses to be tested against the data of history.[18] One virtue (or is it a vice?) of the Virginia Tech program is that it appeals to historical cases as genuine tests rather than as mere illustrations of methodological theses. Another is that it does not rest methodological theses on single cases studied in isolation but produces, in the long run, rough statistical generalizations of the form: In a clear majority of cases taken from several significant scientific debates in different fields, across a considerable timespan, appeal to novel prediction played no significant role in resolving the debate.

The testing program is of a piece with Laudan's *normative naturalism* (a pragmatic attempt to combine naturalism and historicism), according to which the ultimate justification for a methodological claim, as for a first-order scientific claim, is how well it *works* in scientific practice. But how is it possible to test normative, methodological claims against descriptive, historical facts? The key is Laudan's claim that methodological rules can be restated as practical conditionals of the form: If you want to achieve goal G, then procedure P is a reliable (effective, more effective than Q, etc.) way to do so. The consequent of the conditional makes a factual claim that can be checked empirically.

Undoubtedly, evaluating and explaining the "history" of our experience with a practice—its successes and failures—is one way in which to improve upon that

[15] Stephen Toulmin, "The Historicization of Natural Science: Its Implications for Theology," in *Paradigm Change in Theology: A Symposium for the Future,* ed. Hans Küng and David Tracy (New York: Crossroad, 1989), pp. 233–241, on pp. 233–234 (his emphasis).

[16] Alasdair MacIntyre, "Epistemological Crises, Dramatic Narrative, and the Philosophy of Science," in *Paradigms and Revolutions: Applications and Appraisals of Thomas Kuhn's Philosophy of Science,* ed. Gary Gutting (Notre Dame, Ind.: Univ. Notre Dame Press, 1977), pp. 54–74, on p. 66.

[17] See, e.g., Harold Brown, *Perception, Theory, and Commitment: The New Philosophy of Science* (Chicago: Univ. Chicago Press, 1979), intro. and Part II.

[18] Arthur Donovan, Larry Laudan, and Rachel Laudan, eds., *Scrutinizing Science: Empirical Studies of Scientific Change* (Dordrecht: Kluwer, 1988).

practice. However, both philosophers and historians have criticized the Virginia Tech project. It arguably remains too closely tied to hypothetico-deductive methodology, which has come under severe philosophical attack in recent years. It treats history as more a positivistic than a hermeneutic discipline, as providing solid, empirical facts. More radically, many historians and perhaps most contemporary science studies practitioners reject the project on the grounds that it ultimately aims to explain scientific progress and that historical practice refutes *all* the rules—although it may be retorted that this radical view *also* employs history to test methodological claims. Another response is that scientific decisions are like moral and legal decisions, which cover too inhomogeneous a domain of cases to admit of precise, general rules or even statistical generalizations of any normative force. If Kuhn is right that scientific research itself is case based (modeled on previous exemplars) rather than rule based, then why should methodologists expect to find general rules? A related reason for dismissing as pointless the Virginia Tech and similar philosophical inquiries is the increasingly popular view that methodological claims never really guide research but only serve the rhetorical function of rationalizing, post hoc, claims arrived at by less systematic routes. This move instantly converts the status of methodological rules from "internal" to "external," to employ the old terminology—an ironic twist. I consider this objection in Section VI. A variation on this claim is that methods and methodologies are weapons used to attack opponents more than positive guides to research.[19]

Whatever the virtues and defects of the Virginia Tech program, virtually all science studies practitioners sometimes appeal to historical cases to challenge or confirm a methodological thesis. So clearly "provider of information on scientific practice that is useful to test descriptive and normative claims about inquiry" expresses one relation of history to philosophy of science. We shall meet others later. Few philosophers would say that these cases carry special *authority* because they are historical. The appeal to history is not an appeal to traditional authority. Rather, history is a medium by which scientific precedents and test cases are made available to philosophy. History is at least a repository of received wisdom in the form of cases. Surely history does teach by example in this sense. The question is whether there is anything beyond this for *philosophy* to teach.

By no means every philosopher identified with science studies agrees that history has been a boon to philosophy of science. Ronald Giere claimed that the important thing about the appeal to historical practice was the *practice* and not the history: case studies from contemporary science could serve as well or better than appeals to earlier stages of scientific development. Others such as Ernan McMullin answered that the historical perspective was essential to understanding the dynamics of scientific change. Steve Fuller asserted that HPS is "withering on the vine" because philosophers have mistakenly privileged historical evidence over that available from the social sciences. "After all, who else spends their time so unabashedly extracting patterns from history and divining their significance?"[20] But, rhetoric aside, how

---

[19] Laudan himself makes this claim throughout his *Science and Hypothesis: Historical Essays on Scientific Methodology* (Dordrecht: Reidel, 1981).

[20] Giere, "History and Philosophy of Science" (cit. n. 3); McMullin, "Ambiguity of Historicism" (cit. n. 14); and Fuller, "Withering on the Vine" (cit. n. 10).

does this point cut against HPS more than it cuts against social science itself or *any* empirical science?

III. WHAT CAN PHILOSOPHY OF SCIENCE DO FOR HISTORY OF SCIENCE?

While philosophy of science has become more historical since 1957, history of science has become less philosophical. However, philosophy in a wider sense has had a strong impact on history of science in three ways. First, mature history of science was born in the antipositivist insight of Alexandre Koyré, I. Bernard Cohen, Kuhn, and others that the Scientific Revolution owed as much to philosophical (especially metaphysical) ideas as to new facts. Second, recent social historians and sociologists of science have stressed the importance of philosophical ideas of a moral and socio-political character. However, the strong philosophical tone of these works owes more to nineteenth-century German thought than to contemporary philosophy of science and is, in any case, more characteristic of social studies of science than of "straight" historical writing.

Third, history of science owed its rise in popularity and sense of interdisciplinary importance to its perceived role in the debate among the great theories of scientific change advanced by Popper, Kuhn, Lakatos, Toulmin, and other philosophers. History of science was important because one's model of scientific development, one's philosophy of science, was important. Today, totalizing theories of change are passé. And just as the mature historians of yesterday largely rejected the kind of work done by the first generation of professional historians, who were mostly trained as positivistic scientists, so the historians of today often reject the internalist historical work that was particularly useful to the "great systems" debate. "History" is not a fixed abstraction, for, as noted above, history of science itself is undergoing significant historical change that undermines earlier conceptions of the field. "Straight" historians now claim to have found their own problems and prefer to confine themselves to that problematic. Many of them resent being in the service of nonhistorians. Several commentators have noted that nearly everyone doing HPS has been trained more as a philosopher than as a historian.

"History of science for its own sake"? The benefits of disciplinary autonomy, maturity, and professionalism are typically accompanied by a narrowing of focus to more arcane problems and a narrowing of audience to one's professional peers. Irrelevance to a wider audience was indeed the fate of the first professional philosophers of science—the logical positivists. Can professional historians of science escape the same fate? Historians normally avoid the arcane symbolism and other jargon of professionals, but they are today struggling with the problem of making accessible highly technical and esoteric scientific and technological developments. And there are other dangers. Historical work written with an obsession for professional purity tends to become narrow, even anti-intellectual, to shun the big questions that transcend disciplines. Perhaps in Germany everything is too philosophical, but in America everything, including history, is too antiphilosophical. Indeed, it is possible for philosophers of science to feel overrun by, and even "used" by, history of science and the other disciplines. For, again, it was Kuhn's linkage of history to some major "philosophical" theses that made history of science seem so important in the first place and that later opened up the science studies field.

In the current explosion of interest, science and technology studies are headed in all directions and becoming ever more particularistic, with closer and closer attention to the local and the situated.[21] HPS used to be at the center but now history and philosophy have largely broken apart and have themselves fragmented. Perhaps social studies of science, in alliance with relativist social history, form a new center, but the centrifugal tendency envelops those fields as well. People in all areas are beginning to ask how far this fragmenting, particularizing tendency can continue before the pieces are too small to be interesting.

Here philosophers, despite their notorious propensity to disagree with one another, can have a braking effect. To be sure, the later Ludwig Wittgenstein emphasized the significance of local practices, and recent historical philosophy of science has become more field-, content-, and even problem-specific in its claims. But the tendency of most philosophers is to generalize. That many philosophers have generalized too quickly has invited science studies practitioners simply to ignore their work, but once the current, antitheoretical, particularizing tendencies go far enough, controlled attempts to generalize may be considered healthy. Of course, sociologists and even historians also strive for a degree of general understanding. After all, narrative is an attempt to weave a seemingly disparate collection of particulars into a meaningful whole and thereby to tell a story that is of fairly general significance.

My own view is that generalist intellectuals are still needed. Both history and philosophy have become overspecialized. Both used to be of wide intellectual and even public interest. A philosophical role that I find attractive is that of a generalist who can help specialists and local disputants to view their work from different perspectives, to make connections to new precedents and exemplars. In such an exercise rhetoric—metaphor, analogy, and modeling—is as important as logic, for rhetoric can both familiarize the unfamiliar and defamiliarize the familiar. Such a person is a *generalist* but not a *generalizer* in the grand sense. And this sort of generalist may at the same time have special knowledge of a scientific discipline. Today many philosophers contribute to the literature of the biological and cognitive sciences as well as to "foundations" of physics and mathematics.

As Section I suggests, philosophers over the centuries often have dismissed historical work as irrelevant to their goals. However, historians have returned the favor by locating and marginalizing philosophers. It is not unusual for science studies critics to abstract a "rationalist" position (for example) from two or three philosophers whom they happen to dislike, and, before long, to characterize all philosophers, by definition, regardless of period and intellectual context, as rationalists in just this sense. (Philosophers are not the only perpetrators of whig fallacies; they have been victims as well!) On the other hand, it is not unusual for a critic to bash "philosophy" and then immediately turn to other philosophers for inspiration, the most often-cited being Nietzsche, Wittgenstein, Kuhn (as philosopher), Feyerabend, Hesse, and Michel Foucault. Not even the positivists are as terrible as they frequently are depicted by alleged historicists who commit the blatantly whiggish mistake of refusing to consider the positivists in their own problem and cultural contexts. In some of those contexts, after all, to adopt a favorable attitude toward science and scientists was to

---

[21] At the 1991 Stanford conference on "Disunity and Contextualism in HPS," it was pointed out that particularism could be a Butterfieldian anti-whiggish retreat into the specialist monograph. Mario Biagioli added that particularism within a single discipline could be domesticating: "Each of you do your local studies and mind your own business!"

take a moral and political stance that was courageous and progressive. And to avoid stereotyping, it may be helpful to remember that although different philosophical traditions have dominated various times and places, philosophy is not a closed intellectual *discipline* or *specialty* on a par with others. In philosophy a debate over fundamentals is always appropriate. It would be easy to list a dozen quite diverse conceptions or models of philosophical practice down through the centuries, several of which are in active competition during most periods.

Since philosophy is neither a specialist discipline nor a single project, and since people who were recognizably philosophical (e.g., Hegel, Marx, Dewey, Kuhn, hermeneutic philosophers) have been among the first of their eras to impose new standards of historical and social adequacy upon accounts of human development, including scientific development, there is no reason to reject philosophy as intrinsically transcendent, antihistorical, or antisocial.[22] Nor can historians afford to be too smugly judgmental. Since 1800, as we have seen, history has, at various times, enjoyed its own transcendental pretensions, its metaphysical status as God's replacement.

When asked why they do not take note of philosophical debates and distinctions or find them helpful to their own work, historians of science have a standard reply: the result is bound to be whiggish history. Many historians voice these same whiggish fears about the use of their own work by philosophers (or politicians or anyone else). For is it not a whiggish error to extract "lessons" from historical work and then apply them to other contexts? This is precisely what scientists do (according to Kuhn and others) in deforming old exemplars so as to fit new problems at the frontier of research. Evidently, good science is bad history.[23] The point is that a strong antiwhiggism *both* prevents historians from tuning in intellectual debates that ought to interest them *and* discourages others from making use of historical work for their own purposes. For it becomes a whig error to seek historical (in)validation for any enterprise, or to transform historical descriptions into norms of research or guides to public policy. Philosophers sense that historians would like to restrict their audience to other professional historians and to retain complete control over how their results are used. Such behavior is typical of specialists in any field, including philosophy, but it runs counter to both the increasing necessity of interdisciplinary work and current theories of interpretation, which deny authors complete control over the meaning and use of their work. To accuse of whiggism all who appropriate historically produced exemplars is simply to confuse historical work proper with the uses that others might make of it. History is not a logically closed domain in the sense that all uses of historical work must themselves belong to history. To sum up: antiwhiggism has become an isolating principle that insulates historians from

[22] The revolt against "theory" of many areas of contemporary culture can be interpreted as an attack on philosophy. This movement has no doubt been antiphilosophical in some respects, since philosophy is commonly identified with general theory. Yet not all philosophy is highly general, abstract, or speculative. Some philosophers have been antitheoretical, just as some have been positively disposed toward history. Indeed, there is an antitheoretical strain in the current attacks on general methodology in favor of a more discipline-, content-, and problem-specific approach to methodology, as reflected in the emergence of philosophy of biology and philosophy of psychology as separate specialties.

[23] I discuss these matters more fully in Thomas Nickles, "Good Science as Bad History: From Order of Knowing to Order of Being," in *The Social Dimensions of Science,* ed. Ernan McMullin (Notre Dame, Ind.: Univ. Notre Dame Press, 1992), pp. 85–129.

philosophers who had hoped to learn from history and, perhaps, to teach it something in return.

IV. ANTIWHIGGISM REASSESSED

In the years since the 1957 Critical Problems conference, the maturation of professional history of science has been characterized by an ever-increasing antiwhiggism or antipresentism, as scholars uncover more subtle ways in which we have imposed our own intellectual categories and values on the past. Antiwhiggism is also the hallmark of other science studies disciplines that have flowered over the past decade or two, each claiming to be more antiwhig and hence more faithful to the case in question than their predecessors. Andy Pickering accused historians as well as philosophers of taking at face value "the scientists' account" of historical developments, which invariably "[puts] the phenomena first." [24] One advantage of studying ongoing contemporary controversies, noted Harry Collins, is that we cannot be contaminated by the historical knowledge of who won. [25] While full of praise for Martin Rudwick's *The Great Devonian Controversy,* a meticulous reconstruction of a historical controversy, Trevor Pinch could not resist adding that Rudwick's knowledge of how the controversy ended rather spoiled the book's conclusion. [26]

On the whole, this antiwhiggism has been healthy. Indeed, it has been essential to our developing a genuinely historical understanding of other cultures, past and present; and as a doctrine it has effectively countered powerful ahistorical forces in our culture. But now that maturity has been achieved and positivism defeated, we may predict that antiwhiggism has largely run its course. Today it does as much harm as good. It is doctrinaire antiwhiggism as much as anything that harms relations between history, philosophy, and other disciplines and that encourages professional insularity. Doctrinaire because Herbert Butterfield's antiwhiggism has become a kind of founder myth of history of science (far more so than of political history, where the term *whig* originated), as if the field has an essence to protect.

Nor am I alone in making these claims. Not surprisingly, other philosophers chafe at the antiwhiggism bit, but other science studies experts, including historians, increasingly raise similar questions. Let us begin, however, with a philosopher— David Hull's "In Defense of Presentism," a response to Pearce Williams. [27] Hull contends that there are many occasions and purposes for which a whiff of presentism is permissible, necessary, or at least unavoidable. For one thing, historians are writing for a contemporary audience. Where meanings have shifted over time, or common neologisms have been introduced, it may be better for present-day historians to employ our words than those of past actors. Thus it would only confuse the reader to withhold the term *evolution* from discussion of Darwin's theory, on the ground that he did not use the term in *The Origin of Species.* This point is bigger than may

[24] Andy Pickering, "Against Putting the Phenomena First," *Studies in History and Philosophy of Science,* 1984, *15*:85–117, on pp. 81–86.

[25] Harry Collins, "Stages in the Empirical Programme of Relativism," *Social Studies of Science,* 1981, *11*:3–10, introducing a special issue on "knowledge and controversy."

[26] Martin Rudwick, *The Great Devonian Controversy: The Shaping of Scientific Knowledge among Gentlemanly Specialists* (Chicago: Univ. Chicago Press, 1985); and Trevor Pinch, "Strata Various," *Stud. Hist. Phil. Sci.,* 1986, *16*:705–713.

[27] David Hull, "In Defense of Presentism," *History and Theory,* 1979, *18*:1–15; and Williams, "Should Philosophers Be Allowed" (cit. n. 4).

appear, given today's concern for audience and how its members choose to receive and appropriate materials for their own uses. It is worth remembering that historians' audiences are not the people and institutions they study but contemporary and future readers.

Moreover, Hull continues, the good historian had better use the best available scholarly standards, and it is most unlikely that these were the standards of the period studied. Hull cautions against parlaying a decent amount of historical relativity into a full-blown relativism. Obviously, the fact that a scientific debate could not be resolved during the period in question does not imply that it is unresolvable, full stop. In his review of Steven Shapin and Simon Schaffer's *Leviathan and the Air-Pump,* David Oldroyd makes a similar plea for a (reasonably) whiggish social studies of science.[28]

It is not only philosophers who complain about excessive antiwhiggism and who look to history for lessons and for tests of their ideas. Historians themselves and historical sociologists do so as well. As already noted, *Leviathan and the Air-Pump* is an extended seventeenth-century case study widely used in just these ways by sociologists of scientific knowledge.[29] Roy Porter observes that rejecting all forms of whig history invites "over-specialisation, narrowness and fragmentation" and a new mystification of the success of science. Geoffrey Bowker and Bruno Latour point out that an overly robust antiwhiggism is peculiar to Anglo-American historians and that one can be *anti*-antiwhig, as was Canguilhem, without being crudely whiggish. R. A. Jones notes that antiwhigs are easily trapped by the reflexive exercise of providing a progressive history of the professionalization of their own discipline—a triumphant, whig history of the victory of antiwhiggism, as it were. Literary scholars and their "new historicism" have run into many of the problems familiar to historians of science. Dominick LaCapra, for instance, suggests a Freudian "transference" model of the historian's relation (and other contributions to the conference on which it is based) to the past, as an alternative to both historicism and presentism.[30] Other articles in this volume (and other contributions to the conference on which it is based), among them Stephen Brush's "Should Scientists Write History of Science?" confirm my sense that doctrinaire antiwhiggism has run its course.

Even internalist historians such as Rupert Hall have pointed out the serious problems in Butterfield's *Whig Interpretation of History.*[31] One cannot be a pure, antiwhig historian in Butterfield's sense even if one wants to be—retreating so completely into the archival research and the writing of technical monographs that the documents speak for themselves without, apparently, any selection or interpretation.

[28] David Oldroyd, "Why Not a Whiggish Social Studies of Science?" *Social Epistemology,* 1989, *3:*355–372.

[29] See also Barry Barnes, "Sociological Theories of Scientific Knowledge," in *Companion to the History of Modern Science,* ed. R. C. Olby, G. N. Cantor, J. R. R. Christie, and M. J. S. Hodge (London: Routledge, 1990), pp. 60–73, on p. 71.

[30] Roy Porter, "The History of Science and the History of Society," *ibid.,* pp. 32–46, on p. 43; Geoffrey Bowker and Bruno Latour, "A Booming Discipline Short of Discipline: (Social) Studies of Science in France," *Soc. Stud. Sci.,* 1987, *17:*715–748; R. A. Jones, "On Merton's 'History' and 'Systematics' of Sociological Theory," in *Functions and Uses of Disciplinary Histories,* ed. Loren Graham, Wolf Lepenies, and Peter Weingart (Dordrecht: Reidel, 1983), pp. 121–142, on p. 121; and Dominick LaCapra, "Rethinking Intellectual History and Reading Texts," *Rethinking Intellectual History: Texts, Contexts, Language* (Ithaca, N.Y.: Cornell Univ. Press, 1983), pp. 23–71.

[31] Herbert Butterfield, *The Whig Interpretation of History* (London: Bell, 1931); and A. R. Hall, "On Whiggism," *History of Science,* 1983, *21:*45–59.

(What, then, do historians *do*? Only conduct raw archival searches?) Such an account presents the documents as if their meaning is transparent, *given,* thus undercutting that greatest lesson of historical study: that we can assume no givens, no self-evidencing fixities. One difficulty in getting a crisp understanding of whiggism and presentism is that there is not just one, big, whig fallacy but a whole cluster of distinguishable moves, some of which may be well suited for certain purposes. For example, there are many ways of attributing success or failure to historical enterprises without thereby celebrating or denigrating them.

Pure antiwhiggism is one of the shibboleths increasingly challenged by historians and other science studies practitioners working out the implications of a *reflexive* "relativist history" and "relativist social studies of science." Although relativism presents itself as a politically correct dismissal of traditional, "philosophical" attempts to play God, it itself conceals the "god-trick" of pretending to be "nowhere while claiming to be everywhere equally."[32] Yet the sort of reflexivity in which authors attempt to situate themselves fully must also fail. No one can know oneself (one's historical situation) fully. Strong historicism in the sense of strong antiwhiggism, understood as strong antipresentism—the requirement that historians must efface every trace of their own, present historical position—is incompatible with strong historicism understood as the thesis that everything is historically situated, including the historian—that there can be no neutral, ideal observer.

Consider Kuhn's standpoint in *Structure*. His is the view from nowhere, that of the omniscient historian-philosopher who can see that all those normal scientists are doomed to fail, sooner or later, while *their* motivation depends crucially on their conviction that they *are* learning something fundamental about nature. A certain amount of false consciousness (absence of self-knowledge and knowledge of historical situation) seems necessary to their being productive scientists.

It remains to be seen how fast the movement against antiwhiggism will develop, since historians freely admit that they are among the least methodologically self-conscious of science studies practitioners.[33] My remarks, however, are not a call to turn back the clock to, say, George Sarton's vision of the field. There is no question that it is important to understand historical developments in their own terms. My claim is that an important phase in the development of history of science is over, that it is now time to move on. Antiwhiggism works well for the inductivist-idealist intellectual history that Butterfield, with his sympathy for R. G. Collingwood and Michael Oakeshott, championed, and for the internal history of science that flourished in 1957. However, it is not well motivated for those historians turning from "emic" to more "etic" methods of historical investigation, those that put a premium on our distancing ourselves in various ways from the actors, that impose our own analytical categories rather than "naively" understanding things in the actors' own (usually very limited) way.[34]

[32] See Donna Haraway, "Situated Knowledges," *Simians, Cyborgs, and Women: The Reinvention of Nature* (New York: Routledge, 1991), pp. 183–201, on pp. 189, 191.

[33] See, e.g., A. R. Hall, "On Whiggism" (cit. n. 31); A. Wilson and T. G. Ashplant, "Whig History and Present-Centred History," *Historical Journal,* 1988, *31*:1–16; and Ashplant and Wilson, "Present-Centred History and the Problem of Historical Knowledge," *ibid.,* pp. 253–274.

[34] But then there is today's countertendency to allow women, minorities, and historical populations to speak for themselves; yet feminist and minority movements tend to be quite whiggish, as they view their "progress" from the past through the lens of their own, newly gained sensitivities and articulations of problems.

A second reason for thinking that the antiwhig phase of history of science may be over is that science studies have evolved beyond the descriptive stage and entered a more critical and evaluative stage. Third, if history of science is to be of any use to audiences other than professional historians interested in purely academic questions, then historians will have to permit a hint of whiggism here and there. For if these other audiences (including philosophers, politicians, environmentalists, feminists, and policymakers but also students and sovereign citizens) are to find historical knowledge valuable in facing *their* problems, this means that talk of learning lessons from history and of transferring historical knowledge to other problems cannot be wholly illegitimate. If historians want other people to notice their work, they, too, must risk untended appropriations of that work by other actors. Policymakers will use historians' work whether or not they intend it to be so used.[35]

"The historicity of all things" never meant that all things are *produced* by historical research, a view that confuses history as the unfolding of events with history as a scholarly discipline. Clearly, the most urgent problems of life, including the scientific life, are not problems of understanding historical developments. Roughly speaking, they are problems of how to deal with present difficulties and how to build the future, at both individual and social levels. Yet the only resources available to a generation, besides its own ingenuity, are those bequeathed by its past. Whether or not those resources are employed by historically sensitive individuals, they must be used whiggishly, at least in the sense that the problems these individuals face will rarely be identical with the older problems from which the resources derive. Even the application of traditional lifeways to current situations is at least slightly whiggish in this respect: the new cases to which a tradition is applied are not identical with the old; and insofar as the new applications become exemplary of the tradition, they alter that tradition. Traditional practice evolves. Butterfield himself later recognized this: "We are all of us exultant and unrepentant whigs. Those who, perhaps in the misguided austerity of youth, wish to drive out that whig interpretation . . . are sweeping a room which humanly speaking cannot long remain empty. They are opening the door for seven devils which, precisely because they are newcomers, are bound to be worse than the first."[36]

Nor can Butterfield's change of heart on whiggism be entirely written off as an expression of wartime sentiments. His main point in 1944 was that whiggism has been politically and culturally valuable as the chief mechanism for wedding past to future, for legitimating future change by reference to past policies, traditions, and cultural forms that reformers whiggishly adapt to current problems. Thus whiggism helps to solve the major problems *we* face as we *make* history, as we extend and alter our former lifeways. This same point is trivially true of the scientists and technologists whom we study. What else can they do but (often unconsciously) deform problems, solutions, and practices they take as exemplary so as to fit their present situation on the research frontier?[37]

My conclusion to this section is that we science studies scholars must stop being

---

[35] See Bruno Latour, *Science in Action: How to Follow Scientists and Engineers through Society* (Cambridge: Harvard Univ. Press, 1987). On policy and history as teaching lessons see John Heilbron, "Applied History of Science," *Isis,* 1987, *78:*552–563.

[36] Herbert Butterfield, *The Englishman and His History* (Cambridge: Cambridge Univ. Press, 1944), pp. 3–4.

[37] Consult Nickles, "Good Science" (cit. n. 23) for more detail.

automatic antiwhigs and ask ourselves when and under what conditions we must or must not be whiggish. One task for a practically relevant science studies is to find a position between saying that history discloses the meaning and goal of the universe and saying that history has no lessons to teach whatsoever, that it merely provides spectacles of past civilizations for our idle, intellectual amusement. The latter position smacks of a return to the Greek ideal of intellectual contemplation, detached from the workaday world of doing and making, of historical knowledge as mere spectator's or beholder's knowledge, a separation of "theory" (in the etymological sense) from practice. On the other hand, abandoning doctrinaire antiwhiggism for a more judicious choice of topics and tools is easier said than done. We must whiggishly adapt antiwhiggism to our current circumstances!

V. BOOTSTRAP EPISTEMOLOGY; OR,
HISTORICAL (AND NATURALISTIC) METHODOLOGY OF SCIENCE

According to Hall "the most obvious of all historical questions is: 'How did we arrive at the condition we are now in?'" While historians certainly should not "praise or aggrandize scientific achievement," neither should they deny it.[38] Despite the qualification, many historians today would reject this as precisely the wrong kind of question to ask, not to mention the "internalist" answer that Hall would offer. Writes John Schuster:

> Historians of science have been unable to achieve consensus about any of the historiographical issues central to understanding the Scientific Revolution. They cannot agree on what is to be explained. . . . On a deeper level, no consensus has emerged about what would constitute an adequate explanation of either revolutionary or more continuous change. Much of the discussion of this problem has been bogged down in the debate between internalist and externalist approaches to the history of science.[39]

Here we come to the set of issues that, more than anything else, divides most philosophers from those in science studies who practice strong relativism. The assertion and denial of progress in understanding and manipulating nature is surely the single issue that most exercises the two sides. In the past, philosophers, historians, and nearly everyone else took for granted the absolute progress of science as a special process relatively uncontaminated by the social shenanigans that drag down politics, religion, and business life. Moreover, they celebrated science and its creators ("men of genius") for their eternal achievements in bettering human existence, and as the only men to achieve escape velocity from the powerful gravitational pull of "history" and "society," from the situatedness of the here and now. On this view, the Scientific Revolution and its consequences constitute the most important "event" in human history. How to explain this leap forward, this special historical (or antihistorical?) process, is therefore for them *the* important epistemological and historical task.

When relativists first rejected this view, many simply negated it, denying that the sciences were in any way special, that there is any meaningful sense in which we can speak of progress. This denial provoked the aforementioned group

---

[38] Hall, "On Whiggism" (cit. n. 31), p. 54.
[39] John Schuster, "The Scientific Revolution," in *Companion*, ed. Olby *et al.* (cit. n. 29), pp. 217–242, on p. 218.

of philosophers to counter relativism with strong realism. According to the strongest realists, not only is science successful, but the only explanation of this success is that science provides a steadily improving representation of reality.

My own view is that one extreme is as bad as the other, that both lump together too many issues. A first counter to the tendency to lump science as a whole together with the rest of life is the current interest in diversity—in recognizing the distinct "ethnic" or "community" strands within and across the various specialties. Second, we surely can and do evaluate science in many ways for various purposes. To say that something is an achievement is not necessarily to celebrate it or to condemn it in a moral sense. (Nor, in making such a statement, do I commit myself to a sharp distinction between fact and value.) It need only mean that we now have the expertise to produce X routinely by doing Y, whereas no one could do that before. Science studies now seem to be in a period of critical denigration rather than celebration of science and technology, but, as noted above, the one can be just the inversion of the other. Some experts even violate a kind of symmetry principle here by thinking that negative evaluation is fine, while positive evaluation is inappropriate in principle, that praise is whiggish while criticism is somehow antiwhiggish! On the other side, the strong realism of the philosophical reaction to relativism seems strangely out of touch with the fallibilistic and (more recently) the historicist and naturalistic tendencies of twentieth-century epistemology. The logical positivists did not go nearly so far and would have rejected the strong realist position as "metaphysical." On this score it is worth noting that the severest philosophical critic of strong realism, Larry Laudan, is a historical, pragmatist philosopher of science, although one hardly known for his relativist and externalist sympathies.[40]

Surely it is just as indefensible to assume a priori, as a disciplinary requirement, that the sciences have not progressed in any sense as to assume that the sciences tell us the final truth about the world and will provide every aid and comfort in our old age. Is there any more evidence for the former view than the latter? Roy Porter strikes a nice balance:

> Moreover, there is a further pitfall if the social history of science becomes too closely identified with the priorities of demystification, with exposing the hidden authoritarian, sexist and elitist ideologies encoded into science. It is important to de-privilege science in this way. But we may be in danger of creating a new mystification in its place, by rendering profoundly mysterious how and why science has proved such a dramatically successful and powerful enterprise in the West. The science that can put a man on the Moon or destroy our civilisation in a flash . . . has (to return to Butterfield) 'outshone' similar endeavours—organised religion or philosophy come to mind—which, just five centuries ago, were much more imposing activities.[41]

Whether reaching the moon and attaining the capability to destroy the world are morally good, bad, or indifferent, they surely do represent achievements in the sense of gaining power over nature of a kind never approached by other peoples at other times or places. We may express the point in the form of a provocative challenge: If we have not gained tremendous power over nature, partial and ill-considered though it may be, then why are environmentalists and other critics of science and technology

---

[40] Larry Laudan, "A Confutation of Convergent Realism," *Phil. Sci.,* 1981, *48:*1–49.
[41] Porter, "History of Science" (cit. n. 30), p. 44.

so worried? Why do we all worry about the bomb? Nor can this increase in knowl-edge and power be explained entirely in terms of changed social accounting proce-dures or a new rhetoric. To grossly oversimplify: all the talk and mutual back-patting in the world will not get us to the moon. Philosophy cannot bring back to life a Juliet, but perhaps modern medicine can.

As John Schuster notes, there has been little agreement about what an explanation of the scientific revolution is supposed to explain. Surely it will help (at least for now) to stop talking about knowledge and truth in some "high philosophical" sense and to bring the discussion down to earth. Something akin to the seventeenth-century reconceptualization of knowledge is needed. Too many historicisms and relativisms are premised on a near-foundational conception of knowledge, which makes their critical initiatives tantamount to "straw man" fallacies. Equally facile is the relativist move of labeling "knowledge" whatever the actors in question consider knowledge. Noting the branching of ever more specialties over time, Kuhn, at the end of *Structure,* proposes to reconceptualize the development of scientific knowl-edge in terms of a ramifying evolution rather than a progressive evolution. As the evolution metaphor suggests, a truly naturalistic account cannot stop with mere "be-holder's" or "spectator's" knowledge of the universe. Modern science and technol-ogy have given us an exponentially increasing power over nature, an ability to con-trol, to redirect, and to produce specific effects, often unintended effects and often on a worldwide scale. On the other hand, simply to redefine knowledge in terms of power will not work either, since some forms of power have little to do with knowl-edge. Moreover, the notion of power is as obscure as the notion of knowledge.

However we define it, our knowledge of nature and power over it has enjoyed dramatic spurts, differentially over time and across the domains of knowledge, for a variety of reasons that we need to understand better. Surely there is something here that calls for explanation. Over the past few centuries, we have gone from very little knowledge in certain areas to a great deal, often at an exponentially increasing rate. (This claim is perfectly consistent with our having forgotten most of what we previously knew in those and other areas, e.g., craft knowledge.) Moreover, histori-cal studies suggest that this increase is no mere piling up of independent knowledge atoms or bricks. Later work builds on earlier work but transforms it. And knowledge of our naturalistic and historical limits is part of that very knowledge increase. Some of this new knowledge—especially biological, psychological, historical, and socio-logical knowledge of our cognitive faculties, communities, and institutions—has severely *diminished* the power and scope of human cognition relative to that claimed for us by such writers as Bacon, Descartes, and Kant. We are left with the ordinary capacities recognized by naturalistic epistemology and historicist accounts of cogni-tion. And it is historical knowledge more than anything else that has undermined philosophical claims for eternal essences and for epistemic "givens" of traditional empiricists and rationalists—transparent apprehension of the world by means of perceptual or intellectual intuition, and the existence of a faculty of universal reason. We possess no godlike power of cognition, no prescience, no clairvoyance.[42]

[42] See Donald Campbell, "Evolutionary Epistemology," *Methodology and Epistemology for Social Science* (Chicago: Univ. Chicago Press, 1988), pp. 393–434. On exponential increase see Derek Price, *Little Science, Big Science* (New York: Columbia Univ. Press, 1963). For a more cautious discussion of progress see Nicholas Rescher, *Scientific Progress: A Philosophical Essay on the Eco-nomics of Research in Natural Science* (Pittsburgh: Univ. Pittsburgh Press, 1978).

In sum, we have gained more knowledge than Bacon and Descartes ever dreamed we could with far poorer cognitive faculties than they attributed to us. How is that possible?

In my view historicism and naturalism together imply the need for a "bootstrap" or self-amplifying account of the development of new knowledge (as well as an account of the loss or deconstruction or destruction of old knowledge). Somehow, a little knowledge gets parlayed into a lot, within a remarkably short time. In terms of its overall historical development, the knowledge enterprise is, to a significant degree, self-generating and self-transforming. It picked itself up by its bootstraps and continues to do so as it continues to transform or reconstruct itself. Crude tools were used to fashion better tools, and so on and on. This is a project that Dudley Shapere calls *learning to learn*. My own approach is broader-based than Shapere's, which makes internal, technical science self-propelling and self-amplifying, without sufficient attention to the social and contextual features of cognition that sociologists have been studying and the wider cultural resources on which science depends. "Methodology" in my broadened sense devotes much attention to the economy of research, to such matters as the efficient organization of research and of research communities. The pragmatists Charles Peirce and Herbert Simon (in his work on administrative organization and on "satisficing"—making do with good enough rather than optimizing) have addressed such questions.[43]

Whether broadly or narrowly conceived, most relativists will quickly dismiss such a methodological project as a relic of a bygone age. (My account may seem as conservative in the year 2025 as Clark's 1957 report looks to us, but today's relativisms may appear to be equally dated overreactions to an older conception of knowledge and inquiry.) Meanwhile, those of us who disagree with the relativists can happily use their results as key components of the explanation. I refer, for example, to studies of how the experimental laboratory and other workspaces emerged in the seventeenth century and after; studies of how the new ideas of experiment, evidence, probability, and the like were constituted; and studies that identify various "givens" as "black-boxed" social constructions. Deconstruction is not the same as destruction and can help us understand how the sciences "work" in those cases in which the black-boxed construction does indeed work, pragmatically speaking. Those of us who believe in such a project find it ironic that relativists deny themselves this opportunity when they themselves are furnishing tools and results necessary to tackle it.

Such a project (or series of projects, for the sciences and technologies are many and various) should also be of interest to policymakers. It was naive of philosophers to think that traditional methodology, devoid of any reference to social context, could provide a basis for policy judgments. But strong relativists are in no better position, for relativism evidently forswears evaluative and normative judgments of better or worse, on antiwhig and other grounds.[44] Historical science studies today finds itself

---

[43] See Dudley Shapere, *Reason and the Search for Knowledge: Investigations in the Philosophy of Science* (Dordrecht: Reidel, 1984); and Herbert Simon, *Administrative Behavior,* 3rd ed. (New York: Free Press, 1976). I myself do not endorse Simon's rule-based, computerized approach to scientific methodology. For Peirce see Nicholas Rescher, *Peirce's Philosophy of Science: Critical Studies in His Theory of Induction and Scientific Method* (Notre Dame, Ind.: Univ. Notre Dame Press, 1978).

[44] The best critique of philosophers, historians, and other science studies experts on policy grounds is Steve Fuller, *Social Epistemology* (Bloomington: Indiana Univ. Press, 1988).

in the awkward position of being critical of science and technology without being able to offer positive advice for change.

VI. DOES HPS HAVE A FUTURE?

My answer to the question whether HPS has a future is a cautious affirmative, provided that we understand both history and philosophy of science to be participants in a wider science studies. The previous section indicated some fertile areas for joint investigation. Here I must confine myself, after a few general remarks, to a single new example of a domain in which cooperation is needed.

Few philosophers of science any longer believe in a General Method of Science, applicable to all sciences at all times and places and offering genuine insight into and even guidance to scientific work. Most agree that there is no general method of *discovery*. Many still retain the idea of a general method or "logic" of *justification* (e.g., Bayesianism), but it is increasingly challenged. For example, the old idea that every competing theory that is proposed to solve a problem remains alive until explicitly rejected on logical grounds does not sit well with current concern for economy of research or with the view that ideas must be embodied in real, human advocates if they are to operate in the causal nexus of scientific investigation.[45] Many philosophers would agree with John Schuster and Richard Yeo that there is no "single, transferable method responsible for the progress of scientific knowledge," no portable method that is both formal and efficacious in directing and explaining scientific research.[46] But other critics of methodology would go beyond these fairly moderate claims to say that every single bit of research is so local and situated, so context-specific, that no general guidelines could possibly be of any use; or that, at best, methodology dissolves into a rhetoric of post hoc rationalization and justification.

At this point a philosopher is likely to demur. To be sure, studies of "deep practice" suggest that expertise cannot be fully reduced to rules, so that to transfer expertise one usually must transfer the experts themselves; and other studies suggest that methodological rules are negotiated to have a variety of meanings in different contexts. Nonetheless, in point of fact, scientific and technological research practices, procedures, techniques, and "methods" *are* actually "exported" around the world more successfully than some other social practices. (Again, some of these terms are value-loaded: the very fact that a technology is "successfully" transferred to another culture may harm that culture, as such.)

Transferability or portability is an important topic to which philosophers should give more attention. Traditionally, most philosophers have taken for granted the possibility of general knowledge, but only because they have assumed the existence of nonnatural faculties for achieving it. Today the problem is to understand in more

---

[45] See, e.g., David Hull, *Science as a Process: An Evolutionary Account of the Social and Conceptual Development of Science* (Chicago: Univ. Chicago Press, 1988).

[46] John Schuster and Richard Yeo, eds., *The Politics and Rhetoric of Scientific Method: Historical Studies* (Dordrecht: Reidel, 1986), p. ix. Schuster and Yeo say that such conclusions are based on historical evidence—another instance in which respected science studies experts appeal to history to evaluate methodological claims. They do point out that the traditional methodological project was blatantly whiggish but do not seem to notice that their own account of progress in the *historiography* of methodology is itself whiggish. I do not much mind this myself since I already have declared my whiggish tendency to recognize progress when I (by present lights) see it!

historicist, social, and naturalistic terms how any general knowledge is possible, and this typically involves understanding how what begins life as a local "truth" can be generalized to other domains, and how general speculations can be trimmed and deformed to fit specific domains. We need to distinguish various types of synchronic transfer made under subtly different social and epistemic conditions, including from one distinct problem to another within the same specialty; across distinct research strategies or program boundaries within the same field; across specialties and fields; across distinct research centers within the same local and national boundaries; and across international and cultural boundaries, from laboratory to nonlaboratory setting. But we also need to consider diachronic transfers, made from one or a few persons to a colleague; from one generation to the next; from one review of the literature to the next; and even that involved in the development of a single individual's understanding of a technique from one day to the next. We cannot assume constancy just because we are dealing with the same individuals or groups or the same words, nor should we think of transfer as a sort of passive diffusion. Everett Rogers's *Diffusion of Innovations* already dismisses that idea, and recent work on social construction carries the topic much further. For example, Latour explores the extension of the laboratory into the world and develops an actor-network theory. He notes an unhappy inversion in recent science studies. HPSers used to assume that since the products of science are universal, the processes that produced them must also be universal. Now science studies operates under the reverse assumption that since the processes are local, the products must be local also.[47]

Another dimension of portability is standardization. As Ted Porter aptly observes:

> One of the most important and fruitful issues being studied now in history and sociology of science is precisely the question of how local skills are made into general, or public, knowledge. . . . Really successful science eventually becomes reproducible at will, or at least without the need to journey to the site where it originated and imbibe the wisdom of a master craftsman. The separation of knowledge from a particular place and from an association with particular skills and intentions is, in my view, the most interesting meaning of objectivity. In this sense, objectivity is not the same thing as realism. . . . It is more nearly synonymous with standardization, which may soon become one of the key concepts of the history, sociology, and, one may hope, philosophy of science.[48]

It is tempting to think of mathematical techniques such as the differential and integral calculus, statistical inference, and computer programs as highly portable; and indeed they are. Mathematicians can often, with only a little expert assistance, set up and solve problems in domains in which they themselves are not expert. Yet the nineteenth-century British discussion of "embodied mathematics" should give us pause. It was not obvious how to interpret and apply mathematical ideas to heat, electrical, and magnetic phenomena. On the other hand, it only mystifies the subject to think that each context is so unique as to be incommensurable with any other. And even when groups must negotiate the meaning of a rule or procedure, that does not reduce the rule to a completely arbitrary placeholder in the discussion.

---

[47] Everett Rogers, *Diffusion of Innovations,* 3rd ed. (New York: Free Press, 1983); Bruno Latour, *The Pasteurization of France* (Cambridge, Mass.: Harvard Univ. Press, 1988); Latour, *Science in Action* (cit. n. 35); and Latour, "One More Turn after the Social Turn," in *Social Dimensions,* ed. McMullin (cit. n. 23), pp. 272–294. Cf. Nickles, "Good Science" (cit. n. 23).

[48] Theodore M. Porter, "The Uses of Humanistic History" (reply to Fuller, "Withering on the Vine" [cit. n. 10]), *Phil. Soc. Sci.,* 1991, *21*:214–222, on pp. 218–219.

Methodological or procedural rules are not self-explanatory to neophytes and do not capture well the knowledge of experts. Nonetheless, even Hubert and Stuart Dreyfus, in their attack on rule-based artificial intelligence, allow a considerable pedagogical role for rules.[49] Their idea is that, together with some tutorial guidance, applying rules jump-starts students so that they may fairly quickly gain the kind of practical experience they need to become experts, by which time their performance surpasses that of the very rule-based systems on which they learned. The rules play a bootstrap role.

VII. SOME CONCLUDING THOUGHTS

Those readers seeking here a simple statement of *the* relation between history of and philosophy of science will be disappointed. The old idea that the philosopher proposes methodological models or theories that the historian-*cum*-experimentalist subsequently tests against the data of history contains only a grain of truth, as does the idea that philosophy discloses to historians the true significance of their work. These grains of truth become a bit larger if we do not restrict philosophy and history to their respective professions but allow that historians may engage in philosophical thinking and philosophers in historical inquiry, broadly understood. History can help philosophers remain in touch with real science, and history provides a longer-term perspective essential to an adequate understanding of the development of human knowledge. I therefore disagree with those who say that philosophy can do away with history as the "middle man" or "medium" and go directly to social scientific studies of science (e.g., Fuller: but why should such studies exclude history?) or go still more directly to the contemporary sciences themselves (e.g., Giere). Study of contemporary science is, of course, crucial for many philosophical purposes, but absence of a deep historical perspective dangerously moves us back toward the shallow, "time slice" and "instant justification" approaches of yesteryear and the foundational impulse of "starting from zero." How can an adequate account of inquiry ignore the history of inquiry? Inquiry is not instantaneous, and any account that does not consider our long-term experience of past efforts is bound to be deficient. While philosophers should certainly not limit themselves to the record of actual research activity, those who ignore that record, as best historians can provide it, are skating on thin ice. While philosophers have (sometimes legitimately) dreamt of more things than exist on heaven and earth, historians have disclosed to eager generations of historical philosophers of science that more things also exist in heaven and earth than previous philosophy had dreamt of. Knowledge of the history of science and the rest of culture should humble philosophers. How many times have philosophers claimed to have canvassed all conceivable possibilities only to have later science and culture show them wrong? How many scientific revolutions have been self-contradictory and hence inconceivable, from the old perspective? In my view history of science has greatly enriched philosophy of science by raising new problems and suggesting new solutions, so that it can now address a wider range of problems and issues than in 1957. But history has also revealed the unworkability in human practice of many schemes, including scientific methodologies, that philosophers have thought possible.

[49] Hubert L. Dreyfus and Stuart E. Dreyfus, *Mind over Machine: The Power of Human Intuition and Expertise in the Age of the Computer* (New York: Free Press, 1986).

Historicism constantly reminds us of the situatedness of the knowledge enterprise and problematizes all transcendental reaching for the fixed and eternal. It is history more than any science studies discipline that shapes philosophical thinking by showing time and again that what we once took as given and transparent, part of the metaphysical structure of the world, turns out to be a complex, sociocognitive construction that emerged over long historical time. It is of course possible to escape from the here and now to some extent (that, after all, is one achievement of historical study itself), but not nearly so far as traditional philosophies assumed. Ironically, this situatedness is precisely where excessive antiwhiggism begins to undermine itself. For antiwhiggism as antipresentism—understanding the past entirely in its own terms—is incompatible with antiwhiggism now understood as the strong historicist view that every enterprise and its audience are historically situated, including history of science itself. Reflexive antipresentism is self-undermining. Not even historians can escape history!

Thankfully, the mutual bashing of relativist historians, sociologists, and philosophers is dying out, as each "side" finds that it can learn something from the other. I have tried to indicate that methodology today is a wider, more diverse, and livelier field than the traditional "logic of science" stereotype would suggest. That seventeenth-century Baconian, Cartesian, and Newtonian methodologies and twentieth-century positivist and Popperian methodologies have largely failed does not mean that the field is moribund. After all, both history and sociology of science have changed mightily just since 1957. Is there a field alive today that would happily embrace its self-characterization of fifty or a hundred years ago?

There are periods of consolidation and periods of diversification. Since 1957 we have increasingly appreciated the diversity of the sciences and their cultural relations. Oddly, some historians (and other science studies experts) give philosophy's own fragmented state as a reason why they can ignore philosophical discussions—as if they need to cite a single, neat, monolithic authority in the manner recommended in 1957 by Joseph Clark. The world is not so simple. I should have thought that the breakup of the received view in philosophy of science was a healthy thing! No grand new orthodoxy is likely to replace the old in the near future. The battle of the big systems that has raged since 1962 is over, and all of them have lost to history.

At present little work in philosophy of science and epistemology is linear, foundational, nonnatural, or transcendent in the old ways. Individualistic, Cartesian theories of cognition are today losing ground to accounts that are more connectivist and social. Partly because they now realize that logical arguments are not automatically transparent to human cognitive systems, philosophers are beginning to take rhetoric more seriously (e.g., Maurice Finocchiaro, Steve Fuller, Howard Margolis, Marcello Pera). It may be that rhetoric is better tuned than logic to the way human cognition works. During the seventeenth century philosophers replaced metaphysics with the new epistemology as "first philosophy" and proceeded to widen the scope of knowledge to include probable opinion. Today we have rejected altogether the idea of a first philosophy, be it metaphysics or foundational epistemology. Whether a second major reconceptualization of knowledge (a transvaluation of the values of knowledge) is now under way remains to be seen. Philosophers are also currently exploring new sorts of models, such as evolutionary, economic, and engineering models of scientific enterprises (e.g., Giere, Hull, Philip Kitcher, William Wimsatt), but this time the models may not be "rational" in the old senses. Today American philosophers are more in touch with Continental developments (e.g., Gary Gutting, Joseph Rouse, Mary Tiles). In the future philosophers and historians alike will surely be more sensitive to feminist, environmental, and other critical and policy concerns about science and technology and their power implications. Pragmatic philosophies remain attractive and are cited often enough by scholars in a variety of science studies fields to suggest that a serious continuation of the work of Peirce, James, Dewey, and George Herbert Mead—work interrupted by logical positivism—could be in the offing.

The loose association of the science studies disciplines and the open character of science studies conferences invite crossdisciplinary cooperation or at least discussion in a way that requires no formal, administrative linkage of history, philosophy, sociology—or any other study—of science. Other sorts of cooperation are possible without elaborate administrative apparatus—team teaching and joint authorship, for example, or simply soliciting interdisciplinary comment on one's work.

Let us conclude with the observations of three participants at a recent Stanford conference on the future of HPS. Latour made the point that the end of positivism in science studies should not signal the end of philosophy. On the contrary, positivism cut us off from our philosophical past, just as recent, extreme realism cut us off from nature. The fall of positivism and strong realism enables science studies to connect again with philosophy and with the world. In a similar vein, Arnold Davidson warned that, in trying to avoid the perceived errors of the positivists, by simply obliterating or reversing all of the old distinctions, science studies is in danger of reproducing itself as the mirror image or photographic negative of positivism, thereby either committing the same mistakes from the opposite side or falling into the false dilemma of saying that since positivism is wrong, its negation must be right. Finally, Andy Pickering made a remark that could serve as a motto for this report: "In science studies, disciplinary purity is anathema."

The History of Science and
the Sciences of Medicine

By John Harley Warner[*]

W HEN THE HISTORY OF SCIENCE was first taking shape, the history of
medicine was already an active field—a part of medicine rather than a
branch of history. Indeed, in the United States the founding of the Institute of the
History of Medicine at Johns Hopkins led George Sarton to complain bitterly in
1934 that the history of science could never hope for the kind of popularity enjoyed
by the history of medicine. "I do not begrudge the historian of medicine his popularity," he affirmed (with scant conviction), but went on to console himself with the
Comtian reminder that history of science stood far higher on the scale of things than
history of medicine. "We shall not discuss to what extent medicine deserves to be
called a science," he wrote, but added that "however scientific it be or may become,
it is also and will always remain an art." The project of history of science was to
study "a gigantic tree," while historians of medicine would have to content themselves with studying "the few medical twigs which belong to it."[1]

Sarton was being silly, Henry Sigerist, director of the new institute at Johns Hopkins, replied in an open letter. "I am sorry to say that you are brandishing your pen
against windmills, against a conception of medical history that may have existed
decades ago but that nobody thinks of sharing today." Medical history was "political
history, social history, economic history, history of religion and what not" every bit
as much as it was history of science.[2]

Yet Sarton's underlying misconception of the relationship between science and
medicine—that insofar as medicine draws upon the basic sciences it is a science, but
that the practice of medicine is an art—must be taken seriously precisely because it
has been remarkably durable. Such a division suggests that if we were to draw circles
representing history of science and history of medicine and fit them together to form
a Venn diagram, the overlap would be neatly delineated. This way of thinking has
long-standing roots in distinctions between *scientia* and *ars,* theory and practice,

[*] Section of the History of Medicine, School of Medicine, Yale University, 333 Cedar Street, New
Haven, Connecticut 06510–8015.

For their generosity and patience in helping me think through this essay, I wish to thank Rima D.
Apple, Thomas Broman, Harold J. Cook, William Cronon, Alvan R. Feinstein, Georgina Feldberg,
Faye M. Getz, Frederic L. Holmes, Bonnie Kaplan, Ann La Berge, Judith Walzer Leavitt, Mary
Linderman, Harry M. Marks, Lynn Nyhart, Charles E. Rosenberg, Christopher Sellers, Carolyn G.
Shapiro, Stuart Strickland, and especially Ronald L. Numbers, Naomi Rogers, and Nancy G. Siraisi.

[1] George Sarton, "The History of Science versus the History of Medicine," *Isis,* 1935, *23:*315–320,
p. 315.
[2] Henry E. Sigerist, "The History of Medicine *and* the History of Science," *Bulletin of the Institute
of the History of Medicine,* 1936, *4:*1–13, on p. 6.

hand and mind. But its persistent misuse by historians as a way to organize the past conceptually is also rooted in sibling jealousy and the quest for separate identities by fields that long shared their marginality to the historical mainstream.

In more recent years growth in both fields has encouraged specialization, and to historians of science eager to promote their own subspecialties and laboring hard to keep abreast of its literature, defining history of medicine as the Other and duly trimming it off may seem an appealing option. Intellectual changes have pulled the fields apart in more profound ways. As the history of medicine moved from being part of medicine to being part of history, many of its practitioners found they had less to talk about with historians of science than with other social and cultural historians, including those pursuing urban, environmental, business, women's, immigrant, or political history. Historians of medicine are less likely than historians of science to define the world in terms of disciplinary communities. Many of them are decreasingly likely to study the processes of biomedical investigation or the internal logic of change in the biomedical sciences. And as an increasing proportion of contributors regard themselves as historians of illness, disease, suffering, and healing—not of medicine—a growing number have never seriously considered the proposition that they might share substantial intellectual territory with the history of science (a field often assumed to look the way it did in Sarton's day).[3]

And yet, as the concerns of both fields have become integrated with the broader agendas of the humanities and social sciences, they are bound together by their shared exploration of how natural knowledge is produced, organized, and deployed in concrete historical settings. The notion that medical practice, for example, or professional organizations and health care delivery systems, are properly part of history of medicine but not history of science is no longer tenable, for the various institutions that mediate between the sciences and society have become a central preoccupation of the history of science. It is no longer helpful to divide medicine neatly down the middle into science and art, and assert that the former but not the latter overlaps with the history of science. Just as historians of science now work with a much expanded notion of what constitutes science, so historians of medicine have come to recognize the multiple sciences of medicine that coexisted at any

---

[3] The specific examples I use in this essay are intended to illustrate broader patterns, but the selections I have made—from Western medicine principally since the seventeenth century—are also sometimes personal and arbitrary. Recent reviews of the broader reshaping of the history of medicine include Olivier Faure, "The Social History of Health in France: A Survey of Recent Developments," *Social History of Medicine*, 1990, *3*:437–451; Toby Gelfand, "The *Annales* and Medical Historiography: *Bilan et Perspectives*," in *Problems and Methods in the History of Medicine*, ed. Roy Porter and Andrew Wear (London: Croom Helm, 1987), pp. 15–39; Caroline Hannaway, "Historiographical Trends in the History of Medicine: An Editor's Perspective," in *New Perspectives on the History of Medicine: First National Conference of the Australian Society of the History of Medicine, 1989*, ed. H. Attwood, R. Gillespie, and M. Lewis (Melbourne: Univ. Melbourne, 1990), pp. 75–84; Kenneth M. Ludmerer, "Methodological Issues in the History of Medicine: Achievements and Challenges," *Proceedings of the American Philosophical Society*, 1990, *134*:367–386; Nelli-Elena Vanzan Marchini, "Italian Scholars and the Social History of Medicine, 1960–1990," *Soc. Hist. Med.*, 1991, *4*:103–115; Charles E. Rosenberg, "Why Care about the History of Medicine?" *Explaining Epidemics and Other Studies in the History of Medicine* (Cambridge: Cambridge Univ. Press, 1992), pp. 1–6; Nancy G. Siraisi, "Some Current Trends in the Study of Renaissance Medicine," *Renaissance Quarterly*, 1984, *37*:585–600; Nathan Sivin, "Science and Medicine in Imperial China—The State of the Field," *Journal of Asian Studies*, 1988, *47*:41–90; John Harley Warner, "Science in Medicine," *Osiris*, 1985, *1*:37–58; and Paul Weindling, "Medicine and Modernization: The Social History of German Health and Medicine," *History of Science*, 1986, *24*:277–301.

historical moment. Indeed, as I want to suggest in what follows, changes in both fields mean that the history of medicine has come to speak more, not less, to the history of science, albeit less univocally than it once did.

I. FRAGMENTING THE MEANINGS OF "SCIENCE"

One subtle but telling change in the history of medicine literature in recent years is that increasingly the term *science* appears in quotation marks. This might be dismissed as a trivial matter of style, but in fact it offers one marker of the shifting ways science is regarded. The reasons that historians have elected this convention are varied, as are the messages they convey to readers. Sometimes, the historian has in mind a quite specific meaning of science, which he or she then goes on to analyze and explain. More frequently, "science" is placed in quotation marks to suggest that it refers not to authentic science but rather to claims made in its name; the quotation marks imply that science is more than a cumulative body of knowledge about nature. Equally, quotation marks often indicate the historian's skepticism about the term and any claims to truth value that might accompany it. "Science," that is, can refer to the ideal of science or language of science rather than to any substantive content in particular or even can designate pseudoscience. Most often, though, placing "science" in quotation marks protects historians from the charge of naively using a problematic term while sparing them the real work of defining just what they mean by it.

What all these usages of "science" share is a tacit recognition that it is much less clear than it once seemed just what is meant by science in medicine, for it holds divergent meanings at any historical moment. The same historiographic currents that drew attention away from the progressive accumulation of biomedical knowledge simultaneously directed historians to such themes as the social construction of medical ideas, their use as sources of professional and cultural authority, and the power of ideals of medical science sometimes quite unrelated to any particular body of biomedical knowledge or technique. As the focus of medical history expanded, our perspectives on science multiplied. This has led to fragmentation, and all too often "science" appears in quotation marks by default. Yet this fragmentation has given us a more complex depiction of the meanings of science in medicine and a broader understanding of its multiple functions.

Probably the most fundamental transformation is the growing acceptance of the notion that what constituted science in medicine shifted over time. The task of the historian is not so much to trace the accumulation of scientific knowledge as to understand the organization of knowledge and the modes of thinking that fostered change within specific historical contexts. In an older historiographic tradition, for example, William Harvey was celebrated above all as harbinger of modern experimentalism. Studies by Jerome Bylebyl, by contrast, have imbedded Harvey and his work in the intellectual and social context of his time, underscoring the fact that Harvey's science was that of the seventeenth century, not the twentieth.[4] Bylebyl's Harvey appears less modern, though no less important and far more interesting. Indeed, the reasons why later biomedical figures took pains to portray Harvey in

---

[4] Cf. Gweneth Whitteridge, *William Harvey and the Circulation of the Blood* (New York: American Elsevier, 1971); and Jerome J. Bylebyl, ed., *William Harvey and His Age: The Professional and Social Context of the Discovery of the Circulation* (Baltimore/London: Johns Hopkins Univ. Press, 1979).

modern guise have themselves proved revealing, as in Christopher Lawrence's analysis of the way the film *William Harvey and the Circulation of the Blood* was produced by the Royal College of Physicians as a tool to help twentieth-century experimental physiologists fend off antivivisectionist assaults.[5]

Some of the historical rethinking of science in medicine was instigated by the concerted program to deprivilege science that arose in the 1960s and 1970s. The Civil Rights movement, the war in Vietnam, the women's movement, and environmentalism combined to encourage protests against established authority, including authority claimed in the name of scientific expertise. At the same time, the rising vogue of historical sociology of scientific knowledge went far toward dissipating an image of scientific knowledge as objective, impersonal, value free, and autonomous. And while the barriers separating science from other ways of knowing were being pulled down, Michel Foucault and his proselytes drew attention to how elites invoked the authority of the human sciences for exercising power and control.[6]

The recasting of science as ideology found especially powerful expression in matters of health and medicine. At a time when modern reductionist medicine and its apparatus were coming under sharp criticism (by medical ethicists, proponents of preventive health care, activists in the women's and self-help movements, and clinicians who felt their judgment threatened by the hegemony of the laboratory), the depiction of medical science as an ideological tool rather than a symbol of progress was enormously appealing. Activists demonized science as a tool by which medical elites and the interests they served exercised control over the way people led their lives and particularly over the body.[7]

During the 1970s many scholars sought to disclose in the past the abuses of biomedical authority they deplored in the present. Sometimes the lines between admirable activism and shoddy historical practice were blurred in this early work, and some historical studies took on the character of exposés, evoking crude models of social control or victimization that later historians have condemned as theoretically simplistic. Studies on medical theories of insanity, the diseases of women, and definitions of race particularly revealed science as a tool employed by an affluent, white male elite to maintain its hegemony. Yet taken collectively, this work supported the contention—which now appears obvious—that biomedical science is partly an ideological construct that often serves the interests of those dominant in the culture in which it is deployed.[8]

Reflecting and feeding this skepticism toward the authority of science were

[5] Christopher Lawrence, "Cinema Vérité? The Image of William Harvey's Experiments in 1928," in *Vivisection in Historical Perspective,* ed. Nicholas Rupke (London: Croom Helm, 1987), pp. 295–313.

[6] Helpful reviews of this theme include John C. Burnham, "American Medicine's Golden Age: What Happened to It?" *Science,* 1982, *215:*1474–1479; Charles E. Rosenberg, "Afterword, 1987," *The Cholera Years: The United States in 1832, 1849, and 1866* (Chicago: Univ. Chicago Press, 1987), pp. 235–242; and Rosenberg, "Science in American Society: A Generation of Historical Debate," *Isis,* 1983, *74:*356–367.

[7] The most visible assault of the 1970s probably was Ivan Illich, *Medical Nemesis: The Expropriation of Health* (London: Calder & Boyars, 1975), though it is doubtful Illich had much direct influence on historical practice.

[8] Polemical examples include Barbara Ehrenreich and Deirdre English, *For Her Own Good: 150 Years of the Experts' Advice to Women* (Garden City, N.Y.: Doubleday, Anchor Press, 1978); and Ann Douglas Wood, "'The Fashionable Diseases': Women's Complaints and Their Treatment in Nineteenth-Century America," *Journal of Interdisciplinary History,* 1973, *4:*25–52.

historical challenges to the idea that advances in medical science necessarily made physicians better healers. The most influential came from Thomas McKeown, who in 1976 used studies of England and Wales to make the claim that before the 1930s and 1940s advances in medicine did little to increase life expectancy or to diminish mortality. The rise of Europe's population since the eighteenth century, he insisted, had more to do with improvements in hygiene and diet. Accordingly, what had stood as a leading testimonial to the benefits scientific change brought Western civilization seemed groundless. Several years later Gerald Geison suggested that perhaps clinicians skeptical of the value of basic science for their practice had a point. Perhaps, that is, it was the ideological rather than the technical utility of laboratory science that accounts for the remarkable way it was taken up as a rallying point by the medical profession in the late nineteenth and early twentieth centuries. Like McKeown's work, Geison's speculation urged historians to ask what it was, if not efficacy, that physicians gained from a proclaimed allegiance to science.[9]

Studies such as these redoubled attention to the ideological role of science in medicine. As late as 1972, in what one survey identified as the most influential book on the history of American medicine of the 1970s, William Rothstein could depict the advance of science as the force that consolidated the nineteenth-century medical profession and elevated its prestige by merit of the enhanced efficacy it brought.[10] By the early 1980s, though, many historians found such a positivistic treatment of science and professionalization plainly untenable. Instead, less interested in the truth value of science and less convinced by claims that better science translated into greater medical efficacy, they turned their attention to the ways particular models of science were used as tools for elevating the medical profession.

In 1986 Harold Cook, for example, explored how competing claims to the new science functioned professionally in the competitive medical marketplace of seventeenth-century London. For elite physicians attached to the College of Physicians, rival healers who claimed public support by merit of their allegiance to empiricism posed a serious problem. The new philosophy provided one answer, and a battle ensued for control over the new science and the prestige believed to come with it. "In a period in which 'science' was being invented, both sides were claiming empirical evidence as support for their view on which was the superior medicine," Cook noted. The physicians "promoted a new kind of 'scientific' medical learning that incidentally maintained their place at the top of the medical hierarchy as learned men."[11]

---

[9] Thomas McKeown, *The Modern Rise of Population* (New York: Academic Press, 1976); McKeown, *The Role of Medicine: Dream, Mirage, or Nemesis?* (Princeton, N.J.: Princeton Univ. Press, 1979); and Gerald L. Geison, "'Divided We Stand': Physiologists and Clinicians in the American Context," in *The Therapeutic Revolution: Essays in the Social History of American Medicine,* ed. Morris J. Vogel and Charles E. Rosenberg (Philadelphia: Univ. Pennsylvania Press, 1979), pp. 67–90. The perception that historians of disease are "writing in McKeown's shadow" (p. 470) is discussed in Nancy Tomes, "The White Plague Revisited," *Bulletin of the History of Medicine,* 1989, *63:*467–480.

[10] William G. Rothstein, *American Physicians in the Nineteenth Century: From Sects to Science* (Baltimore: Johns Hopkins Univ. Press, 1972); and Ronald L. Numbers, "The History of American Medicine: A Field in Ferment," *Reviews in American History,* 1982, *10:*245–263, on p. 248.

[11] Harold J. Cook, *The Decline of the Old Medical Regime in Stuart London* (Ithaca/London: Cornell University Press, 1986), pp. 106, 126. See also Cook, "The New Philosophy and Medicine in Seventeenth-Century England," in *Reappraisals of the Scientific Revolution,* ed. David C. Lindberg and Robert S. Westman (Cambridge: Cambridge Univ. Press, 1990), pp. 397–436; and Jerome J. Bylebyl, "The Medical Meaning of *Physica,*" *Osiris,* 1990, *6:*16–41.

The growing sophistication of work on the ideological nature of science as used to augment the medical profession's status also marked work on the use of science to expand that profession's authority over daily life. During the 1970s studies of medicalization, spurred by Foucault's writings, often told conspiratorial tales in which the medical profession, depicted as a calculating elite, set out to use the name of science to enhance physicians' authority and income. More recent studies of medicalization, however, tend to depict a more complex historical process, one in which a variety of players actively took part and made choices. Judith Walzer Leavitt, for example, in her 1986 study of childbirth in America, told a partly familiar story about how obstetricians displaced midwives: "'Science' enters the birthing room," as she put it. But instead of casting women as the passive substratum in a process over which they had no control, she stressed how and why women actively made choices about their childbirth attendants. Colin Jones, assessing the role of lay pressure in the reform of medical education in late-eighteenth-century Montepellier, similarly presented what he termed "a 'demand-oriented' account of the 'medicalisation' process." It was the idea of science more than any concrete benefits it brought to the consumer that drove medicalization in both cases.[12]

As some historians of medicine came to assume that, whatever else it might be, science was an ideology, historiographic emphasis gradually shifted from revealing how the authority of science was abused to a more complex analysis of what made science authoritative in the first place. One strain, radical in the 1970s but now mainstream, has been to explicate how medical ideas purporting to be scientific were imbued with the social and moral values of the culture in which they arose. Medical theories about the diseases of women, or the creation of masturbation as a disease, clearly drew upon the authority of science to reinforce and impose certain social norms. At the same time, by embodying the values and expectations of social elites, medical theories and their scientific foundations gained in cultural power and authority.[13]

A more recent preoccupation has been with the language of science. Like the broader field of cultural history, history of medicine showed signs of taking a literary turn. This has been expressed as attention to the ways medical narratives were constructed and conveyed meaning, and by close analysis of the rhetorical strategies used to make particular biomedical propositions persuasive.

These analyses have focused mostly on how in the late nineteenth century an orthodox medical profession set on collective uplift used imagery and language drawn from the medical sciences. Yet other groups as well could deploy the rhetoric of science. In seventeenth-century England both academic physicians and the

---

[12] Judith Walzer Leavitt, *Brought to Bed: Childbearing in America 1750–1950* (New York: Oxford Univ. Press, 1986), esp. pp. 36–63; Leavitt, "The Medicalization of Childbirth in the Twentieth Century," *Transactions and Studies of the College of Physicians of Philadelphia*, ser. 5, 1989, *11*:299–319; Colin Jones, "Montpellier Medical Students and the Medicalisation of Eighteenth-Century France," in *Problems and Methods*, ed. Porter and Wear (cit. n. 3), pp. 57–80, on p. 58. See also Claudia Huerkamp, "The Making of the Modern Medical Profession, 1800–1914: Prussian Doctors in the Nineteenth Century," in *German Professions, 1800–1950*, ed. Geoffrey Cocks and Konrad H. Jarausch (New York/Oxford: Oxford Univ. Press, 1990), pp. 67–84.

[13] Strict constructionist analyses remained surprisingly few, the most cited being Karl Figlio, "Chlorosis and Chronic Disease in Nineteenth-Century Britain: The Social Construction of Somatic Illness in a Capitalist Society," *Social History*, 1978, *3*:167–197. For recent perspectives see Jens Lachmund and Gunner Stollberg, eds., *The Social Construction of Illness: Illness and Medical Knowledge in Past and Present* (Stuttgart: Franz Steiner, 1992).

virtuosi and chemists who challenged them sought support in the language of empiricism associated with the new experimental philosophy. Similarly, in early-nineteenth-century America both elite physicians committed to sensual empiricism in the tradition of the French idéologues and their sectarian assailants, who proclaimed an aggressively democratic medical epistemology, deployed the language of empiricism to sustain their position and invalidate the authority of their opponents.[14]

The language of science, then, can be pressed into service by a variety of interests. In England during the first half of the nineteenth century radical science functioned as an ideological tool wielded to challenge the established medical order rather than to maintain it. General practitioners, those often termed apothecaries or surgeon apothecaries, borrowed the name of science in calling for fundamental reforms in the system of medical education, certification, and professional rewards. Asserting that an allegiance to science should be the new hallmark of professional identity, they sought to uplift their position vis-à-vis the privileged physicians and surgeons whose status stemmed more from social than scientific qualifications. To the reformers, science stood less for any body of knowledge or technique in particular than for a value system—a blueprint for meritocratic reforms that would make mastery of science the leading basis for claims to professional status. Although elite physicians and surgeons did not resort to this particular rhetoric, as Stephen Jacyna has shown, during the same period elite surgeons found in another language of science—that of a Hunterian tradition of learned surgery—a vehicle for distancing themselves from lingering craft associations and for affirming their own status as learned professionals. Different factions within the medical profession understood science and its meaning for their world in divergent ways.[15]

Historians have looked increasingly beyond the confines of the established medical profession to explore the meanings of science for other groups, medical and lay. The practitioners and clients of alternative medical systems, marginalized or neglected in traditional histories, have especially attracted scrutiny since the rise of the new social history in the 1970s, with its historiographic preoccupation with the authority of science and with groups situated on the fringes. Further, mounting skepticism about reductionist medicine and the concomitant surge of interest in alternative approaches to illness and wellness—ranging from homeopathy to New Age medicine—prompted historians to seek out in the past patterns of alternative belief and behavior that might shed light on the present.

The literature on alternative medicine underscored the depriviling of science during the 1970s. Some contributors were willing to take seriously only alternative

---

[14] Cook, *Decline of the Old Medical Regime* (cit. n. 11); and John Harley Warner, "The Fall and Rise of Professional Mystery: Epistemology, Authority, and the Emergence of Laboratory Medicine in Nineteenth-Century America," in *The Laboratory Revolution in Medicine,* ed. Andrew Cunningham and Perry Williams (Cambridge: Cambridge Univ. Press, 1992), pp. 310–341. Of a rapidly growing literature exemplifying the sensibility to language, see also Christopher Hamlin, *A Science of Impurity: Water Analysis in Nineteenth-Century Britain* (Berkeley/Los Angeles: Univ. California Press, 1990).

[15] John Harley Warner, "The Idea of Science in English Medicine: The 'Decline of Science' and the Rhetoric of Reform, 1815–1845," in *British Medicine in an Age of Reform,* ed. Roger French and Andrew Wear (London: Routledge, 1991), pp. 136–164; and L. S. Jacyna, "Images of John Hunter in the Nineteenth Century," *Hist. Sci.,* 1983, *21:*85–108. See also Adrian Desmond, *The Politics of Evolution: Morphology, Medicine, and Reform in Radical London* (Chicago/London: Univ. Chicago Press, 1989).

systems that flourished prior to the late nineteenth century, repeating the positivist myth that the rise of experimental science undermined medical sectarianism. Others composed their histories as partisan briefs for the enduring truth value of systems such as homeopathy.[16] Much of the literature, though, tried to place orthodox and unorthodox medicine on an equal footing: orthodox medical science was deemed no better (and perhaps worse) than alternative systems. As evidence most studies cited the therapeutic shortcomings of orthodox medicine, oftentimes recounting gruesome contemporary accounts of the therapeutic misdeeds of regular physicians. Their point—that orthodox science did not necessarily inform better patient care— was reinforced by McKeown. Still, underlying most of this work was a residual positivism, rooted in a fundamental division between medical science before and after a perceived late-nineteenth-century watershed.

Yet, as Ronald Numbers pointed out in 1985, the persistence of nonorthodox medical sects into the twentieth century should put to rest the notion that the rise of experimental medicine undermined alternative healing.[17] The pluralism that is a hallmark of postmodern medical culture underscores this point. In the 1980s studies of both orthodox and unorthodox practice began to emphasize that judging past therapeutics in terms of twentieth-century appraisals of efficacy, rather than in terms of what made them meaningful at the time, is not helpful historically. The new studies, adopting a relativist perspective, were less likely to judge the medical systems of the past and more likely to see them as useful contexts for exploring the links between social interests and ideological allegiances.[18]

These works often analyzed the reasons why some people constructed certain alternative systems of natural knowledge and entrepreneurial practice and why others chose to place trust in them. Recent studies of hydropathy in mid-nineteenth-century America, for example, note how it recruited followers from among the adherents of other contemporary reform movements, such as antislavery, feminism, temperance, dress reform, and vegetarianism. Hydropathy offered not only a cure for disease but also a plan of natural living designed to counterbalance the evils of civilization. Historians have established convincing correlations between other medical creeds and particular social classes, political parties, and religious faiths. These studies further suggest that the science of orthodox doctors was often the chief target of those who assailed the established medical profession, as in hydropathic

---

[16] For a positivist approach see Rothstein, *American Physicians in the Nineteenth Century* (cit. n. 10); and for a partisan approach see Harris L. Coulter, *Divided Legacy: A History of the Schism in Medical Thought* (Washington, D.C.: McGrath, 1973), Vol. III.

[17] Ronald L. Numbers, "The Fall and Rise of the American Medical Profession," *Sickness and Health in America: Readings in the History of Medicine and Public Health* (Madison/London: Univ. Wisconsin Press, 1985), ed. Judith Walzer Leavitt and Numbers, pp. 185–196.

[18] For the rethinking of efficacy see Charles E. Rosenberg, "The Therapeutic Revolution: Medicine, Meaning, and Social Change in Nineteenth-Century America," in *The Therapeutic Revolution,* ed. Vogel and Rosenberg (cit. n. 9), pp. 3–25; and John Harley Warner, *The Therapeutic Perspective: Medical Practice, Knowledge, and Identity in America, 1820–1885* (Cambridge, Mass.: Harvard Univ. Press, 1986). Recent studies on unorthodox medicine include W. F. Bynum and Roy Porter, eds., *Medical Fringe and Medical Orthodoxy, 1750–1850* (London: Croom Helm, 1987); Roger Cooter, ed., *Studies in the History of Alternative Medicine* (Houndmills, Hants./London: Macmillan, with St. Antony's College, Oxford, 1988); Robert C. Fuller, *Alternative Medicine and American Religious Life* (New York/Oxford: Oxford Univ. Press, 1989); Norman Gevitz, ed., *Other Healers: Unorthodox Medicine in America* (Baltimore/London: Johns Hopkins Univ. Press, 1988); and Phillip A. Nicholls, *Homoeopathy and the Medical Profession* (London: Croom Helm, 1988).

railings against science as an oppressor of humanity in general and women in particular. Such studies present the voices of critics of established medical science and
the values some laypeople attached to that science.[19]

Much the same impulse to rewrite the history of medicine in ways that give voice
to those silent in the canon has directed attention to the experiences and perceptions
of such historiographically marginal healers as midwives, patent medicine vendors,
women physicians, and advice book authors. During the past decade this movement
has found its fullest expression in calls to heed the layperson's perspective. "Traditional history of medicine simply ignored the patient," Roy Porter, the most articulate proselytizer of this program, noted in 1985. "He or she was of no interest." Yet
the sufferer has always been a key participant in medical encounters. So far, the
context most densely studied has been seventeenth- and eighteenth-century Britain,
where scholars have begun to mine diaries for what they reveal about the experience
of being ill in preindustrial society, lay conceptions of health and disease, patterns
of self-medication, and the relationship of sufferers to the variety of sources of healing advice and treatment.[20]

Science, of course, was only one among a whole host of elements that patients
took into account in assessing and selecting their medical attendants. Yet especially
as historians pay more and more attention to medical use of the languages of science,
it becomes increasingly important to understand how this language was heard and
interpreted by the laypeople to whom it was directed, for it was they, after all, who
held the power to translate that language into cultural authority.

Attention to science in the history of medicine, then, has by no means faded, nor
have the intellectual agendas of the field wrenched it apart from the history of science. Shifts in focus have widened the scope of the field at the same time that they
have turned attention away from some themes central in an earlier historiography,
while new modes of analysis have multiplied the meanings of science we recognize
in the past. Indeed, in many respects those movements in history of medicine that
might most betoken a move away from history of science are the very ones that have
done the most to expand our knowledge of how natural knowledge in medical contexts was produced, became authoritative, and informed action. Despite the persis-

---

[19] See Susan E. Cayleff, *Wash and Be Healed: The Water-Cure and Women's Health* (Philadelphia:
Temple Univ. Press, 1987), and Jane B. Donegan, *"Hydropathic Highway to Health": Women and
Water-Cure in Antebellum America* (New York/Westport, Conn.: Greenwood Press, 1986).

[20] Roy Porter, "Introduction," in *Patients and Practitioners: Lay Perceptions of Medicine in Pre-
Industrial Society* (Cambridge: Cambridge Univ. Press, 1985), pp. 1–22, on p. 2. See also Porter,
"The Patient's View: Doing Medical History from Below," *Theory and Society,* 1985, *14:*175–198;
Barbara Duden, *The Woman beneath the Skin: A Doctor's Patients in Eighteenth-Century Germany,*
trans. Thomas Dunlap (Cambridge, Mass.: Harvard Univ. Press, 1991); Mary E. Fissell, *Patients,
Power and the Poor in Eighteenth-Century Bristol* (Cambridge: Cambridge Univ. Press, 1991); and
Roy Porter and Dorothy Porter, *In Sickness and in Health: The British Experience, 1650–1850* (Lon-
don: Fourth Estate, 1988). Recent studies on other contexts include Barbara Bates, *Bargaining for
Life: A Social History of Tuberculosis, 1876–1938* (Philadelphia: Univ. Pennsylvania Press, 1992);
Michael R. McVaugh, *Medicine before the Plague: Practitioners and Their Patients in the Crown of
Aragon, 1285–1345* (Cambridge: Cambridge Univ. Press, 1993); and Sheila M. Rothman, *Living in
the Shadow of Death: Tuberculosis and the Social Experience of Illness in American History* (New
York: Basic Books, 1994). Something of the same historiographic impulse is seen in attention to
popular conceptions of health, as in Martha H. Verbrugge, *Able-Bodied Womanhood: Personal
Health and Social Change in Nineteenth-Century Boston* (New York/Oxford: Oxford Univ. Press,
1988), and in the study of the dissected rather than the dissectors, as in Ruth Richardson, *Death,
Dissection and the Destitute* (London/New York: Routledge & Kegan Paul, 1987).

tent risk of historiographic reductionism—the inclination to fix upon one among the many meanings of science to the exclusion of all others—what has emerged is a much more complex and variegated depiction of science in medicine.

II. REVISIONS, GAINS, AND LOSSES

This expanded view of science in medicine has not been purchased without a price. Drawing attention to the importance of neglected topics has tended to privilege them, and not only to deprivilege but to delegitimize the study of longer-established themes that are equally critical to our understanding of the medical past: the ideas of medical elites, the technical content of medical knowledge and practice, and the dynamics of conceptual change in the biomedical sciences.

No impulse has brought more welcome changes to the history of medicine than the new social history that arose during the 1970s. Yet some of its practitioners reconstructed an image of traditional medical history that was remarkably whiggish—more a tool of self-legitimation than a depiction of historiographic reality. The image of an older field consecrated to the celebration of great doctors and their ideas, unrepentantly internalist and positivist, rightly targeted some of its most glaring shortfalls. Yet it hardly represented the complexity and enduring value of work by such early laborers as Ludwig Edelstein, Erwin Ackerknecht, Oswei Temkin, and Henry Sigerist.

No less than physicians or scientists writing about the past of their disciplines, historians had good reasons for their whiggery. The history of medicine was undergoing rapid professional changes during the 1970s, and the construction of a straw figure had important symbolic value for a group trying to define its separate identity, just as it served as a rallying point for establishing a fresh historiographic program. Authors decreasingly likely to have read work from the field's early years often caricatured and assailed it in order to enhance the importance of their own contributions. Predictably enough, this polemic engendered a counterpolemic, and the 1970s ended with lamentations about "medical history without medicine" that echoed complaints about history of science without science.[21]

By 1981 Ronald Numbers could reflect on the complaint that historians of American medicine placed "too much emphasis on heroic physicians and medical milestones and too little on the context in which healing takes place" and conclude: "If these criticisms fairly reflect American medical historiography before the 1970s— and I suspect they do not—they certainly are no longer valid." Yet the jeremiad continued. Thus in 1988 the editors of the new journal *Social History of Medicine* chose to define its program in opposition to what had become an anachronistic and hackneyed stereotype. "What distinguished the social history of medicine from the history of medicine," they proclaimed, "is the approach to the subject—the belief that topics within the history of medicine can only be understood in the context of the society of which they are a part." Their program, they continued, "rejects the history of medicine which focuses on a 'progressive' view of medicine as emanating from a disinterested science, the view that the history of medicine is the culmination

---

[21] Lloyd G. Stevenson, "A Second Opinion," *Bull. Hist. Med.,* 1980, *54:*134–140; and [Leonard Wilson], "Medical History without Medicine," *Journal of the History of Medicine and Allied Sciences,* 1980, *35:*5–7.

of great discoveries and technological advances or achievements of great men (and occasionally women) of the past." At best, they were rousing the troops to battle against an enemy long since put to rout, magnifying the danger of stragglers in order to sustain the momentum of a crusade no longer novel.[22]

The production of such polemics is understandable, mirroring as it does broader competitions for the limited supply of academic turf. What is disturbing, though, is the way programs of caricature and delegitimation sustain an either-or mentality that is historiographically divisive and intellectually simplistic. It perpetuates unhelpful dichotomies between medical science as ideology and medical science as a body of knowledge and technique; between science as discourse and science as social practice; between science as an agent of power and science as an agent of healing; and, indeed, between science and society. What should be heartening, however, is a mounting call in recent years to shun reductionism and embrace complexity—to recognize that no single perspective on science in medicine will even begin to exhaust historical understanding.

One lightning rod for either-or thinking and protests against it has been relativism. Radical social constructionism has had much less influence on the history of medicine than on the history of science, in part because it has always been hard to ignore how society shapes explanations of disease, experiences of illness, and the ways health care is managed. By the 1970s, instead of judging past medical theories in the light of modern scientific knowledge, historians increasingly set out to understand them in the context of their own times and to assess what made them meaningful, and in this endeavor methodological relativism often proved helpful. For some historians, a relativist stance was part of the radical program to deprivilege medical science and authority. But during the 1980s this approach was assimilated into the mainstream; indeed, as it became a common tool in medical historiography, some early proponents seemed distressed that it had lost its radical fervor. Thus, in reviewing a study that acknowledged its relativist approach to nineteenth-century therapeutics, Steven Shapin commented that it was "an important and radical book." Yet he went on to note critically that the author was "regrettably timid on the question of relativism," lamenting the fact that the book was "written in a way that masks its radicalism."[23]

In polemical exchanges (more over seminar tables than in print), ordinarily sober historians sometimes acted as if they were being asked to attribute the construction of disease to *either* social *or* pathophysiological processes. Yet most historians of medicine never shed a guarded leaning to biological realism. This reflects the paradox of McKeownism: while it drew attention away from biomedical science as a factor in population growth and thereby to the social agendas embedded in claims to its importance, it simultaneously reminded scholars of the biological realities of nutrition, physical environment, disease, epidemics, and death. Studies on the history of public health, like those in the emerging field of environmental history, have paid persistent attention to the material conditions of life and death, including the biological environment. It may be that those historians of health and illness who are

---

[22] Numbers, "History of American Medicine" (cit. n. 10), pp. 252–253; and Linda Bryder and Richard Smith, "Editorial Introduction," *Soc. Hist. Med.* 1988, *1*:v–vii, p. v.

[23] Steven Shapin, "Practical Healers," review of Warner, *Therapeutic Perspective* (cit. n. 18), *Times* (London) *Higher Education Supplement,* 31 July 1987, p. 19.

readiest to cast biological reality as an actor in their accounts tend to be situated farthest from the history of science, while those most squeamish about recognizing the material reality of disease have the greatest affinity to history of science.[24] In any event, a historical reality rooted in existential biological factors has increasingly reentered studies on the history of medicine.

The appearance of AIDS early in the 1980s furthered this shift. No observer can reasonably deny that AIDS is socially constructed, the product of cultural values, social prejudices, and power relationships that prevail toward the end of the twentieth century. Nevertheless, as Charles Rosenberg recently noted, "the biomedical aspects of AIDS can hardly be ignored; it is difficult to ignore a disease with a fatality rate approaching 100 percent." Seeing it as merely a social product has seemed to many activists irresponsible, insensitive, and dangerous. Among historians, Rosenberg suggested, "AIDS has, in fact, helped create a new consensus in regard to disease, one that finds a place for both biological and social factors and emphasizes their interaction. Students of the relationships between medicine and society live in a necessarily postrelativist decade."[25]

What Rosenberg has observed in the historiography of disease is true of the broader field of history of medicine. We may all agree that medical interpretation, explanation, and definition of the pathological and the normal are fundamentally shaped by social and cultural values; but many of us simultaneously insist that historical understanding grounded in these factors alone is simply not enough. It may be precisely because there is no longer any need to stridently assert that medical knowledge, like disease, is partly socially constructed that we may look to other factors without fear of lapsing into crude biologism. It seems doubtful that most historians of medicine are prepared to endorse Rosenberg's suggestion that we have entered a postrelativist era by openly embracing a postrelativist historiography: the specter of positivism in our field's past still looms too menacingly large. But as Joan Jacobs Brumberg put it in her exemplary 1988 study of anorexia nervosa, "It is clearly time to drop the either/or approach and begin to think about the reciprocity of biology and culture."[26]

---

[24] See, e.g., Gwyn Prins, "But What Was the Disease? The Present State of Health and Healing in African Studies," *Past and Present*, 1989, *124*:159–179; and Donald Worster, "History as Natural History: An Essay on Theory and Method," *Pacific Historical Review*, 1984, *54*:1–19. "If we had our way," the environmental historian William Cronon noted in a recent review, "historians would be no more willing to ignore questions about ecological context—about nature—than they would questions about gender or class or race": Cronon, "Modes of Prophecy and Production: Placing Nature in History," *Journal of American History*, 1990, *76*:1122–1131, on p. 1131. For studies that take for granted the material reality of disease see, e.g., Alfred W. Crosby, *Ecological Imperialism: The Biological Expansion of Europe, 900–1900* (Cambridge: Cambridge Univ. Press, 1986); Philip D. Curtin, *Death by Migration: Europe's Encounter with the Tropical World in the Nineteenth Century* (Cambridge: Cambridge Univ. Press, 1989); Mirko D. Grmek, *History of AIDS: Emergence and Origin of a Modern Pandemic*, trans. Russell C. Maulitz and Jacalyn Duffin (Princeton: Princeton Univ. Press, 1990); and James C. Riley, *The Eighteenth-Century Campaign to Avoid Disease* (New York: St. Martin's Press, 1987).

[25] Charles E. Rosenberg, "Disease and Social Order in America: Perceptions and Expectations," in *AIDS: The Burdens of History*, ed. Elizabeth Fee and Daniel M. Fox (Berkeley/Los Angeles: Univ. California Press, 1988), pp. 12–32, on p. 14. See also Virginia Berridge and Philip Strong, "AIDS and the Relevance of History," *Soc. Hist. Med.*, 1991, *4*:129–138; and Fee and Fox, "The Contemporary Historiography of AIDS," *Journal of Social History*, 1989, *23*:303–314.

[26] Joan Jacobs Brumberg, *Fasting Girls: The Emergence of Anorexia Nervosa as a Modern Disease* (Cambridge, Mass.: Harvard Univ. Press, 1988), p. 7. See also Randall McGowen, "Identifying Themes in the Social History of Medicine," *Journal of Modern History*, 1991, *63*:81–90.

The move toward pluralism is also apparent in the units of analysis historians now choose, for in recent years splitting has taken the upper hand over lumping. Book reviewers continue to call for greater attention to either context or content, external or internal factors, the scientific or the social. But critics are most likely to fault authors for the voices they have excluded. The 1990s opened with a rising chorus of *mea culpa* among some historians of medicine—public lamentations over past omissions and calls to pay attention to the complex interaction of gender, race, class, religion, and ethnicity.[27] Certainly different groups in any population did not necessarily interpret their experiences of health, illness, and medical care in the same way. And as we learn more about the perceptions and expectations of different social groups, the multiple meanings of science in medicine to them will be even further fragmented.

The aim of course should be not a steady stockpiling of perspectives, but more complex historical narratives; and with calls for complexity have come intensifying calls for synthesis and integration. "It may be too much to ask any individual historian to integrate adequately all approaches and all peoples within a single study," Judith Walzer Leavitt noted in a 1990 review essay, "but it is essential that we learn to look broadly at our own work and the work of others and begin the long-term task of fitting together the complex pieces of the puzzle of health care history."[28]

Other historians share this impulse to integration and have acted on it. Nancy Siraisi's *Medieval and Early Renaissance Medicine* (1990), for example, focused on a learned medical elite and their intellectual and technical acquirements. But Siraisi was sensitive throughout not only to the social context of literate medicine but also to the interconnections between it and the worlds of patients, folk practitioners, and religious healing. "Syntheses always run the risk of being premature," she noted, but her own work makes it clear that the intellectual and scientific activities of the academically trained medical community and those of popular medical culture need not be topics of separate historical narratives.[29]

Leavitt selected a narrower focus for her study *Brought to Bed: Childbearing in America, 1750 to 1950* (1986), but, like Siraisi's, her work wove together threads spun out by a generation of historical writing. Medical science as ideology is a theme running throughout the text; yet Leavitt did not depict science as a tool used by obstetricians solely out of a will for dominance in the marketplace or at the bedside, and she took seriously the conviction of many doctors that scientific and technical change would bring better care. Moreover, she did not locate power on one side of the doctor-patient relationship alone, and by looking at the perceptions and choices women expressed in their diaries and letters, she went far toward demonstrating why the historian interested in the changing place of science in medicine *must* care about the patient's perspective.

---

[27] As at, e.g., the workshop "Women and Difference," annual meeting, Organization of American Historians, Washington, D.C., 24 March 1990; and "All the Men Are Doctors, All the Women Are Sick, but Some of Us Are Getting Better: A Roundtable on Gender and Race in American Women's and Health Care History," Eighth Berkshire Conference on the History of Women, New Brunswick, N.J., 9 June 1990.

[28] Judith Walzer Leavitt, "Medicine in Context: A Review Essay of the History of Medicine," *American Historical Review,* 1990, *95:*1471–1484, on p. 1480.

[29] Nancy G. Siraisi, *Medieval and Early Renaissance Medicine: An Introduction to Knowledge and Practice* (Chicago/London: Univ. Chicago Press, 1990), p. xii.

Charles Rosenberg's *The Care of Strangers: The Rise of America's Hospital System* (1987) also represents a model integration of the changing perspectives that have characterized recent medical historiography. Rosenberg did not dismiss the technical yield scientific change brought to patient care, but emphasized that the image of American hospitals as "temples of science" had as much to do with the idea of science as with its content.

> Not everyone agreed as to what that science actually was or should be—or how it might best be related to the clinician's mundane tasks. But such imprecision did little to moderate the impact of appeals to the laboratory or undercut the role of science in the forging of a new style of medical identity. Quite the contrary, a lack of precise meaning has rarely interfered with the efficacy of appeals to science and the promise of its application.

Rosenberg drew on important studies from the early 1980s that had moved beyond the assumption that the hospitalization of the middle classes came as an inevitable consequence of the rise of science. Yet so successful had these studies been in asserting that social rather than scientific changes had empowered this transformation, and in shifting the focus of hospital history away from an iatrocentric narrative, that Rosenberg's assertion "that the history of American hospitals can best be made comprehensible by focusing on the physician's role in eliciting and shaping change" itself came as bold and revisionist.[30]

What these three studies exemplify is a shared willingness to construct more complex, less reductionist historical narratives. The problem of integration is no longer productively conceptualized in terms of a dichotomy between internalist and externalist, not only because that is an inbuilt and probably creative tension rather than an obstacle to be overcome, but also because it no longer captures the realities of the multiple agendas now fruitfully being pursued. Science, after all, is no more internal to medicine than culture is. What we stand to gain from historiographic pluralism is not an integration of history of medicine with history of science, but a more integrated historical understanding of the multifaceted meanings of science in medicine.

III. NEEDS AND POSSIBILITIES

Most studies are not going to be wide-ranging syntheses, and should not be. We need closely focused case studies, but, equally, as we intensify demands for complexity, we need as authors to specify the limits to our particular focus and as readers to accept that authors make their studies manageable by focusing on one or another aspect of a larger problem. The task is to supplant an either-or approach with a readiness to acknowledge that we are looking *either* at one aspect *or* at another of a more complex issue.

---

[30] Charles E. Rosenberg, *The Care of Strangers: The Rise of America's Hospital System* (New York: Basic Books, 1987), pp. 9, 150, 333. Chief among the earlier studies are David K. Rosner, *A Once Charitable Enterprise: Hospitals and Health Care in Brooklyn and New York, 1885–1915* (Cambridge: Cambridge Univ. Press, 1982); and Morris J. Vogel, *The Invention of the Modern Hospital: Boston, 1870–1930* (Chicago: Univ. Chicago Press, 1980). See also Rosemary Stevens, *In Sickness and in Wealth: American Hospitals in the Twentieth Century* (New York: Basic Books, 1989).

Once we are free from looking for the *one* meaning of science in medicine, then the work of the past two decades appears not merely as a fragmentation of perspectives in need of integration, but as a springboard to posing more refined questions. To illustrate some of the possibilities, I will first take here the example of American medicine between the 1870s and 1920s. This is the context most densely investigated in recent studies, but I offer it only as one illustration of the kinds of problems that should be explored in other equally significant times and places. I will then point briefly to some broader comparative topics that seem especially promising in deepening our understanding of science in medicine.

Few themes in the late-nineteenth- and early-twentieth-century transformation of American medicine have received closer attention than the rising status and authority of the medical profession, while the role accorded to scientific change in this transformation has been subjected to more intense revisionism than any other component in this historical process. The idea that advances in science brought increased medical efficacy and that this in turn elevated the standing of the medical profession—once largely taken for granted—has been soundly challenged by the proposition that it was the cultural more than the technical value of science that propelled change.

One of the most effective prods to rethinking the relationship between science and professional status was an essay published by Gerald Geison in 1979, which underscored the cultural utility of experimental science. From the late nineteenth century onward, Geison argued, "the experimental sciences, like Latin in an earlier era, have given medicine a new and now culturally compelling basis for consolidating its status as an autonomous and learned profession." Several years later S. E. D. Shortt expanded on this point, drawing attention to the extent to which physicians uplifted their profession's status by using "not the content but the rhetoric of science."[31] The key here was rising public faith in the kind of special learning science represented, and with it the perception by physicians that claims to scientific expertise offered a promising source of status. What particularly made the profession's claims to mastery of science convincing was the thoroughgoing reform of medical education, for the elevation of standards both limited professional ranks to those capable of pursuing a rigorous course of study and certification and gave those who remained plausible credentials as medical scientists.

The sociologist Paul Starr, in his Pulitzer Prize–winning *The Social Transformation of American Medicine* (1982), reinforced the point that cultural authority need not rest upon competence. The new and convincing bid to authority the medical profession made in the name of laboratory science during the Progressive Era, he proposed, can be understood as "the renewal of legitimate complexity." As "the American faith in democratic simplicity and common sense yielded to a celebration of science and efficiency," doctors were able to claim a new measure of cultural

---

[31] Geison, "Divided We Stand," p. 85; and S. E. D. Shortt, "Physicians, Science, and Status: Issues in the Professionalization of Anglo-American Medicine in the Nineteenth Century," *Medical History*, 1983, *27*:51–68, on p. 60. See, however, Christopher Lawrence's alternative perspective in "Incommunicable Knowledge: Science, Technology and the Clinical Art in Britain 1850–1914," *Journal of Contemporary History*, 1985, *20*:503–520. And on the ideal of science as a platform for elevating the nursing profession see Susan M. Reverby, *Ordered to Care: The Dilemma of American Nursing, 1850–1945* (Cambridge: Cambridge Univ. Press, 1987), p. 158.

authority over a public impressed by the promise of laboratory medicine and ready to accept that it was complex beyond their mastery.[32]

Much as this perspective added to historical understanding, it also raised fundamental questions that remain largely unanswered. Whose conception of science mattered here? And whose aspirations? The American medical profession, to be sure; but who within that large and highly variegated body? Kenneth Ludmerer has convincingly argued that the reformation of medical education was driven by the aspirations of medical academics, particularly basic scientists eager to carve out careers for themselves. "Entrance requirements," he noted as an example, "were introduced not by the profession as a whole but by medical educators, whose interests lay with the development of academic medicine as a career rather than with how much money the average doctor in practice made."[33]

How is this interpretation of the reform of American medical education to be reconciled with the position, voiced by Shortt among others, that the medical profession employed science as a socially acceptable means of restricting access to the profession and enhancing the cultural authority of those who remained in ways that would bring payoff in the marketplace? Was the elevation of the profession's status merely a side effect of academic physicians' attempts to attain their own career goals? Or if the wider medical profession did have a hand in educational reforms, how were its desires enforced? Ludmerer rightly criticized "the common error of viewing the medical profession as a monolith."[34] To speak of how the medical profession saw science, or of how American physicians used it in attaining their ends, is too simple, glossing over the occupational, social, and intellectual diversity among individual physicians. We must begin by disaggregating such constructs as "the American medical profession." Put simply, one group within it saw the greater infusion of experimental science into American medicine as a vehicle for scientific career making, while another saw it as a vehicle for augmenting cultural authority and income. If, as historians have suggested, members of both groups were active players, then we should pay close attention not merely to motivation and results, but also to the process by which each group pursued its objectives and to the interaction of these processes.

Having recognized differences in the meaning science held for academic physicians and nonacademic practitioners, we should also expect that science had different resonances for diverse social groups within these aggregates—urban specialists and rural general practitioners, for example—yet the texture of these differences has been explored only lightly. Similarly, those who shared membership in the regular medical profession were sometimes divided by their gender, ethnicity, race, and religion. A sophisticated literature on women physicians gives us every reason to think such factors could be important in shaping attitudes toward science. As Regina Morantz-Sanchez and others have shown, some women physicians, especially those of the first generation, believed that their gender gave them not only a special mission but also a special relationship to science, while others played down the

[32] Paul Starr, *The Social Transformation of American Medicine* (New York: Basic Books, 1982), p. 73.
[33] Kenneth M. Ludmerer, *Learning to Heal: The Development of American Medical Education* (New York: Basic Books, 1985), p. 118.
[34] *Ibid.*

significance of gender. This kind of diversity, like differences between women and men physicians, merits much closer attention. How were the allegiances of women physicians divided by race or sectarian medical education, for example?[35]

Historians have little explored transformations of the meaning of science in alternative medicine from the late nineteenth century onward. Naomi Rogers has pointed out that toward the end of the century sectarian medical colleges, like their regular counterparts, came to portray themselves as scientific to appeal to students who might be attracted to a self-consciously alternative education but who also demanded the accoutrements of scientific training.[36] What was the relationship between medical knowledge and rhetoric in such situations? Did sectarian educators share the language of science deployed by regular physicians, even though they subscribed to a substantially different belief system?

Even if we grasp why various physicians staked their claims in the name of science, to understand how this elevated the authority of the medical profession we also need to understand just why the American public bought it. And at this juncture the patient's perspective becomes indispensable to understanding the transforming power of science. "The renewal of legitimate complexity," Starr's phrase, aptly describes what happened without explaining why or how. After all, it was in the eyes of the American public that scientific medicine was judged to be (or not to be) legitimately complex, beyond the ken of the common man or woman.

Historians have long recognized the rising esteem for science during the Progressive Era, and studies on the medical context have suggested that it was an image of science—slippery when it came to precise definition—that the public revered. "The exact meaning of the term 'science' remained extremely vague," Rima Apple noted of the "scientific motherhood" that led women to depend on the advice of medical experts; "its definition remained elusive." Scientific motherhood gave mothers status, just as choosing to give birth in a hospital—the most prominent icon of science in public view—could serve the same function. As Leavitt noted, what attracted birthing mothers to the hospital was the "aura surrounding 'Science,' in their minds spelled with a capital S."[37] Still, we know remarkably little about the particular ways the American public witnessed medical science, became convinced of its power, and chose to give it their trust—the processes, that is, through which the rhetoric of science came to be translated into authority, and confidence into status.

We also need to know more about *which* American public, for not all Americans looked upon medical science through the same eyes. Here again, we need to understand not an abstract patient, but concrete ones bounded by class, gender, ethnicity, religion, and residence, factors that shaped individual experience and percep-

[35] Regina Morantz-Sanchez, "Feminist Theory and Historical Practice: Rereading Elizabeth Blackwell," *History and Theory,* 1992, *31:*51–69, and Morantz-Sanchez, "Physicians," in *Women, Health, and Medicine in America: A Historical Handbook,* ed. Rima D. Apple (New York/London: Garland Publishing, 1990), pp. 477–495.

[36] Naomi Rogers, "The Proper Place of Homeopathy: Hahnemann Medical College and Hospital in an Age of Scientific Medicine," *Pennsylvania Magazine of History and Biography,* 1984, *108:* 159–201; and Rogers, "Women and Sectarian Medicine," in *Women, Health, and Medicine,* ed. Apple (cit. n. 35), pp. 281–310. Christian Science healing might also be considered in this context.

[37] Rima D. Apple, *Mothers and Medicine: A Social History of Infant Feeding, 1890–1950* (Madison/London: Univ. Wisconsin Press, 1987), p. 100; and Leavitt, *Brought to Bed* (cit. n. 12), p. 174.

tion. The cultural authority of science must not be equated with cross-cultural authority, an important limitation if we recognize the variety of cultures that coexisted within American society: thus we cannot assume that the middle-class and native-born shared their attitude toward the authority of science (or authority in general) with working-class immigrants. Work such as that of Charlotte Borst, which explores how physicians in the German and Polish communities of Progressive Era Milwaukee translated scientific medicine to immigrant subgroups, begins to elucidate the complexities of patient perception.[38] Others have suggested that publicly embracing scientific medicine could serve as a way of affirming a modern value system and middle-class status, a vehicle of Americanization for upwardly mobile immigrants. But until we know much more than we now do, statements about how the American public saw science should be regarded as working generalizations at best.

Function is not meaning, though, or at any rate does not exhaust it. And if we cast medical science as an ideological commodity alone, we lose much of the texture of contemporary perception. Some American women contemplating their options in the management of childbirth or infant feeding may well have sought status in relying on scientific medicine, but as studies by Leavitt and Apple convincingly show, those who turned to scientific childbirth or scientific motherhood were also making decisions about the welfare of their children and themselves. Certainly *they* often saw science as a technical, not just a cultural, resource. If we take seriously the depth of their cognitive and emotional lives, we must recognize that more was involved in their proclaimed faith in science than a single-minded quest for status.

So too, we must assume that physicians were looking for more than status and its material rewards in their turn to a new ideal of science, and unless we decline to take their lives seriously as well, it is too simple to seek the meanings science held for them solely in the marketplace. It is clear, for example, that many physicians attached wide-ranging social and moral meaning to the new ideal of science rooted in the experimental laboratory. In their eyes it posited a new relationship not only between science and practice, but also between science and professional identity and between science and moral legitimacy. Not all welcomed this new conception of science, moreover, and some regarded it as subversive to the established moral order of things. Their vocal disputes should not be dismissed as some intellectual front to cover up *real* issues of status and income, for redefining the place of science in medicine was for many a disturbing process that inseparably bound together issues of medical epistemology with issues of integrity. In fact, the new ideal of science was divisive enough that in the 1880s it cleaved apart the leadership of the regular medical profession, creating intellectual and organizational rifts that lasted into the twentieth century.[39]

[38] Charlotte G. Borst, *Catching Babies: From Midwife to Physician-Assisted Childbirth in Wisconsin, 1870–1920* (Cambridge, Mass.: Harvard Univ. Press, in press).

[39] John Harley Warner, "Ideals of Science and Their Discontents in Late-Nineteenth-Century American Medicine," *Isis*, 1991, *82*:454–478. See also Regina Markell Morantz-Sanchez, *Sympathy and Science: Women Physicians in American Medicine* (New York/Oxford: Oxford Univ. Press, 1985), pp. 184–202; and Barbara Gutmann Rosenkrantz, "The Search for Professional Order in Nineteenth-Century American Medicine," in *Sickness and Health in America*, ed. Leavitt and Numbers (cit. n. 17), pp. 219–232.

What remains less clear is precisely why some physicians welcomed the new science as a source of social, moral, and technical betterment, while others saw it as a source of corruption. What accounts for the differences between the faithful and the heretics? Differences between them cannot be reduced neatly to generational divisions, though that was one line of demarcation. Prosopographical studies would be valuable here. We also need close readings of public rhetoric, such as Susan Lederer has given to antivivisectionist protests against the new science and the counterarguments of its defenders. And we should explore closely doctors' letters and diaries for more private discourse and reflection, just as we are exploring those of patients.[40]

We also need to know more about how science—as an ideal and as a body of knowledge—entered the physician's workaday world. In repeating the once-revisionist insight that the rise of experimental science did not initially win allegiance by any demonstrated power to increase medical efficacy, we have tended to gloss over the ways physicians at the time believed science gave them immediate payoff at the bedside. William F. Bynum's *Science and the Practice of Medicine in the Nineteenth Century,* a welcome corrective that displays close interaction, should serve as a blueprint for further exploration.[41] Changes in surgical treatment and medical diagnosis were impressive, and the few medical treatments that emanated from the laboratory loomed larger at the time than in retrospect. Machines became prominent fixtures in clinical settings, gleaming emblems of science to the eyes of contemporary witnesses if not necessarily products of science in the appraisal of later historians. And medical explanation, something most agreed was enhanced by the new sciences, was one form of intervention.[42] Saying that the ideal of science was more important than the reality—however warranted—begs the question of just how science altered clinical cognition and activity. We are now well acquainted with the rhetoric of educational reformers who argued that giving medical students experience in the laboratory would train them in scientific modes of thinking. What we know little about, though, is how such instruction *did* change their patterns of thought and behavior on the wards.

The meanings of science to biomedical researchers also need to be further explored rather than assumed. We know much about the transport of German ideals of experimental research to America and about the efforts of those committed to these ideals to create flourishing disciplines of, say, physiology and biochemistry, partly

---

[40] Of a large literature on vivisection reform see Susan E. Lederer, "The Controversy over Animal Experimentation in America, 1880–1914," in *Vivisection in Historical Perspective,* ed. Rupke (cit. no. 5), pp. 236–258; and Lederer, *Subjected to Science: Human Experimentation in America before the Second World War* (Baltimore/London: Johns Hopkins Univ. Press, in press).

[41] W. F. Bynum, *Science and the Practice of Medicine in the Nineteenth Century* (Cambridge/New York: Cambridge Univ. Press, 1994).

[42] For an exemplary study of how work in experimental physiology reshaped clinical conceptions of the heart in the British context, see Christopher Lawrence, "Moderns and Ancients: The 'New Cardiology' in Britain 1880–1930," in *The Emergence of Modern Cardiology,* ed. W. F. Bynum, C. Lawrence, and V. Nutton (London: Wellcome Institute for the History of Medicine, 1985), pp. 1–33. On machines, see, e.g., Robert G. Frank, Jr., "The Telltale Heart: Physiological Instruments, Graphic Methods, and Clinical Hopes, 1854–1914," in *The Investigative Enterprise: Experimental Physiology in Nineteenth-Century Medicine,* ed. William Coleman and Frederic L. Holmes (Berkeley/Los Angeles: Univ. California Press, 1988), pp. 211–290, and Joel D. Howell, "Machines and Medicine: Technology Transforms the American Hospital," in *The American General Hospital: Communities and Social Contexts,* ed. Diana Elizabeth Long and Janet Golden (Ithaca, N.Y.: Cornell Univ. Press, 1989), pp. 109–134.

through the reform of medical institutions.[43] Still, we know surprisingly little about what animated these reformers beyond their desire for occupational niches, or about how individuals selecting a career chose between the aesthetic appeal of research at the bench and the normative appeal of practice at the bedside.

As we trace the ways science transformed American medicine, however, we should equally attend to fundamental continuities. There can be no doubt that the products of the bacteriological and physiological laboratories broadly reshaped assumptions about culpability for individual illness just as they reassigned responsibility and accountability for public health. Yet recent research has begun to reveal the extent to which earlier associations between dirt, disease, and marginal social status preserved their importance long after the putative objectification of disease that came with germ theories. Naomi Rogers, using the case of polio in the 1910s, and Judith Leavitt, exploring typhoid carriers, have shown how ethnicity and gender persistently informed medical expectation, while Nancy Tomes has shown significant continuity in domestic hygiene before and after the advent of medical bacteriology.[44] Just how the ideals and practices of the new sciences were assimilated into earlier schemes of expectation and explanation remains a central problem at the interface between history of medicine and history of science.

The singular density of literature on late-nineteenth- and early-twentieth-century American medicine owes much to the integration of medical history into the central currents of American historiography and cultural studies. But this stimulating move has also made it decreasingly likely that American medicine will be viewed within the wider context of medical pursuit in other times and places. History of medicine thus has a less internationalist perspective than the history of science. Comparison, however, offers one of the most promising avenues to better understanding the place of science, both in the American context and in other settings. Deterministic explanations—positing, for example, that with the rise of experimental science particular educational, institutional, or technical changes became inevitable—can be checked by the counterexample of another context. More than this, comparison across national, regional, or class boundaries not only displays the contingency of meaning but also helps the historian identify what it is that needs to be explained.[45]

Viewed in this way, few recent movements in medical historiography appear so promising as the emerging vogue for exploring relationships between medicine and colonialism. The roots of this movement are many, but it stems partly from a preoccupation with the authority of science and medicine. Medicine, and more particularly public health, often functioned as a vehicle for social engineering, with the ideological might of science and the political force of the imperial power sometimes reinforcing each other. Within colonial contexts the ideological dimensions of

---

[43] An especially helpful recent study is Robert G. Frank, Jr., "American Physiologists in German Laboratories, 1865–1914," in *Physiology in the American Context, 1850–1940,* ed. Gerald L. Geison (Bethesda, Md.: American Physiological Society, 1987), pp. 11–46.

[44] Judith Walzer Leavitt, " 'Typhoid Mary' Strikes Back: Bacteriological Theory and Practice in Early Twentieth-Century Public Health," *Isis,* 1992, *83*:608–629; Naomi Rogers, *Dirt and Disease: Polio before FDR* (New Brunswick, N.J./London: Rutgers Univ. Press, 1992); and Nancy Tomes, "The Private Side of Public Health: Sanitary Science, Domestic Hygiene, and the Germ Theory, 1870–1900," *Bull. Hist. Med.,* 1990, *64*:509–539.

[45] For a work that exemplifies the potential of comparison see Ronald L. Numbers, ed., *Medicine in the New World: New Spain, New France, and New England* (Knoxville: Univ. Tennessee Press, 1987).

medical science are often witnessed in bold relief. Historical studies of colonialism are especially appealing because comparison is inbuilt. An ongoing study of the meanings attached to bacterial etiology in the control of plague by British colonial medical officers in early-twentieth-century Calcutta, to cite one example, strongly invites comparison with the meanings attached to it in Britain. The case of Calcutta further invites comparison with management of plague in Cape Town and in Hong Kong during the same period and partly under the supervision of the same medical officers. This particular comparison draws attention to the relationships between medical science, racism, and individuating local political and cultural circumstances. Other occasions for comparison appear as historians look at medical science not just through the eyes of the medical agents of empire, but also through the eyes of indigenous peoples.[46]

The medical marketplace offers another forum for comparison, the opportunity to see an array of practitioners interacting, often in competition, and to compare the particular kinds of knowledge and expertise to which they laid claim. Matthew Ramsey's study of medical practitioners in France, Hilary Marland's study of the medical communities in two Yorkshire towns, and Evelyn Ackerman's study of medical care in the department of Oise-et-Seine display the kind of rewards intensive local studies can bring.[47] All these works cast their nets widely, looking equally at learned physicians, alternative practitioners, folk healers, and (to some eyes) outright quacks, and, just as important, at interactions among these groups. They also explore the place of these healers within their communities and their interactions with patients. In such studies the historian can depict the variety of strategies used by a wide range of healers, functioning within a shared social context, in their efforts to attract clients—as well as examine the criteria patients used in selecting their sources of health advice and assistance. Science appears as only one among many factors in winning confidence and custom. Through focused local studies, moreover, the shifting meanings of science can be related to the social factors that determined change.

Education also offers opportunities for comparative exploration. To begin with, it can point to profound differences in what it meant to be a physician at various times and places. In England during the first half of the nineteenth century, education and the de facto licensing system were crucial in defining professional identity; during the same period in America, however, professional identity was principally defined by practice. Being a regular medical practitioner meant something different, and

[46] Molly Sutphen, "Imperial Hygiene in Calcutta, Cape Town, and Hong Kong: The Early Career of Sir William John Ritchie Simpson (1855–1931)" (Ph.D. diss., Yale University, in progress). Helpful starting points in the rapidly expanding literature on medicine and colonialism include Roy MacLeod and Milton Lewis, eds., *Disease, Medicine, and Empire: Perspectives on Western Medicine and the Experience of European Expansion* (London/New York: Routledge, 1988); David Arnold, ed., *Imperial Medicine and Indigenous Societies* (Manchester: Manchester Univ. Press, 1988); and Arnold, *Colonizing the Body: State Medicine and Disease in Nineteenth-Century India* (Berkeley/London: Univ. California Press, 1993).

[47] Evelyn Bernette Ackerman, *Health Care in the Parisian Countryside, 1800–1914* (New Brunswick, N.J./London: Rutgers Univ. Press, 1990); Hilary Marland, *Medicine and Society in Wakefield and Huddersfield, 1780–1870* (Cambridge: Cambridge Univ. Press, 1987); and Matthew Ramsey, *Professional and Popular Medicine in France, 1770–1830: The Social World of Medical Practice* (Cambridge: Cambridge Univ. Press, 1988). See also Monica Green, "Women's Medical Practice and Health Care in Medieval Europe," *Signs*, 1989, *14*:434–473; and Roy Porter, *Health for Sale: Quackery in England 1660–1850* (Manchester/New York: Manchester Univ. Press, 1989).

comparative studies can clarify the relationship between medical identity and science. It may be true that at any point in time elite physicians sought to present themselves as embodying whatever their society most esteemed—the scientific expert in twentieth-century America, the gentleman in Victorian England, the philosopher in Renaissance Italy. By looking closely at what physicians were being educated to become (healers, to be sure, but always more than that alone), we can begin to understand something of the place of science both in the identity of the physician and in the wider culture. Here, too, students' expectations of their education—what it was that instruction in the basic sciences would give them, for example—constitute a promising but remarkably little explored theme. By and large the history of medical education remains the history of medical teaching rather than of the process of learning, but students have often left behind abundant letters, diaries, class notes, and, in more recent years, scripts from second-year shows that certainly make it possible to trace their perceptions and experiences.[48]

IV. THE SCIENCES OF MEDICINE

So far I have focused on how a generation of historical scholarship has expanded our understanding of the meanings of science in medicine. Yet, there are still deeper ways in which the history of medicine overlaps with the history of science. Focusing solely on the place of science within medicine begs the question of the place of medicine as science. The proposition that medicine is (or is not) a science has a long, checkered, and continuing history, and, while trying to resolve it is not a particularly productive enterprise for the historian, it is certainly worth considering how our historical actors grappled with this issue. More than this, as we widen our notion of what is encompassed by the history of science to include the social sciences and human sciences, it is appropriate to recognize that how people in the past explained illness, organized their thinking about it, and managed it are indeed parts of the history of science.

Perhaps the least problematic element of this proposition rests in recognizing the extent to which the life sciences have developed as a part of medicine. The social and cognitive rooting of what we have come to call biology provides a good example. Reflecting on the state of history of biology in 1981, William Coleman reminded *Isis* readers that "the social, economic, and in large part intellectual life of biology before 1900 or even 1930 was in many respects dominated by the medical profession and its clientele." He particularly stressed the extent to which medical inquiries gave rise to biological investigations, noting that "biological problems are not simply presented to an inquirer by the doings of organisms." Coleman concluded

---

[48] Of a large recent literature, particularly helpful studies include Thomas Broman, "University Reform in Medical Thought at the End of the Eighteenth Century," *Osiris,* 1989, *5:*36–53; Jan Goldstein, *Console and Classify: The French Psychiatric Profession in the Nineteenth Century* (Cambridge: Cambridge Univ. Press, 1987); Christopher Lawrence, "Ornate Physicians and Learned Artisans: Edinburgh Medical Men, 1726–1776," in *William Hunter and the Medical World of the Eighteenth-Century,* ed. W. F. Bynum and Roy Porter (Cambridge: Cambridge Univ. Press, 1985), pp. 153–176; Susan C. Lawrence, *Charitable Knowledge: Hospital Pupils and Practitioners in Eighteenth-Century London* (Cambridge: Cambridge Univ. Press, in press); Margaret Pelling, "Medical Practice in Early Modern England: Trade or Profession?" in *The Professions in Early Modern England,* ed. Wilfrid Prest (London: Croom Helm, 1987); and Arleen Marcia Tuchman, *Science, Medicine and the State in Germany: The Case of Baden, 1815–1871* (New York: Oxford Univ. Press, 1993).

by cautioning against the pitfalls of a separatist history of biology, noting that "whatever we select as our guide from among the shifting fashions in the study of the history of biology, we will always be returned to the medical context," and urged that "we had best, therefore, again pay heed to the history of medicine in general and of the medical sciences in particular."[49]

One recent model study that revealed interaction where others had long assumed separation is John Lesch's *Science and Medicine in France: The Emergence of Experimental Physiology, 1790–1855* (1984). Lesch challenged the canonical view that the dominance of the clinic in post-Revolutionary Parisian medicine had a negative, hindering influence on the development of the basic sciences. He did acknowledge that the Paris clinicians not only rejected the proposition that clinical medicine should be grounded on such accessory sciences as physics, chemistry, physiology, and microscopy, but saw them as dangerous encouragement to speculation and rationalistic system-building. However, Lesch went on to show how the Parisian medical milieu shaped and encouraged experimental work in chemistry, pharmacology, and physiology. This supportive relationship went well beyond the simple fact that most physiologists were medically trained and occupied positions in medical institutions. Physiological practice owed a heavy debt to surgery, for example: with boldness, technique, and localistic ways of seeing informed by surgical training, such experimental physiologists as Xavier Bichat, François Magendie, and Claude Bernard wrested knowledge from the organism by direct surgical interference with life processes. The physiologist also relied on the clinician for pathological information about the natural experiments performed by disease and upon access to living hospital patients to test new chemical compounds. Parisian experimentalists further shared the characteristic empiricist and antitheoretical attitude of their clinical counterparts. Lesch's revisionist account convincingly showed that the development of experimentalism in French physiology cannot be understood unless it is carefully placed within its specific medical context.[50]

And yet even as this study supplanted an image of stark antagonism between clinic and laboratory with one of interaction, it perpetuated and reinforced an even sharper division between science and medicine. It recognized clinical medicine as a resource for science, but did not take seriously the perception of contemporary Parisian physicians that the pursuits of the clinic *were* science. Lesch was sensitive to how the historical actors saw their world, yet he attended closely to the rhetoric of the experimentalists without listening to that of the clinicians. Accordingly he tacitly equated science with experimentalism, and tended to depict Parisian clinical medicine, committed as it was to a strident sensual empiricism, as the founders of the Société de Biologie depicted it: a narrowly practical art rather than a science. Lesch implicitly contrasted those committed to clinical medicine with "the more scientifically

[49] Coleman, in Martin Rudwick, William Coleman, Edith Sylla, and Lorraine Daston, "Retrospective Review: Critical Problems in the History of Science," *Isis*, 1981, *72*:267–283, on pp. 275–276. Recent models of integration include Georgina Feldberg, *Disease and Class: Tuberculosis and the Shaping of Modern North American Society* (New Brunswick, N.J.: Rutgers Univ. Press, forthcoming); Mirko D. Grmek, *La prèmiere revolution biologique: Réflexions sur la physiologie et la médecine du XVIIe siècle* (Paris: Editions Payot, 1990); and Anne Marie Moulin, *Le dernier langage de la médecine: Histoire de l'immunologie de Pasteur au Sida* (Paris: Presses Universitaires de France, 1991).

[50] John E. Lesch, *Science and Medicine in France: The Emergence of Exp␣rimental Physiology, 1790–1855* (Cambridge, Mass.: Harvard Univ. Press, 1984).

minded segment of the Paris medical community." The result was a portrayal of French medicine that ignored the cardinal article of the Paris clinician's creed, that is, that science meant empiricism. This is all the more surprising given the author's own assertion that "science is, first and foremost, knowledge about nature, whether that knowledge aims for theoretical understanding or practical control."[51] It is precisely because Lesch's book is such a splendid contribution to our understanding of the interactions between clinic and laboratory that his failure to see the contemporary perception that clinical medicine *was* science is so revealing.

An important counterpoint is Russell Maulitz's *Morbid Appearances: The Anatomy of Pathology in the Early Nineteenth Century* (1987). Maulitz, like Lesch, set out to explore a science pursued in the medical environment of early-nineteenth-century Paris, but his science—pathology—was firmly rooted in the clinic. "Whether conceived as the theory of disease, as the practice of anatomical pathology, or as clinicopathological correlation, pathology was part of clinical medicine before becoming enshrined as a separate discipline." In this prebacteriological era, before pathology entered the laboratory, "pathology remained an integral part, indeed in France the cornerstone of clinical medicine." Seeing pathology cultivated as a science in no way denies that it had practical as well as intellectual ends. "Medical theory," Maulitz contended, "was not merely a sort of *jeu* for the well-placed elite of French medicine, who had the luxury and time to speculate on the arcane reaches of pathology. It was also, and most significantly, a cognitive product with direct implications both for practice and for the structure of the profession." Regarding pathology as the centerpiece of clinical medicine allows Maulitz to recognize that René Laennec's *Mediate Auscultation* (1819)—best known for introducing his new instrument, the stethoscope—was above all a work of pathological anatomy, a contribution to the science of pathology more than to the practice of physical diagnosis. Maulitz was thus rightly able to conclude that "it is presentist to see it exclusively, or even primarily, as a work designed to popularize an instrument which, as it happened generations later, would end up dangling from virtually every medical neck."[52] Accordingly Laennec, no less the clinician, was recast as a figure squarely within the history of science. The broader significance of Maulitz's perspective, especially contrasted with that of Lesch, is his depiction of a science of pathology pursued in a singularly medical space—not merely science in the clinic, but *clinical medicine as a science*.

The Paris Clinical School has drawn special attention in medical historiography partly because many historians have seen it as the starting point of modern scientific medicine, and partly because the persistent vogue of Foucault's work has drawn attention to his focus on this medical episode in *The Birth of the Clinic* (1963). Lesch's study on experimental physiology has shown that links did exist between the Paris clinic and the laboratory, and the work of other historians—by Ann La Berge on microscopy, for example—is displaying analogous patterns. Yet to the extent that the birth of this particular clinic plausibly can be claimed as the birth of modern scientific medicine, the critical changes were rooted in the clinic itself, not in the reliance of medicine on any more basic sciences. This was a medical science

[51] *Ibid.,* pp. 3, 223.
[52] Russell C. Maulitz, *Morbid Appearances: The Anatomy of Pathology in the Early Nineteenth Century* (Cambridge: Cambridge Univ. Press, 1987), pp. 61, 65, 99, 135.

not of the laboratory bench but of the bedside and autopsy table. The vigorous historical rethinking of Paris medicine now going on is rightly integrating important revisionist perspectives such as Lesch's into a picture at once wider and finer grained, but it stands to contribute most to the history of science by telling us a story not of physiology, chemistry, and microscopy, but of clinical science.[53]

The broader point is not simply that clinical medicine in the narrow, hospital-based sense can speak to the history of science, but that the ways doctors in the past strove to understand and explain illness must be seen as a part of scientific enterprise. Hippocratic observation of the natural history of disease and composition of case histories, textual study and the collecting of *consilia* by learned medieval and Renaissance physicians, generalization of individual case histories into disease histories by physicians in the seventeenth century, the construction of rationalistic systems of nosology in the eighteenth—all of these were at least in part efforts to organize knowledge about nature. In these activities the dual aims of caring for the sick and codifying natural knowledge were intermeshed, but that does not diminish their utility to the historian who would explore the ways they reveal patterns of thinking about the natural world.

One reason to underscore this point is that the notion of *scientific medicine* stands as among the sturdiest bastions of presentism in the field. To a remarkable extent scientific medicine remains equated in the historical literature with medicine rooted in experimental laboratory science. This historical assumption tacitly informs Lesch's distinction between *medicine* (the practice of the clinic) and *science* (the experimentalism of the laboratory), and in less sophisticated works it is often made explicit. From the perspective of the historiography of medicine, this does damage by leading to sometimes-misphrased questions about the relationship between medicine and the something else that is science, an unhelpful polarity that glosses over the varieties of scientific medicine that existed (and sometimes competed with one another) in concrete historical settings. From the perspective of the historiography of science, it is damaging in its tendency to exclude much of the scholarship on medicine that can illuminate the contexts in which natural knowledge was produced, the nature of applied science, the cultural role of science as authority, and the movement of scientific knowledge between societies and among various cultures within a given society.

The identification of scientific medicine with laboratory experimentation is one common assumption in accounts of the rise of the experimental sciences. It affects even the best studies, as W. Bruce Fye's fine book *The Development of American Physiology: Scientific Medicine in the Nineteenth Century* (1987) shows, for the "scientific medicine" of that title is treated by and large as self-evident, monolithic, and exclusionary. Like other historians, Fye argued that "those who wanted to reform American medicine wanted to endow it with the authority of science." But in seeking indicators of experimental physiology's development—its progressive growth toward the goal of reformers "to make medicine more scientific"—the author never acknowledged that the meaning of "scientific medicine" in the nineteenth century

---

[53] See, e.g., Ann La Berge and Mordechai Feingold, eds., *French Medical Culture in the Nineteenth Century* (Amsterdam/Atlanta: Rodopi, 1994); and Caroline Hannaway and Ann La Berge, eds., *Reinterpreting Paris Medicine,* in progress.

was highly problematic.[54] American physiologist-reformers pinned their hopes for creating scientific medicine on the experimental science of the laboratory. Yet other equally conscientious, intellectually elite medical thinkers committed to French empiricism maintained that the only sturdy foundation for an authentically scientific medicine was close clinical observation. Told solely from the perspective of the physiologist-reformers, Fye's narrative ignored that most physicians believed American medicine already was scientific—but for them science meant empiricism. Attention to this alternative conception of scientific medicine would have shown that reformers sought not only "to make medicine more scientific," but, more profoundly, to redefine what constituted scientific medicine.

Part of the problem is a whiggish acceptance of the definition of scientific medicine propagated by the ultimately victorious physiologist-reformers. Indeed, the very idea that clinical medicine is not a science (except insofar as it might be rooted in more basic experimental sciences) is a creation of the nineteenth century. It is, more particularly, an artifact of polemical campaigns mounted in order to win a legitimate and secure place in medicine for the experimental laboratory sciences— and of the subsequent political programs of discipline building. In the present century it has been reinforced by romantic nostalgia and ethically informed remorse for whatever it is that has been lost from an earlier medical world with the rise of high-tech, reductionist medicine—a regret phrased in language that often juxtaposes the art of medicine, clinical activity pursued with the goal of healing, against the science of medicine, imagined as a subversive threat to the doctor-patient relationship. This perspective may do important service for medical ethicists, policy makers, and humanistic clinicians. But historians trying to find their way in the medical worlds of the past cannot depend on a map of the present, most especially not a Michelin guide to scientific medicine that directs us from one monument of experimentalism to the next.

One concrete historiographic payoff of recognizing that physicians who flourished before the rise of the laboratory were convinced they *had* a scientific medicine is that it helps us to see the medicine of the past as it was witnessed by participants. It helps us wrench loose (if never free) from teleological narratives, and draws our attention to alternative forms of medical enterprise unrelated to the laboratory but nonetheless regarded by contemporaries as scientific. The library, lectern, bedside, and field thus all emerge as sites where scientific medicines were pursued; the learned Latin textual tradition in medieval and Renaissance medicine becomes one focus, as do the field studies of environmental conditions that formed the core of epidemiology in the prebacteriological era, and the occupational medicine rooted in investigations of the twentieth-century workplace.[55]

Recognizing that clinical medicine could represent one form of scientific medicine opens up a particularly wide range of possibilities to the historian prepared to

[54] W. Bruce Fye, *The Development of American Physiology: Scientific Medicine in the Nineteenth Century* (Baltimore/London: Johns Hopkins Univ. Press, 1987), pp. 1, 3.

[55] See, e.g., James H. Cassedy, *Medicine and American Growth, 1800–1860* (Madison/London: Univ. Wisconsin Press, 1986); William Coleman, *Yellow Fever in the North: The Methods of Early Epidemiology* (Madison/London: Univ. Wisconsin Press, 1987); and David Rosner and Gerald Markowitz, eds., *Dying for Work: Workers' Safety and Health in Twentieth-Century America* (Bloomington/Indianapolis: Indiana Univ. Press, 1987).

explore the interface between history of science and history of medicine. In both fields a preoccupation with *practice* has recently surfaced, though the focus remains rather different in the two. To move away from theory-dominated accounts of the construction of knowledge in the laboratory, some historians of science have looked self-consciously at the practice of experimentation, focusing on processes more than results. In studies of such laboratory-based biomedical sciences as biochemistry, physiology, and bacteriology, they have looked closely at the activities of scientists at the bench. Some of the most intriguing work has been based on laboratory note-books: Close scrutiny of laboratory operations informs a finer grained narrative of the investigative enterprise than can be produced by studying published records alone. Indeed, a comparative analysis of scientists' published writings with the private record of activities can restructure our understanding of their work, as Geison has displayed elegantly in a study of Louis Pasteur.[56]

Such exploration of the practice of laboratory-based biomedical science is generally accepted as part of the history of science. Less recognized is that studying practice at the bedside might also enhance our understanding of the history of science. In part this may be because some of the most innovative recent works on practice have emphasized routine matters of patient care over the production of new knowledge. Thus in 1990 Laurel Thatcher Ulrich, in her Pulitzer Prize-winning study of an eighteenth-century New England midwife, claimed that the greatest strength of the account was its "very dailiness, the exhaustive, repetitious dailiness." Most studies of physicians' bedside practice have also elected to examine mundane routine rather than the way the exceptional practitioner set out to create new knowledge about disease or its treatment.[57]

Nevertheless, even the practice of routine patient care can enrich our understanding of the history of science, especially when it offers a systematic portrayal of how clinicians translated an organized body of natural knowledge into workaday behavior. Hospital practice—more particularly activity in the clinic, as it emerged from the eighteenth century as an institution where physicians who saw themselves as not only healers but also natural scientists worked at the bedside, autopsy table, and, later, clinical laboratory—has especially rewarded investigation. To reconstruct the actual practice of clinical medicine historians have begun to turn to hospital patient records in much the same way that historians of other sciences have turned to laboratory and field notebooks. As a tool for exploring the nature and meanings

[56] Gerald L. Geison, *The Private Science of Louis Pasteur* (Princeton: Princeton Univ. Press, in press). On laboratory practice see Jan Golinski, "Experiment in Scientific Practice," *Hist. Sci.*, 1990, *28*:203–209; David Gooding, Trevor Pinch, and Simon Schaffer, eds., *The Uses of Experiment in the Natural Sciences* (Cambridge: Cambridge Univ. Press, 1989); Frederic L. Holmes, "Laboratory Notebooks: Can the Daily Record Illuminate the Broader Picture?" *Proc. Amer. Phil. Soc.*, 1990, *134*:349–366; Holmes, "Scientific Writing and Scientific Discovery," *Isis*, 1987, *78*:220–235; and Timothy Lenoir, "Practice, Reason, Context: The Dialogue between Theory and Experiment," *Science in Context*, 1988, *2*(1):3–22.

[57] Laurel Thatcher Ulrich, *A Midwife's Tale: The Life of Martha Ballard, Based on Her Diary, 1785–1812* (New York: Alfred A. Knopf, 1990), p. 9. For reconstructions of medical practice see Jacalyn M. Duffin, *Langstaff: A Nineteenth-Century Medical Life* (Toronto: Univ. Toronto Press, 1993); Martin S. Pernick, *A Calculus of Suffering: Pain, Professionalism, and Anesthesia in Nineteenth-Century America* (New York: Columbia Univ. Press, 1985); Guenter B. Risse, *Hospital Life in Enlightenment Scotland: Care and Teaching at the Royal Infirmary of Edinburgh* (Cambridge: Cambridge Univ. Press, 1986); Ronald C. Sawyer, "Patients, Healers and Disease in the Southeast Midlands, 1597–1634" (Ph.D. diss., Univ. Wisconsin, 1986); and Warner, *The Therapeutic Perspective* (cit. n. 18).

of science in the wards, patient records permit a systematic exploration of the relationship between scientific ideas and clinical activities. Through patient records we can begin to discern the behavior that underlay ideological positions, and to develop operational definitions for the vocabulary of medical science. By comparing clinical behavior with its portrayal in other texts, moreover, the historian can begin to understand what expectations clinicians held about their bedside activities, how they explained particular interventions, and how they assessed the consequences of their actions. What can emerge is a rich portrait of science in context, in a reconstruction that draws together technical content and institutional milieu, ideology and behavior, the professional agendas of physicians and the realities of patient experience.[58]

While patient records have attracted historians chiefly as a source on what physicians actually did diagnostically or therapeutically, they provide an important indicator of other bedside practices as well, most especially the clinical practice of narrative. As we begin to recognize the day-by-day inscription of the patient chart as itself a significant clinical activity—as we begin, that is, to see the practice of writing as one among the practices of medicine—it becomes apparent that the shifting representational conventions physicians employed in these records offer one promising means of charting how experimental science and its ideals entered bedside practice.

Clinical medicine cannot be reduced to a forum for the application of science, however, for in the clinic the context of application and the context of discovery were seldom neatly separated. The practice of patient care was often intertwined with the practices of teaching and investigation. The practice of clinical teaching in particular presents the historian with a unique setting for exploring the pedagogical transmission of science. Recent studies of topics ranging from instruction in Hermann Boerhaave's Leiden clinic at the start of the eighteenth century to the integrated clinical and laboratory training of late-twentieth-century medical students reveal how natural knowledge was ordered and sometimes reordered for the purposes of pedagogy. But many questions recently posed about scientific training in other contexts have been little explored for the clinic. How did clinical instruction both convey tacit knowledge of craft skills and initiate medical students into the processes of science and even actual scientific practice? How did it shape their identities as not just healers but scientific practitioners? We have an ample literature on the history of medical education, but we still know far more about educational philosophy and curricular structure than we do about what went on in the classroom, teaching laboratory, or clinic.[59]

The richest promise that patient records may hold, though, may well be insight into how new scientific knowledge was produced in the clinical context. Yet, except in history of psychiatry, there has been little systematic use of clinical records—as others have used laboratory notebooks—to elucidate the process of clinical creativity; indeed, it remains the least explored dimension of clinical practice.[60] Jacalyn

[58] See Guenter B. Risse and John Harley Warner, "Reconstructing Clinical Activities: Patient Records in Medical History," *Soc. Hist. Med.*, 1992, *5:*183–205.

[59] See esp. "Clinical Teaching, Past and Present," special issue of *Clio Medica,* 1987/88, *21.*

[60] On the relationship between individual case histories and psychiatric theorizing see, e.g., Mark S. Micale, "Charcot and the Idea of Hysteria in the Male: Gender, Mental Science, and Medical Diagnosis in Late Nineteenth-Century France," *Med. Hist.,* 1990, *4:*363–411; Jack Pressman, *Active Treatment: Psychosurgery and the Rise of Scientific Psychiatry in America* (Cambridge: Cambridge Univ. Press, forthcoming); and Frank J. Sulloway, "Reassessing Freud's Case Histories: The Social Construction of Psychoanalysis," *Isis,* 1991, *82:*245–275.

Duffin's study of Laennec's clinical epistemology does rely on his case records in tracing how his ideas grew from his clinical experience.[61] But the fine structure of investigation in clinical science has received scant attention. How, to cite but one example, out of the protean signs and symptoms displayed by patients, did clinicians actually select those deemed significant in making diagnostic distinctions, such as separating typhus from typhoid fever? The published record tells us much, but clinical case histories can give new clues about the production of clinical knowledge.

The point in stressing clinical medicine is not to privilege it as a focus for historical study, much less to imply that clinical medicine prior to the mid-nineteenth century, an era before domination by the experimental laboratory sciences, represents a somehow more authentic science of medicine. Nevertheless, in conventional notions of the intellectual territory encompassed by the history of science, the biomedical sciences pursued in the laboratory still occupy a more acknowledged and less problematic place than the biomedical sciences of the clinic. We need to know much more about the intercalation of the worlds of the clinic and of the laboratory. Intriguing recent studies on how laboratory ways of thinking were assimilated into clinical thinking, and on the emergence of the clinical laboratory in the nineteenth century, have begun to explore that interaction. We can look forward to historical studies on the new kind of clinical science that came into being early in the twentieth century, a clinical science self-consciously rooted equally in the ward and in the laboratory.[62]

What doctors and patients thought about illness is one part of what people in the past thought about the natural world. There is no particular reason for historians of medicine to preach this point, yet it has important implications for the ways historians of science should delineate the intellectual concerns of their field. Medicine, after all, is in some respects singular. People's conceptions of the normal and pathological in their own bodies are central to popular belief about nature and deployment of natural knowledge. Indeed, if clinically based pathology is a part of history of science as well as of history of medicine, perhaps we should ask as well about the natural knowledge that comes into play when a parent is confronted with a sick child.[63] In any event, one of the most fundamental things that the history of medicine stands to offer the history of science in general is a different context for studying scientific investigation, pedagogy, and application—that is, at the bedside of the

[61] Jacalyn M. Duffin, "Laennec: Entre la pathologie et la clinique" (Ph.D. diss., Paris I–Sorbonne, 1985); and Duffin, "The Medical Philosophy of R. T. H. Laennec (1781–1826)," *History and Philosophy of the Life Sciences,* 1986, *8*:195–219.

[62] See, e.g., L. S. Jacyna, "The Laboratory and the Clinic: The Impact of Pathology on Surgical Diagnosis in the Glasgow Western Infirmary, 1875–1910," *Bull. Hist. Med.,* 1988, *62*:384–406; and Edward T. Morman, "Clinical Pathology in America, 1865–1915: Philadelphia as a Test Case," *Bull. Hist. Med.,* 1984, *58*:198–214. On clinical trials as experiment see Harry M. Marks, "Notes from the Underground: The Social Organization of Therapeutic Research," in *Grand Rounds: One Hundred Years of Internal Medicine,* ed. Russell C. Maulitz and Diana E. Long (Philadelphia: Univ. Pennsylvania Press, 1988), pp. 297–336.

[63] For anthropological, literary, and historical studies of popular conceptions of the natural world and the place within it of the healthy and diseased body see, e.g., Piero Camporesi, *The Incorruptible Flesh: Bodily Mutation and Mortification in Religion and Folklore* (1983), trans. Tania Croft-Murray (Cambridge: Cambridge Univ. Press, 1988); Jacques Gélis, *History of Childbirth: Fertility, Pregnancy and Birth in Early Modern Europe* (1983), trans. Rosemary Morris (Cambridge: Polity Press, 1991); and Heinrich von Staden, "Women and Dirt," *Helios,* 1992, *19*:7–30.

sick. At the very least, to our list of forums where science is pursued—the laboratory, museum, field, lectern, and observatory—we should add the clinic. For their part, as historians of medicine embrace historiographic pluralism, they should work to elucidate the multiple meanings of science in medicine rather than merely acknowledging (however rightly) that *science* has always been a freighted term.

Opening the Black Box: Cognitive Science and History of Science

By Nancy J. Nersessian*

A NEW INTERDISCIPLINARY FIELD, cognitive science, has emerged over the last twenty years. Cognitive science is a loose confederation of cognitive psychology, artificial intelligence, cognitive neuroscience, linguistics, philosophy, and cognitive anthropology. Recently a research frontier I call *cognitive history* has emerged within the history of science and is finding its place in this confederation. The emphasis of cognitive history is on the "thinking practices" through which scientists create, change, and communicate their representations of nature. I view the task of this article as largely programmatic; that is, to discuss the theoretical presuppositions of cognitive history and to outline a sample of research projects and problems. What I hope to persuade the reader is that cognitive history of science provides new insights into science as a human enterprise. While my own contributions to cognitive history have been inherently multidisciplinary, I will concentrate the discussion here on issues of concern to historians. My objectives are to introduce historians of science to developments in cognitive science; to argue and demonstrate that in some areas historical interpretation has already been enriched by "cognitive-historical" analysis; to outline potential areas of research within history of science; and to entice historians of science into collaborative research with cognitive scientists on topics of mutual interest.

I. WHAT IS COGNITIVE HISTORY OF SCIENCE?

Cognitive history of science examines the cognitive tools scientists employ and the artifacts they construct in theoretical and experimental thinking practices. It attempts to reconstruct the cognitive dimensions of the processes through which vague speculations get *articulated* into scientific understandings, are *communicated* to other scientists, and come to *replace* existing representations of a domain. These reconstructions employ a "cognitive-historical" method, which joins historical inquiries with those carried out in the sciences of cognition. Dimensions of the history of science that are amenable to cognitive-historical analysis include devising and executing real-world and thought experiments, constructing arguments, inventing and using

* School of Literature, Communication, and Culture, Georgia Institute of Technology, Atlanta, Georgia 30332–0165.

I am grateful for the comments of numerous colleagues, especially those of Anne Harrington and Stephen Kosslyn. Research for this article was supported in part by the National Science Foundation, Scholars Awards DIR 8821422 and DIR 9111779, and the Princeton University Committee on Research in the Humanities and the Social Sciences.

mathematical tools, creating conceptual innovations, devising means of communicating ideas and practices, and training practitioners.

Cognitive history also addresses metatheoretical issues central to the practice of history of science. These include such problems as the status of different kinds of source materials (e.g., notebooks, diaries, and published papers); what counts as a "method" in science (e.g., whether visual representation is ancillary or essential when it is used); how cultural resources within and external to science enter the representational content of a theory; and how representations are transported out of their initial cultural milieu.

As with other forms of history of science, cognitive history views a wide range of historical records as resources for gaining access to the practices through which scientists create, change, and communicate understanding. These include diaries, laboratory notebooks, publications, correspondence, experimental equipment, drawings, diagrams, and pedagogical notes and texts. In cognitive history the temporal and contextual perspective of history provides an integrated picture of a set of achievements—not necessarily "successes"—over a period of time and against a background of community and wider cultural resources.

What makes the method of analysis "cognitive" is that its interpretations create a working synthesis between case studies of historical scientific practices and investigations of human reasoning and representation by the cognitive sciences. Cognitive history assumes science is one product of the interaction of the human mind with the world and with other humans. It presupposes that the cognitive practices scientists have invented and developed over the course of the history of science are sophisticated outgrowths of ordinary thinking. Understanding how science develops and changes requires knowledge both of what the actual practices of scientists are and of how human cognitive abilities and limitations produce and constrain these practices. Neither of these can be determined a priori. Empirical research is needed in both cases. Thus cognitive historians draw on scientific investigations of how humans reason, judge, represent, and come to understand.

At the outset, the cognitive historian needs to grapple with an obvious methodological problem. What epistemological status should be accorded to the "knowledge" drawn upon? Is it legitimate to use scientific knowledge to develop a theory of the production and communication of scientific knowledge? This problem is not unique to cognitive history. Historians who draw on knowledge in other domains, sociology, anthropology, political science, for example, must address the same issue. Although this is not the place to argue the point in detail, the approach taken by cognitive historians is similar to that of philosophers practicing naturalized, antifoundational epistemology, who have argued extensively for its legitimacy.[1] Cognitive historians draw on cognitive theories and methods insofar as they help interpret the historical cases, while at the same time examining to what extent current theories

---

[1] See, e.g., Ronald N. Giere, *Explaining Science: A Cognitive Approach* (Chicago: Univ. Chicago Press, 1988); Alvin Goldman, *Epistemology and Cognition* (Cambridge, Mass.: Harvard Univ. Press, 1986); Nancy J. Nersessian, *Faraday to Einstein: Constructing Meaning in Scientific Theories* (Dordrecht: Martinus Nijhoff, 1984); Thomas Nickles, "Reconstructing Science: Discovery and Experiment," in *Theory and Experiment: Recent Insights and New Perspectives on Their Relation*, ed. Diderick Batens and J. P. van Bendegem (Dordrecht: D. Reidel, 1988), pp. 33–53; Willard V. O. Quine, "Naturalized Epistemology," *Ontological Relativity and Other Essays* (Cambridge, Mass.: Harvard Univ. Press, 1969), pp. 69–91; and Dudley Shapere, *Reason and the Search for Knowledge* (Dordrecht: Kluwer, 1984).

of cognitive processes can be applied to scientific practices and indicating where these theories might need extension, refinement, and revision. Assessments of the fit between the cognitive analyses and the historical practices are fed back into cognitive science. These assessments are evaluated and used in developing richer and more realistic models of cognition, which, in turn, will be applied and evaluated in further historical analyses. The assumptions, methods, and results from both sides are subjected to critical evaluation, with corrective insights moving in both directions. The goal is to bring historical and cognitive interpretations into a state of reflective equilibrium, so as to make the circularity inherent in the approach virtuous rather than vicious.

Although some assumptions must initially be accorded privileged status in order to get the analysis off the ground, the expectation is that at some point further on these assumptions too will be subject to critical scrutiny. Consider, for example, one assumption of cognitive psychology that historians find particularly troubling. Although there has not been much explicit discussion of it, cognitive scientists do presume that for at least the last few millennia, nothing much has changed in how human cognition takes place. Most would agree that *what* people think about is affected by context, but until quite recently most have assumed that *how* people think is not so affected. That is, cognitive scientists have presumed that their models of problem solving, learning, decision making, and so forth do not themselves need to be historicized. While this presupposition served a purpose in the preliminary analyses of cognition, enriched analyses require that it now be subjected to critical scrutiny.

From the historian's perspective it is much more likely that over the course of history contextual assumptions and values and developments such as instrumentation, mapping techniques, mathematical structures—to name just a few—have affected human cognition. There is evolutionary warrant for assuming that the "wiring" of the brain has not changed much in the period for which there are written records of human thinking. But whether and to what extent human cognition is shaped by the historical-cultural context in which it has taken place ought to be seen as an open question. Recent encounters of cognitive psychologists with research from cognitive anthropology and from the history and philosophy of science have put the question on the agenda of some researchers. For example, in their comparative study of how alchemists, modern scientists, and experimental subjects have reasoned by means of analogy, the cognitive psychologists Dedre Gentner and Michael Jeziorski have proposed that only the abilities to recognize similarities, patterns, and structure are basic aspects of cognition, whereas how humans use those abilities has been affected by the historical context in which they have been used.[2] They argue that such contextual factors as metaphysical presuppositions, cognitive values (e.g., what counts as "simplicity"), and cultural assumptions have historically influenced what kinds of similarities are taken to be significant, what counts as a chain of association, and how correspondences are mapped in analogical reasoning. This repre-

[2] Dedre Gentner and Michael Jeziorski, "The Shift from Metaphor to Analogy in Western Science," in *Metaphor and Thought*, ed. A. Ortony, 2nd ed. (Cambridge/New York: Cambridge Univ. Press, in press). For cognitive anthropology see, e.g., Jean Lave, *Cognition in Practice: Mind, Mathematics, and Culture in Everyday Life* (New York: Cambridge Univ. Press, 1988); and Lucy A. Suchman, *Plans and Situated Actions: The Problem of Human-Machine Communication* (Cambridge/New York: Cambridge Univ. Press, 1987).

sents a significant modification to Gentner's "structure mapping" theory of analogy, in which only structural features of the analogical source domain are directing the mapping process. The new analysis suggests that past reasoning practices have implications for current accounts of cognition and lends support to those who argue that cognition is always contextually "situated."

The cognitive-historical method differs from other, well-known approaches that have attempted to integrate psychology, one of the central disciplines of cognitive science, with historical analysis. Attempts at integration have, primarily, taken two forms. First, some analyses have viewed the role of scientists' personalities in their achievements through the lens of Freudian psychoanalytic theory. Second, Jean Piaget and others have tried to fit episodes of scientific change—indeed the entire developmental history of science—into the framework of his theory of cognitive development. Additionally, while not strictly attempts at integration, Thomas Kuhn, Paul Feyerabend, and others have sought to reinforce and to explain their conception of scientific change by adopting theories from the psychology of perception and from Gestalt psychology.

What all of these earlier attempts share is that they fit the history of science to models of thinking and behavior imported from psychology. In so doing they have much in common with the reconstructive approaches of philosophers that historians—and some philosophers—have found so objectionable. As with all interpretation, cognitive history consists of reconstruction. However, it does not fit historical cases to models drawn from the cognitive sciences, but attempts to integrate the cognitive and the historical findings in models that suit the actual scientific practices encountered in the historical record. Cognitive historians strive to create a true synthesis, hoping to avoid the excesses and pitfalls of the earlier approaches.

Finally, my own vision of the research program includes working in active collaboration with cognitive scientists to forge interpretive frameworks that will accommodate scientific change. There has been a long history of mutual influence between history and psychology. I believe that more progress can be made on problems of common interest to both disciplines by having some practitioners from each discipline working together. Most cognitive historians have, at some point, been drawn into collaborations. Nevertheless, historians of science can profit from cognitive science without having to adopt my full agenda. Insights from the cognitive sciences can be used in the context of a larger historical argument to assist reconstruction and can open new avenues for investigation. For example, artificial intelligence (AI) programs modeling scientific discoveries can, as Frederic L. Holmes remarked of an AI model of Hans Krebs's discovery of the ornithine cycle, furnish "us with means to reflect on the degree of contingency and fluidity inherent in the historical discovery processes we seek to reconstruct."[3]

II. WHAT IS COGNITIVE SCIENCE?

Cognitive science employs experimental techniques and computer modeling to investigate how knowledge is produced and is represented by the mind.[4] The central

---

[3] Frederic L. Holmes, "Research Trails and the Creative Spirit," MS (1993), p. 22.

[4] For an engaging popular history of cognitive science see Howard Gardner, *The Mind's New Science: A History of the Cognitive Revolution* (New York: Basic Books, 1985). Barbara Von Eckardt,

assumption of the strong, reductionist version is that the same information processes can be implemented in computer hardware and in the brain. The weaker version holds that it is possible to simulate hypotheses about human reasoning by means of computer modeling in much the same way that astrophysicists use computer simulations to investigate cosmological models. For both "computational assumptions" computer modeling, based on empirical investigations carried out in cognitive psychology, is an essential tool through which to explore and experiment with theories of human reasoning and representation.

Cognitive science has made significant advancements in understanding the mental structures and processes underlying knowledge representation, acquisition, and construction. It has now produced sophisticated cognitive theories of language acquisition, conceptual development, decision making, problem solving, learning, and brain functioning. Cognitive neuroscience, cognitive psychology (including psycholinguistics), and artificial intelligence constitute the core disciplines—and the three levels of analysis—in cognitive science. In the current state of the field, research in cognitive psychology has the most to offer immediately to history of science. Although the findings of neuroscience act as a constraint on any theory of human cognitive activities, the analyses in that field are at too basic a level to make a direct contribution to our understanding of scientific practice.[5] The role of AI in cognitive history is also mainly indirect at present.

AI is central to cognitive science as it is now conceived because all cognitive scientists have adopted one of the two "computational assumptions" discussed above. Yet they disagree on what kind of computational modeling, standard AI or neural-connectionist, is best for modeling human cognition. Still, a significant body of AI research that attempts to model knowledge representation, qualitative reasoning, and analogical reasoning has formed the basis of the work discussed in Section IV. AI programs for modeling scientific discovery could well be used to test historians' hypotheses about the cognitive activities of earlier scientists, to enrich and expand knowledge of the heuristics processes employed in scientific investigations, and to exhibit multiple potential pathways to a scientific discovery. But existing "discovery" programs, such as those implemented by Herbert Simon and his coworkers thus far, although the product of highly sophisticated AI research, have tackled only rather simple problem-solving heuristics employed by scientists.

Within cognitive psychology the areas with the most to contribute to the analyses of cognitive historians are those focused on cognitive development and conceptual change and on problem solving, understanding, and reasoning—specifically, the areas of analogical problem solving, expert and novice reasoning, qualitative reasoning, heterogeneous reasoning, and mental modeling. (I discuss pertinent investigations in Section IV.)

The usefulness of cognitive psychology to cognitive historians stands to increase as it develops. As William Bevan has noted in a recent analysis of the state of cogni-

---

What Is Cognitive Science? (Cambridge, Mass.: MIT/Bradford Books, 1993), presents the first in-depth analysis of the foundations of the field.

[5] For a different assessment see Paul M. Churchland, *A Neurocomputational Perspective: The Nature of Mind and the Structure of Science* (Cambridge, Mass.: MIT Press, 1989).

tive psychology, the discipline does not yet comprise a coherent body of knowledge.[6] On the one hand, most cognitive psychologists are reluctant to create overarching theories that would synthesize the vast body of empirical research they have conducted in specific areas and thus to develop the "unified theory of cognition" that Allen Newell called for and attempted.[7] On the other hand, those who do speculate often present little substantive backing for sweeping claims that border on speculative metaphysics. Although much of the research is sophisticated, too much of the literature in cognitive psychology reports on narrowly focused experiments, reflecting more the empiricist spirit of the nineteenth century than the cognitive aspirations of the late twentieth century.

To be fully adequate, a theory of human cognition has to supply an account of what it means to be a human thinker acting in the world. James Greeno, among others, has suggested that the strong information-processing model of cognition cannot be made to accommodate the fact that human reasoning always takes place in a social context.[8] Having a better understanding of how people create collective understandings is critical to both cognitive science and history of science. Achieving this end will require extensive investigations into cognition in complex, interactive contexts, including the role of personality in cognition. However, that cognition is inherently contextual does not mean that we can learn nothing general about it. As Edward Hutchins, who has worked extensively on cognition in the workplace, has put it,

> There are powerful regularities to be described at a level of analysis that transcends the details of the specific domain. It is not possible to discover these regularities without understanding the details of the domain, but the regularities are not about the domain specific details, they are about the nature of cognition in human activity.[9]

The interdisciplinary nature of the field of cognitive science provides great benefits for those who probe cognition as a sociocultural phenomenon. For example, much of the research on social cognition to date has been carried out by cognitive psychologists and anthropologists working together on science and mathematics education or on work practices. Practitioners of these disciplines were attracted to collaborative projects because they recognized that the cognitive activities involved in learning situations and in the workplace are inherently social and collective activities.

If cognitive science is in an embryonic state, is it perhaps premature even to attempt cognitive history? My own response is no. Creating a working synthesis does not mean importing interpretations, methods, and assumptions of the cognitive sciences wholesale into historical analysis. Rather, the heart of the cognitive-historical method is reflexive. Input from historical investigations is required to

---

[6] William Bevan, "A Tour Inside the Onion," *American Psychologist,* 1991, *46*:475–483.

[7] Allen Newell, *Unified Theories of Cognition* (Cambridge, Mass.: Harvard Univ. Press, 1990).

[8] James G. Greeno, "Situations, Mental Models, and Generative Knowledge," in *Complex Information Processing,* ed. D. Klahr and K. Kotovsky (Hillsdale, N.J.: Lawrence Erlbaum, 1989), pp. 285–318.

[9] As quoted in David D. Woods, "Towards a Theoretical Base for Representation Design in the Computer Medium: Ecological Perception and Aiding Human Cognition," in *The Ecology of Human-Machine Systems,* ed. J. Flach *et al.* (Hillsdale, N.J.: Lawrence Erlbaum, in press), p. 15.

produce comprehensive models of cognition. Cognitive scientists need analyses of the sophisticated reasoning practices of scientists—living and dead, ordinary and genius—to create their models of human reasoning just as cognitive historians need models of human cognition to assist their reconstructions of the historical record. The method employs the kind of bootstrapping procedure commonly used in science. Cognitive science has advanced to the state where productive interaction of this sort is possible. Some of these interactions, which in a few cases are actually collaborative endeavors, will be discussed in Section IV, but a concrete example might be in order here.

As Martin Rudwick has pointed out, the visual representations in scientific papers, notebooks, and the like, although largely ignored by historians, seem to perform more than an ancillary function in communicating scientific research. In his struggle to find a way to talk about the cognitive function of these practices, Rudwick ended by laying a "cognitive grid" on the historical analysis. But as he acknowledged, in order to understand how the new visual modes of representation changed the way people did geology, one must be able to explain the cognitive function of visual representation.[10] Recent research in cognitive science has posed the question of what value pictorial representation might have over linguistic representation in texts for problem solving and reasoning. I have used insights from these analyses in my interpretation of the role of visual representation in the construction of Michael Faraday's and James Clerk Maxwell's field representations of electromagnetic forces.[11] These analyses provide assessments and critiques of cognitive research that are in turn being fed back into cognitive psychology, in part through my collaboration with James Greeno on the possible roles of visual representation in creating a shared scientific understanding.[12]

Before discussing the research program of cognitive history, I want to consider the place of that program within the discipline of history of science.

III. HOW DO COGNITIVE HISTORIANS VIEW THEIR DISCIPLINE?

The conception of the discipline of history of science with which cognitive historians work is pluralist. There are many kinds and levels of historical analysis, and historians have long used the resources, insights, and tools of many other disciplines to deepen historical understanding. The mode of analysis and the resources that historians use in any given analysis will depend on what they want to learn from that analysis. Contemporary history of science draws to a significant degree on the sociology of scientific knowledge, for example. However, the rhetoric of much of "sociohistorical" analysis—the attempt to integrate history and sociology—is reductionist and openly anticognitive in stance. In its most radical form, sociology of

[10] Martin J. S. Rudwick, "The Emergence of a Visual Language for Geological Science, 1760–1840," *History of Science,* 1976, *14:*149–195.

[11] Nancy J. Nersessian, "Reasoning from Imagery and Analogy in Scientific Concept Formation," in *PSA, 1988*(1):41–47; and Nersessian, "How Do Scientists Think? Capturing the Dynamics of Conceptual Change in Science," in *Cognitive Models of Science,* ed. R. N. Giere, Minnesota Studies in the Philosophy of Science, 15 (Minneapolis: Univ. Minnesota Press, 1992), pp. 3–44.

[12] Nancy J. Nersessian and James S. Greeno, "Multiple Abstracted Representations in Problem Solving and Discovery in Physics," *Proceedings of the Cognitive Science Society,* 1990, *11:*77–84; and Nersessian and Greeno, "Constructive Modeling in Creating Scientific Understanding" (in progress).

science attempts to explain the production of scientific knowledge wholly in terms of social factors. Bruno Latour and Steve Woolgar have gone so far as to call for a "ten-year moratorium" on cognitive explanations, claiming that at the end of that time sociological explanations will have established that there is nothing left to explain.[13]

Although sociological insights and methods have enriched the history of science, the concerns of the historian of science do not reduce to those of the sociologist, and none of the work produced by sociohistorical analysis supports the reductionist position. One valuable contribution sociologists of science have made to history is to show how positivist philosophy of science has unduly influenced the historian's conception of the scientific enterprise. This insight is important because one's conception of science does motivate how one frames questions, what problems one works on, what avenues of inquiry one pursues, and what factors one sees as relevant to the explanations that are advanced. Yet because of this very circumstance historians should not end up substituting one monolithic conception for another. Sociology of science advances a specific conception of the nature and production of science. The cognitive historian contends that although it is true that science is a social product, it is also the most cerebral of enterprises and the complex cognitive activities of the individual scientist are equally pertinent to a historical understanding of the product and the processes through which it is produced.

What factors are relevant to historical explanation is not a question that can be answered a priori. Historians of science frequently use the resources of other disciplines—anthropology, economics, politics, literature, sociology—to further their analyses. What resources outside of history one draws upon in any given analysis depends on the questions one is asking. Drawing from the cognitive sciences in the way outlined in Section I can help to make sense of certain theoretical and experimental practices of scientists. But the cognitive is only one dimension to these practices, and questions about scientific cognition are by no means the only questions historians want to ask.

Should historians of science, in the current research climate, view cognitive history as reactionary? As a harbinger of a return to the external-internal distinction in a new guise? Cognitive history is not reactionary because its practitioners do not see themselves in conflict with the sociology of science. Many of the important concerns raised by sociohistorical analysis are concerns for cognitive historians as well. Both cognitive and social historians agree, for example, that "thinking" is an inherently social activity, it is not something that simply goes on "in the head." Where they disagree is over the claim that social factors are all that count in explaining the production of scientific knowledge.

The reductionist claim that knowledge production is purely a product of environmental influences has curious resonances with behaviorist psychology.[14] Indeed, tacit psychological assumptions do underlie sociohistorical analysis. For example,

---

[13] Bruno Latour, *Science in Action: How to Follow Scientists and Engineers through Society* (Cambridge, Mass.: Harvard Univ. Press, 1987), p. 247; and Bruno Latour and Steve Woolgar, *Laboratory Life: The Construction of Scientific Facts*, rev. ed. (Princeton: Princeton Univ. Press, 1986), p. 280.

[14] For this argument see William Shadish, Jr., *et al.*, "The Psychology of Science: An Introduction," in *Psychology of Science: Contributions to Metascience*, ed. B. Gholson, W. R. Shadish, Jr., R. A. Neimeyer, and A. C. Houts (New York: Cambridge Univ. Press, 1989), pp. 1–16; and Giere, *Explaining Science* (cit. n. 1).

in attempting to argue that rationality is a factor in sociohistorical analysis, Steven Shapin notes that "actors" are themselves treated as if their "cognitive wiring" was "in proper working order," as possessing a "natural rationality." [15] But how can one make this assumption without knowing what it means to possess "natural rationality"? Cognitive-historical analysis addresses the questions of what the "cognitive wiring" is and what constitutes "natural rationality." Sociohistorical analysis has "black-boxed" the individual scientist. Cognitive history is attempting to open the box and to show how the cognitive and the social are fused in the scientist's construction of knowledge.

As I noted earlier, the relationship between the social and the cognitive dimensions of thinking and reasoning is territory ripe for exploration. Multiple constraints shape cognition, and social contexts and interactions have to be figured into the account. Taking them into account may lead to substantial changes within cognitive science itself, as some of the cognitive scientists cited earlier contend. But acknowledging that the cognitive activities of scientists are embedded in complex social contexts does not require historians to "black-box" the individual. That is, an accounting is still required of how and what the cognitive activities of the individual scientist contribute in the construction of knowledge.

In sum, cognitive history aims at integrating the social and the cognitive dimensions of knowledge production by showing how they are linked in a scientist's life work and how the individual and group levels of knowledge production mesh. Thus, while the concerns of the cognitive historians do resonate with those of the earlier intellectual historians, their aim is to give a more comprehensive and nuanced analysis of scientific knowledge as the product of cognitive, social, tool-using agents. Traditional history of ideas has tended to treat ideas as having a life and dynamic of their own and to analyze them mainly in the context of other ideas. Cognitive history aims to include the contextual embeddedness of human agents and their knowledge products in its accounts. Just as Peter Burke has argued that *mentalité* history occupies a place in the discipline of history "between an intellectual history with society left out and a social history with thought left out," [16] so too cognitive history occupies a place between a traditional history of scientific ideas and a new sociological history of science that tends to marginalize the cognitive dimension of scientific practices.

IV. WHAT PROBLEMS DO COGNITIVE HISTORIANS ADDRESS?

Cognitive historians address problems customarily treated under the rubrics of scientific revolutions, scientific discovery, and individual creativity. The research produced thus far encompasses work on conceptual change, theoretical and experimental practices, and methodological and technological innovation. Work in the genre of cognitive history has from the outset been interdisciplinary; thus the research discussed here includes that carried out by historians of science, philosophers of

---

[15] Steven Shapin, "History of Science and Its Sociological Reconstructions," *Hist. Sci.*, 1982, *20*:157–211.

[16] Peter Burke, "Strengths and Weaknesses in the History of Mentalities," *History of European Ideas*, 1986, *7*:439–451, p. 440.

science, and cognitive psychologists. A brief overview of projects in the research program of cognitive history and avenues for further research follows.[17]

Creativity in Experimental Practice

Psychologists were the first to begin to use the resources of cognitive psychology in studying scientific practices. Individual creativity has been a focal point of their analyses.[18] The psychologist Ryan Tweney's examination of Faraday's notebook, diaries, and loose notes has made significant contributions to our understanding of his discovery of electromagnetic induction. Tweney's finely detailed analysis provides an integrated model of Faraday's experimental and theoretical reasoning between his first attempts at detecting electromagnetic induction in 1821 and his success at it in 1831.[19] There was already a substantial literature on this discovery. Pearce Williams's extensive examination of Faraday's activities during that ten years succeeded in conveying the impression that the discovery was not simply the product of a flash of creative insight. Although Williams had the important intuition that the work on acoustical figures and crispations just prior to the discovery of electromagnetic induction "must have affected Faraday deeply," he did not reveal the patterns lying beneath Faraday's many activities and thoughts during that period that we get from Tweney's analysis.[20] By utilizing the notions drawn from cognitive psychology of "schemata," "scripts," and "heuristics," Tweney was able to reconstruct these patterns and provide a plausible interpretation of how they led to the 1831 discovery.

David Gooding, too, has focused on the role of Faraday's experimental practices in his conceptual innovations. In wrestling with the problem of how to access and unravel the procedural knowledge that informs experimental practices, Gooding is concerned more generally to develop methods of recovering and representing experimental procedures and to examine the practical and inferential structure of experimental problem solving. His recent book provides detailed analyses of Faraday's experimental procedures that add much to our understanding of Faraday, laboratory records in general, and the cognitive and social dimensions of experimental practice. His latest work extends the analysis to thought experiments and sees them as rooted in the procedural knowledge of how experimental processes work in the real world.[21] One analytical tool that Gooding used to capture the "on-line" dynamics of

---

[17] For a longer version of this article, with extensive references to pertinent literature in the cognitive sciences, write to Nancy Nersessian, at the address on p. 194, for technical report CSL 53.

[18] Howard E. Gruber, *Darwin on Man: A Psychological Study of Scientific Creativity* (New York: Dutton, 1974), stands as the first major cognitive-historical analysis, although it predates much of the literature in cognitive psychology that recent cognitive history draws on.

[19] Ryan D. Tweney, "Faraday's Discovery of Induction: A Cognitive Approach," in *Faraday Rediscovered: Essays on the Life and Work of Michael Faraday,* ed. David Gooding and Frank A. J. L. James (New York: Stockton Press, 1985), pp. 189–210.

[20] L. Pearce Williams, *Michael Faraday: A Biography* (London: Chapman & Hall, 1965), quoting from p. 180. See also Joseph Agassi, *Faraday as a Natural Philosopher* (Chicago: Univ. Chicago Press, 1971); and William Berkson, *Fields of Force: The Development of a World View from Faraday to Einstein* (New York: Wiley, 1974).

[21] David Gooding, *Experiment and the Making of Meaning: Human Agency in Scientific Observation and Experiment* (Dordrecht: Kluwer, 1990); and Gooding, "The Procedural Turn; or, Why Did Faraday's Thought Experiments Work?" in *Cognitive Models of Science,* ed. Giere (cit. n. 11), pp. 45–76.

experimental practice, endeavoring to remove the layers of reconstruction that occur when experiments are packaged into narratives, was adapted from a tool used in the cognitive sciences. Gooding devised "experimental maps" to depict sequences of experimental procedures acting on physical and conceptual objects. The maps exhibit how experimentation is a nonlinear, reflexive process, with multiple possible pathways between goals and solutions. They display possible as well as actual pathways in the experimental record. When compared with maps of later experimental accounts, they exhibit how contingent choices in experimentation become reconstructed as decisions. This process reveals experimental failures to be as important as successes in their informative value.

Gooding's experimental maps elaborate a technique developed by cognitive scientists to analyze "think-aloud protocols."[22] These protocols are the records of every thought experimental subjects verbalize during a problem-solving task in which they have been instructed to record all of their thinking. Tweney took the first step in extending this technique to analysis of Faraday's diaries. He argued that specific features of Faraday's record keeping make it possible to view his diary as approximating a think-aloud protocol.[23] The purpose of mapping a protocol in cognitive psychology is to develop a computer simulation of the reasoning processes exhibited in the protocol. Tweney and Cathy Hoffner mapped 135 experiments conducted after the discovery of electromagnetic induction; some of these were later simulated by a computer program.[24] Although Gooding's original purpose was different, he has recently undertaken a research project with Thomas Addis, an AI researcher, to use the maps in creating real-time, interactive simulations of experimental expertise.[25]

Conceptual Change

From the perspective of cognitive history, the problem of conceptual change appears as follows: How is it that scientists, working alone or in collaboration, combine their human cognitive abilities with the conceptual resources available to them as members of scientific communities and wider social contexts to create and communicate new scientific representations of a domain? This formulation shifts the focus of the problem from the customary preoccupation with how linguistic structures change to the role of human agency in conceptual change.

My own research centers on the theoretical reasoning and representational practices of scientists who created major conceptual innovations. Despite all the attention given to theory in the history and philosophy of science, there is still very little understanding of the theoretical *practices* that scientists employ in general, and in conceptual innovation in particular. Thus, my investigation is complementary to

---

[22] See Alan Newell and Herbert A. Simon, *Human Problem Solving* (Englewood Cliffs, N.J.: Prentice-Hall, 1972); and Karl A. Ericsson and Herbert A. Simon, *Protocol Analysis: Verbal Reports as Data* (Cambridge, Mass.: MIT Press, 1984).

[23] Ryan D. Tweney, "A Framework for the Cognitive Psychology of Science," in *Psychology of Science,* ed. Gholson *et al.* (cit. n. 14), pp. 342–366; and Tweney, "Fields of Enterprise: On Michael Faraday's Thought," in *Creative People at Work,* ed. D. B. Wallace and H. E. Gruber (New York: Oxford Univ. Press, 1989), pp. 91–106.

[24] Ryan D. Tweney and Cathy Hoffner, "Understanding the Microstructure of Science: An Experiment," *Proc. Cog. Sci. Soc.,* 1987, 9:677–681; and Deepak Kulkarni, *The Processes of Scientific Research: The Strategy of Experimentation* (Ph.D. diss., Carnegie-Mellon University, 1987).

[25] Thomas R. Addis, David Gooding, and Judith J. Townsend, "A Functional Description of Faraday's Discovery of the Electric Motor" (draft, 1991).

those on experimental practice. Unlike some work in cognitive history, mine makes extensive use of published as well as unpublished sources. I concur with Holmes that writing a scientific paper is part of the creative process and that much can be learned from published work about the constructive and communicative practices of scientists.[26] In many cases a scientist's thinking about a problem is still going on and important transformations take place in the "final" paper. This historiographical assumption finds support in the cognitive literature on writing.[27]

Historians have long been divided on the issue of how new conceptual structures emerge in scientific revolutions and come to replace existing structures. Some have viewed conceptual change as occurring suddenly and, following Kuhn, have likened the process to a psychological gestalt switch. Others, such as Holmes, Gerald Holton, and R. S. Westfall, have viewed it as extending over a long period.[28] This tension, reflected in the work of Kuhn himself, can be characterized as that between a short-term cognitive approach and a long-term historical one. It seemed to me that many of the problems of "incommensurability" plaguing both the history and philosophy of science were artifacts of the tools and the implicit presuppositions with which each approached the problem of scientific change. The cognitive-historical method specifically achieves a deeper and more refined understanding of conceptual change than is possible with either approach.

Numerous metatheoretical presuppositions have left their imprint on historical writing. I have discussed several from a cognitive perspective elsewhere.[29] Here I will only mention one: the presupposition that scientific reasoning is limited to deductive and inductive algorithms. This notion of scientific method has led many to ignore the constructive practices recorded by scientists in their papers, diaries, and notebooks. Many scientists use analogies, visual representations, and thought experiments extensively when creating and communicating new concepts. Although these thinking and communicative practices are widespread, none except analogy has received more than scant attention in the philosophical literature on scientific method. The main problem philosophers have had in even countenancing these as methods is that they are not algorithmic in application, and thus even if used correctly, they may lead to the wrong solution or to no solution. This very feature, however, makes them much more realistic historically. Historians too need to understand the constructive practices scientists use in their research irrespective of whether the

[26] Frederic L. Holmes, "The Fine Structure of Scientific Creativity," *Hist. Sci.*, 1981, *19:*60–70; Holmes, "Scientific Writing and Scientific Discovery," *Isis*, 1987, *78:*220–235; and Holmes, "Argument and Narrative in Scientific Writing," in *The Literary Structure of Scientific Argument: Historical Studies,* ed. Peter Dear (Philadelphia: Univ. Pennsylvania Press, 1991), pp. 164–181.

[27] See, e.g., Vernon A. Howard and J. H. Barton, *Thinking on Paper* (New York: William Morrow, 1986).

[28] Thomas S. Kuhn, *The Structure of Scientific Revolutions* (Chicago: Univ. Chicago Press, 1962); Frederic L. Holmes, *Lavoisier and the Chemistry of Life: An Exploration of Scientific Creativity* (Madison: Univ. Wisconsin Press, 1985); Holmes, "Patterns of Scientific Creativity," *Bulletin of the History of Medicine,* 1986, *60:*19–35; Gerald Holton, *Thematic Origins of Scientific Thought: Kepler to Einstein* (Cambridge, Mass.: Harvard Univ. Press, 1973); Holton, *The Scientific Imagination: Case Studies* (New York: Cambridge Univ. Press, 1978); and Richard S. Westfall, *Force in Newton's Physics: The Science of Dynamics in the Seventeenth Century* (New York: American Elsevier, 1971).

[29] Nancy J. Nersessian, "Faraday's Field Concept," in *Faraday Rediscovered,* ed. Gooding and James (cit. n. 19), pp. 175–188; Nersessian, "A Cognitive-Historical Approach to Meaning in Scientific Theories," in *The Process of Science: Contemporary Philosophical Approaches to Understanding Scientific Practice,* ed. Nersessian (Dordrecht: Martinus Nijhoff, 1987), pp. 161–178; and Nersessian, "How Do Scientists Think?" (cit. n. 11).

scientists in question went down dead ends, contributed to "winning" science, or employed different strategies in making a "simultaneous discovery." Cognitive research on ordinary reasoning practices provides support for their salience and insight into how they function in scientific reasoning.

Though the topic of scientific method might seem more appropriate for a philosophical audience, historians do make implicit assumptions about method that affect their analyses significantly. As is well known, the "whig" history of the past was heavily influenced by the positivist conception of method, and that conception continues to influence historical analyses and plays a critical role in some historical controversies. Witness the case of Maxwell. From Pierre Duhem to the present historians have argued whether the analogical model Maxwell presented in his second paper on electromagnetism was critical to his derivation of the electromagnetic field equations. Those who see the analogical model as off to the side, while "the results are known by some other means before use of the model," have at least tacitly bought the philosopher's assumption that reasoning is only by means of inductive or deductive algorithms.[30]

Research in cognitive psychology forces the conclusion that analogies are not "merely suggestive" or an "unproductive digression" but are fundamental in the reasoning process. An extensive psychological literature shows how prominent a role analogy plays in problem solving in general; these investigations seek to determine how analogy generates solutions and to differentiate between productive and nonproductive uses. Viewing Maxwell as reasoning *through* the analogy fits the historical record better and obviates the need to throw away inconvenient pieces, such as his "errors." Indeed, this view better fits Maxwell's own insights into how what he called "the method of physical analogy" functions in the mathematization process.

From the perspective of cognitive history the analogies, visual representations, and thought experiments scientists employ in their research are not ancillary, but play a crucial role in the creative process. They all involve creating, manipulating, and reasoning about intermediary models.[31] They are also the principal means through which a scientist transports conceptual resources drawn from the wider cultural milieu into science and transmits novel representations to the scientific community.

Technological Innovation

A relatively new undertaking is to apply cognitive-historical analysis in studies of technological innovation. Michael Gorman, a psychologist, and Bernard Carlson, a

---

[30] Pierre Duhem, *The Aim and Structure of Physical Theory* (New York: Atheneum, 1962), quoting p. 68. See also Duhem, *Les théories électriques de J. Clerk Maxwell: Etude historique et critique* (Paris: A. Hermann, 1902); and for recent studies, Berkson, *Fields of Force* (cit. n. 20); Joan Bromberg, "Maxwell's Displacement Current and His Theory of Light," *Archive for the History of the Exact Sciences*, 1968, *4*:218–234; Alan F. Chalmers, "The Heuristic Role of Maxwell's Mechanical Model of Electromagnetic Phenomena," *Studies in the History and Philosophy of Science,* 1986, *17*:415–427; Peter Heimann, "Maxwell and the Modes of Consistent Representation," *Arch. Hist. Exact Sci.*, 1970, *6*:171–213; Nersessian, *Faraday to Einstein* (cit. n. 1); Nersessian, "How Do Scientists Think?" (cit. n. 11); and Daniel Siegel, "The Origin of the Displacement Current," *Historical Studies in the Physical Sciences,* 1986, *17*:99–145.

[31] Nancy J. Nersessian, "Why Do Thought Experiments Work?" *Proc. Cog. Sci. Soc.,* 1991, *13*:430–438; Nersessian, "Abstraction via Generic Modeling in Concept Formation in Science," in *Idealization and Science,* ed. N. Cartwright and M. R. Jones (Amsterdam: Rodophi, in press); and

historian, joined forces to compare the invention of the telephone by Thomas Edison and by Alexander Graham Bell.[32] Their comparative analysis examines the interaction between mental models, mechanical representations, and heuristics in technological innovation. Thus far they have demonstrated the complex relation between mental modeling and conducting experiments and have revealed significant differences in the cognitive styles of Bell and Edison. Gary Bradshaw, a cognitive psychologist, has applied Herbert Simon's problem-solving model of discovery to the invention of the airplane to see how well it can be extended to technological innovation.[33]

Communicative Practices

Many contemporary historians are examining the form and rhetoric of presentations of experimental and observational activities and results. They usually frame their studies in terms of how scientists adopt certain modes and conventions of writing in order to gain authority and thus to persuade others of their inferences. What tends to be overlooked is that there is a significant cognitive dimension to persuasion. In the process of persuading one has to get one's colleagues to comprehend the new ideas, and, again, in order to negotiate one has to comprehend what is being proposed. These are interconnected, though not necessarily sequential, steps. Success at communication, of course, does not entail success at persuasion.

A cognitive historian takes the public communications of scientists presenting new representations as attempts to create a shared understanding. That is, these communications are devised to bridge the gap between doing, on the one hand, and comprehending and evaluating, on the other. Although Steven Shapin and Peter Dear have presented insightful analyses of Robert Boyle's contributions to shaping the form of research papers,[34] there is more to be told in this story. For example, Shapin claims that Boyle created his literary techniques to gain authority for his conclusions, devising a way of "witnessing" an experiment while not being present. But how and why does creating an "impression of verisimilitude" convey authority and "compel assent"? My analysis of thought experiments points to a possible answer. Boyle's texts, including both the narrative and the visual parts, assist readers in constructing their own mental simulation, thereby creating an understanding of what they have not actually witnessed themselves.

Holmes's analysis of the relation between argument and narrative in the scientific paper argues that the difference in style between research papers produced at the Royal Academy and at the Académie des Sciences stems, in part, from the practice of French academicians of carrying out communal investigations. In their case authority was not in question.[35] A cognitive perspective would add that when creating

---

Nersessian, "In the Theoretician's Laboratory: Thought Experimenting as Mental Modeling," *PSA,* 1992, 2:291–301.

[32] Michael E. Gorman and W. Bernard Carlson, "Interpreting Invention as a Cognitive Process: The Case of Alexander Graham Bell, Thomas Edison, and the Telephone," *Science, Technology, and Human Values* (forthcoming).

[33] Gary Bradshaw, "The Airplane and the Logic of Invention," in *Cognitive Models of Science,* ed. Giere (cit. n. 11), pp. 239–250.

[34] Steven Shapin, "Pump and Circumstance: Robert Boyle's Literary Technology," *Social Studies of Science,* 1984, *14:*481–520; and Peter Dear, "*Totius in verba:* Rhetoric and Authority in the Early Royal Society," *Isis,* 1985, *76:*145–161.

[35] Holmes, "Argument and Narrative" (cit. n. 26).

a shared understanding through mental modeling, the French could also cut the narrative short because they could presume tacit and explicit knowledge of experimental procedures and apparatus.

One last significant dimension of how scientists communicate novel ideas is visual representation. Scientific texts contain many different kinds, among them diagrams, schematic representations, paintings, and photographs. A cognitive historian would try to determine just how they create understanding within a community, keeping in mind that they may serve different cognitive functions. The visual representations in physics and geology often appear in abstracted, idealized form, displaying what the representer deems essential to the analysis or for understanding it. Such schematic renderings and diagrams seek to reify solutions to problems and to convey quantitative understandings. Depictions in biological texts or in early astronomical treatises may include artistic embellishments, to heighten the senses and fix attention on specific features of phenomena. Medical and astronomical texts may present faithful renderings or photographic images. Quite recent texts in several areas now include images reconstructed from computer data collected from such observational devices as satellites and CAT scanners. Again, these all provide ways of fixing the phenomena, as the presenter has thought about them, in the mind of the reader.

A conspicuous role these visual representations all play in communication is to provide a stable embodiment that is public. Latour has called such embodiments "immutable and combinable mobiles." However, in the process of communicating, they often mutate, creating an enhanced or a different understanding. For example, James Griesmer and William Wimsatt use modifications in "Weisman diagrams" to document conceptual variation within a community of understanding over a period of time.[36] My analysis of schematic representations in physics finds that they help readers construct mental models, much as they do in thought experiments.

Training Practitioners

The role of training in scientific change and in science more generally has only recently attracted the attention of historians. The history of pedagogical practices offers potentially fertile ground for examining how cognitive and social factors interact in scientific change.

Training, whether as a theoretician or as an experimentalist, comprises more than learning through texts and working problem exemplars. Kathryn Olesko's study of Franz Neumann's seminar attempts to capture the explicit and implicit pedagogical practices that formed "the activity of learning" within the seminar.[37] Her analysis reconstructs various methods Neumann devised to teach measuring procedures, theoretical techniques, mathematical skills, and evaluative strategies. She argues that the work of individual students exhibited techniques they learned in the seminar and thus shows how these educational practices shaped the context and content of the local scientific practices.

[36] James R. Griesemer and William Wimsatt, "Picturing Weismannism: A Case Study of Conceptual Evolution," in *What the Philosophy of Biology Is: Essays for David Hull,* ed. Michael Ruse (Dordrecht: Kluwer, 1989), pp. 75–137.

[37] Kathryn M. Olesko, *Physics as a Calling: Discipline and Practice in the Königsberg Seminar for Physics* (Ithaca, N.Y.: Cornell Univ. Press, 1991). See also Nersessian, "Theoretician's Laboratory" (cit. n. 31); Nersessian, "How Do Scientists Think?" (cit. n. 11); and Nersessian and Greeno, "Dynamic Mental Modeling" (cit. n. 12).

A cognitive historian would take this analysis further by examining how historical and psychological inquiries into pedagogy could enhance one another. For example, of central importance to both is how a community's implicit knowledge is made explicit for pedagogical purposes and then becomes tacit for the new practitioners. The trainee in physics, in learning modeling techniques, must first learn the generative principles and constraints for physical models in a domain. These then transform into the tacit assumptions of the thinking and communicative practices of the community of expert physicists.

V. APPLIED HISTORY OF SCIENCE

The bulk of this article examines how cognitive science could enrich the history of science. What does the history of science have to offer cognitive scientists in turn? John Heilbron has called "applied history of science" that which is done with the "client's purposes" in mind.[38] Although applied history most likely differs from history done with historians' professional interests in mind, it is unlikely to dilute work produced for the profession and very likely to make a valuable contribution elsewhere. I will indicate fertile areas for such collaboration in cognitive psychology, AI, and science education.

Cognitive Psychology

The history of scientific change could well create more sophisticated psychological models that see conceptual change, creativity, and organization of knowledge as intertwined wholes. Many cognitive psychologists recognize this, and several have undertaken historical studies on their own. Since they are not trained historians, though, psychologists would prefer that historians carry out the analytical work necessary. Historical analysis, unlike their customary interviews and protocol analyses, provides psychologists with an integrated picture of a significant set of achievements over an extended period of time—one that can enrich their data. Used in combination with interviews, historical analyses can clarify issues and prevent misleading interpretations, such as those that arose from Wertheimer's interviews of Einstein.[39] These analyses can inform the design of laboratory experiments and challenge interpretations based on psychological studies alone.[40] Additionally, as scientists have reflected throughout history on how to learn about and represent nature, they have refined and extended their representational and communicative practices. These "metacognitive" reflections are a valuable resource for cognitive psychologists concerned with learning.

One prominent question in the contemporary psychological literature is, Are the

---

[38] John L. Heilbron, "Applied History of Science," *Isis*, 1987, *78:*552–563.

[39] Max Wertheimer, *Productive Thinking* (New York: Harper & Row, 1945). See also, e.g., Gentner and Jeziorski, "Shift from Metaphor" (cit. n. 2): and Marianne Wiser and Susan Carey, "When Heat and Temperature Were One," in *Mental Models*, ed. Dedre Gentner and Alan L. Stevens (Hillsdale, N.J.: Erlbaum, 1983).

[40] Nancy J. Nersessian, "Conceptual Change in Science and Science Education," *Synthese*, 1989, *80:*163–183; Nersessian and Greeno, "Multiple Abstracted Representations in Physics" (cit. n. 12); and Nersessian and Lauren B. Resnick, "Comparing Historical and Intuitive Explanations of Motion: Does 'Naive Physics' Have a Structure?" *Proceedings of the Eleventh Annual Conference of the Cognitive Science Society* (Ann Arbor: Univ. Michigan Press, 1989), pp. 412–420.

conceptual changes (or "restructurings") that take place in cognitive development or learning like those in "scientific revolutions"? Several psychologists argue that they are indeed.[41] The main support for this hypothesis comes from research that compares the initial state of the child's or student's representation of a domain with the desired final state. That research shows that the kinds of changes necessary to get from one state to the other resemble those that have taken place in scientific revolutions, as characterized by Kuhn. However, as with Kuhn's analysis, the nature of the processes through which conceptual change occurs has not been explored in any depth in psychology, and it is widely acknowledged that this problem now needs to be placed on the agenda. My analysis of the constructive practices of Faraday and Maxwell indicated how important analogy and mental modeling are in conceptual change and urged the need for a dynamical notion of "mental model" when analyzing change. Cognitive psychologists could well also explore the possible role of analogy in conceptual change in learning.

Artificial Intelligence

Computer simulations of scientific reasoning processes are a valuable resource for those AI researchers who use computer modeling in studying human intelligence and for those who wish to model expert systems on human problem solving. Analyses by historians have become more important as this latter project has unfolded. Computer "discovery programs" have evolved from the early BACON program, which was concerned not with historical accuracy but rather with examining heuristics through which Kepler's laws could in principle have been generated from the empirical data. The more realistic STAHL program attempts to model eighteenth-century styles of reasoning about the components of substances, and the KEKADA program attempts to model the actual heuristics employed by Hans Krebs in his discovery of the ornithine cycle. Simon and his colleagues make explicit use of the work of historians, especially Holmes, in constructing these later simulations.[42] Whether computer simulations will be successful in capturing the complexity of actual discovery processes in past science and in extracting general problem-solving heuristics from them depends largely on the development of more sophisticated computer-modeling techniques. What the historian can contribute to this project is a deeper understanding of the nature of the discovery processes that need to be modeled.

Science Education

Historians of science have long recognized that they can make a significant contribution to science education. Earlier attempts to apply the fruits of historical research

---

[41] See, e.g., Susan Carey, *Conceptual Change in Childhood* (Cambridge, Mass.: MIT Press, 1985); Richard A. Duschl, *Restructuring Science Education* (New York: Teachers College Press, 1990); Frank C. Keil, *Semantic and Conceptual Development: An Ontological Perspective* (Cambridge, Mass.: Harvard Univ. Press, 1979); and Stella Vosniadou and William Brewer, "Mental Models of the Earth: A Study of Conceptual Change in Childhood," *Cognitive Psychology,* in press.

[42] Patrick Langley *et al., Scientific Discovery: Computational Explorations of the Creative Processes* (Cambridge, Mass.: MIT Press, 1987); Jan W. Zytkow and Herbert A. Simon, "A Theory of Historical Discovery: The Construction of Componential Models," *Machine Learning,* 1986, *1:*107–136; and Deepak Kulkarni and Herbert Simon, "The Processes of Scientific Discovery: The Strategy

to science teaching turned on the beliefs that exposing students to the historical roots of scientific ideas will facilitate learning the ideas themselves by placing them in a context, and that providing examples of scientific discovery enables students to understand scientific methods. However, though there are good reasons for teaching science students some history of science, the historical examples cannot by themselves help students learn scientific theories.

As I have argued recently, a new way of incorporating the history of science into science pedagogy does further that aim.[43] A new interdisciplinary research area, "cognition and instruction," has developed at the intersection of cognitive psychology and research in science and mathematics education. A hallmark of the research in the field of cognition and instruction is a "constructivist" approach to learning, based on the belief that as part of the process of learning a science, students need to construct their own representations of the extant conceptual structure of the science. Cognitive historians too conceive of what is customarily called "scientific discovery" as a process of construction. Scientists actively construct representations by employing problem-solving procedures not unlike those we employ in ordinary problem solving. On a constructivist account, the cognitive activity of the scientist is pertinent to learning. Thus the historical cognitive processes can provide a model for the learning activity itself. For that purpose the history of science would provide a repository of strategic knowledge of how to go about constructing, changing, and communicating scientific representations. Historians, working in conjunction with cognitive scientists, can "mine" historical records for these strategies and then devise ways of integrating and transforming the more realistic exemplars of scientific problem solving into instructional procedures.

This article has argued for the significance to history of science of cognitive-historical analysis and has established the areas in which history of science and cognitive science can be mutually enriching. Even so brief a survey demonstrates that, although a relatively new undertaking, cognitive history has already added significant insights to the historical literature on scientific change, and that collaborations with historians, though few thus far, have proven beneficial for cognitive scientists as well.

---

of Experimentation," *Cognitive Science,* 1988, *12:*139–175. See also Frederic L. Holmes, "Hans Krebs and the Discovery of the Ornithine Cycle," *Federation Proceedings,* 1980, *39:*216–225.

[43] Nancy J. Nersessian, "Constructing and Instructing: The Role of 'Abstraction Techniques' in Creating and Learning Physics," in *Philosophy of Science, Cognitive Psychology, and Educational Theory and Practice,* ed. R. Duschl and R. Hamilton (Albany: SUNY Press, 1992), pp. 48–68; and Nersessian, "Should Physicists Preach What They Practice? Constructing Modeling in Doing and Learning Physics," *Science and Education,* 1995, *4* (forthcoming).

KNOWLEDGE AND AUTHORSHIP

Engraved for the Univerſal Magazine.

THE HON.ᵇˡᵉ ROBERT BOYLE

F.M.laCave ſculp.

For J. Hinton at the King's Arms in Newgate Street.

One can be present-minded without being whiggish. One need not avoid noting that Robert Boyle, for example, misunderstood the physics of his vacuum apparatus. See page 220. Photograph courtesy of the Chemical Heritage Foundation.

Scientists as Historians

By Stephen G. Brush[*]

"SHOULD SCIENTISTS write history of science?"[1] Surely this question would sound bizarre to those who are not professional historians of science. They would be more likely to ask: "Should anyone who is *not* a scientist write history of science?" The idea that anyone would write about the history of science without having at least acquired advanced training in science, if not significant research experience, strikes most people—and especially scientists—as absurd.

The short answer to both questions is yes. Scientists should write history of science if they are willing to acquire the skills and background knowledge of the historian of science; and nonscientist historians should write history of science if they are willing to learn enough science to understand what they are going to write about.

The real issue is the proposition that historians and scientists, even if they share the same knowledge and skills, have distinctively different ways of writing about the history of science. History written by scientists has been called, by Allen Debus and others, "science-history."[2] Is it different from, or an identifiable version of, "history of science"? Is there any good reason for imposing a single style or viewpoint on all writings about the history of science?

During the 1960s and 1970s, as historians of science struggled to establish an academic discipline within history rather than as an adjunct to science, the most popular way to stigmatize science-history was the disdainful phrase "the whig interpretation of the history of science."[3] Did historians of science strive so hard to be antiwhig that they slipped into a dogmatic "contextualism" that has been outgrown

[*] Department of History and Institute for Physical Science & Technology, University of Maryland, College Park, Maryland 20742.

Research for this article was supported in part by the History and Philosophy of Science Program of the National Science Foundation, grant number DIR-9011872. It was completed while I was a member of the School of Social Science, Institute for Advanced Study, Princeton, New Jersey, with support from the Andrew W. Mellon Foundation. I thank Diana Barkan, Charles Gillispie, and George Stocking for useful suggestions.

[1] This was the assigned topic for the paper presented at the Conference on Critical Problems and Research Frontiers.

[2] Allen G. Debus, "The Relationship of Science-History to the History of Science," *Journal of Chemical Education,* 1971, *48:*804–805.

[3] For accounts of the development of the history of science discipline in the United States, portrayed in part as liberation from the domination of scientists and as an assertion of the professional independence of the historian, see Arnold Thackray and Robert K. Merton, "On Discipline Building: The Paradoxes of George Sarton," *Isis,* 1972, *63:*473–495; Thackray, "The Pre-History of an Academic Discipline: The Study of the History of Science in the United States, 1891–1941," *Minerva,* 1980, *18:*448–473; Frederick Gregory, "The Historical Investigation of Science in North America," *Zeitschrift für Allgemeine Wissenschaftstheorie,* 1985, *16:*151–165; and Nathan Reingold, "History of Science Today, 1: Uniformity as Hidden Diversity: History of Science in the United States, 1920–1940," *British Journal for the History of Science,* 1986, *19:*243–262.

by other historians? Have they committed themselves to an untenable belief in the objective reality and knowability of past events? Conversely, has antiwhiggism led historians to exclude themselves from any role in contemporary affairs where they might be able to use their professional knowledge?

While it is easy enough to find contemporary examples of the whig interpretation, that label is no longer a very useful club to beat the scientists with, for reasons to be examined below. Instead, the current fashion is to stress the social construction of scientific theories and concepts, and to deny that scientists are actually discovering the truth about the world, or that their efforts to do so have any moral or epistemological superiority to those of pseudoscientists, humanists, and theologians. The historian no longer assumes that scientific research is an admirable activity and scientific progress a benefit to society, whereas the scientist, by definition, thinks science is worth doing.

Readers unfamiliar with recent publications in science studies may not believe that any reputable scholar actually holds such extreme views, and might even dismiss the abstract "historian" in the last sentence as a straw person. Fortunately, Paul Forman has published an essay that beautifully describes the distinction between the putative historian's and scientist's views, and has clearly declared his allegiance to the former. He also identifies a specific document—Max Dresden's biography of H. A. Kramers—as an example of science-history, although we do not of course have to accept that document as being representative of the genre.[4] Nevertheless, Forman has greatly advanced the discussion of the issue. In his terms, historians of science should aim at *independence*—should be neutral or skeptical observers rather than partisans of science, in contrast to the scientists, whose emotional attachment to the *transcendence* of science disqualifies them from being suitably objective about it.

Whatever may be its defects from Forman's viewpoint, Dresden's book certainly illustrates some of the *advantages* of science-history. In particular, Dresden's technical knowledge allows him to make a detailed (and not always favorable) evaluation of Kramers's contributions to science, while at the same time his own experience in doing research enables him to understand and empathize with the emotional state of a researcher. Neither is impossible for nonscientist historians, but both are increasingly rare. To what extent can our understanding of the history of science be advanced by putting ourselves in the place of and reliving, physically and psychologically, the experiences of a scientist? Apparently Forman would say, not at all.

One of the critical problems, not just in the history of science but in the philosophy and sociology of science, is the extent to which scientific facts are socially constructed rather than discovered. Unfortunately, social construction has all too often been simply asserted, not demonstrated. Stephen Cole stated in a recent book that scholars who claim that social processes influence the cognitive content of science have failed to produce a single example in which this alleged causal influence is demonstrated by the standards of sociological research. In reviewing this book

---

[4] Paul Forman, "Independence, Not Transcendence, for the Historian of Science," *Isis,* 1991, *82:* 71–86; and Max Dresden, *H. A. Kramers: Between Tradition and Revolution* (New York: Springer-Verlag, 1987).

Steven Shapin, a leader of the social construction school, quoted that statement but could not refute it with any specific examples.[5]

One way to test an assertion about social construction would be to try to reproduce the calculation or experiment, to see if nature or mathematics yields the reported result. In doing so, one is behaving as a scientist—or at least one must enlist the expertise of a scientist to perform the test. Of course it cannot be assumed that the modern replication yields a true fact about nature, and that the original result must be wrong if it is different; I suggest only that replication can provide additional relevant evidence. *not relevant – not address pts of entry*

I. FROM WHIG TO PRIG?

The phrase "whig interpretation of the history of science" was apparently introduced by George W. Stocking, Jr., as an explicit application of Herbert Butterfield's characterization of writings on British political history. Stocking quoted Butterfield's definition: "The tendency in many historians to write on the side of Protestants and Whigs, to praise revolutions provided they have been successful, to emphasize certain principles of progress in the past and to produce a story which is the ratification if not the glorification of the present."[6] *1965*

Stocking identified the whig interpretation as a version of *presentism*—the term then being used by historians for the general tendency to see the past in terms of the present—and contrasted it to *historicism,* which he defined as "commitment to the understanding of the past for its own sake." While admitting that even professional historians cannot completely avoid presentism, Stocking suggested that scientists are more likely to judge the history of science in terms of present-day science:

> There is a sort of implicit whiggish presentism virtually built into the history of science. . . . However disillusioned we may be with the idea of progress in other areas, however sophisticated in the newer philosophy of science, most of us take it for granted that the development of science is a cumulative ever-upward progress in rationality. Indeed, George Sarton, long-time doyen of historians of science, described his study as "the only history which can illustrate the progress of mankind" because "the acquisition and systematization of positive knowledge are the only human activities which are truly cumulative and progressive."

Stocking also asserted that Thomas S. Kuhn's *Structure of Scientific Revolutions* marked the beginning of a trend away from whiggism in the history of science, even though it was "imperfectly historicist in its focus on the inner development of science" to the neglect of external factors.[7]

[5] Stephen Cole, *Making Science: Between Nature and Society* (Cambridge, Mass.: Harvard Univ. Press, 1992), p. 81: and Steven Shapin, "Mertonian Concessions" (rev. of Cole), *Science,* 1993, *259*:839–841.

[6] George W. Stocking, Jr., "On the Limits of 'Presentism' and 'Historicism' in the Historiography of the Behavioral Sciences," *Journal of the History of the Behavioral Sciences,* 1965, *1*:211–218; and Herbert Butterfield, *The Whig Interpretation of History* (London: Bell, 1931), p. v. See also W. F. Bynum, E. J. Browne, and Roy Porter, eds., *Dictionary of the History of Science* (Princeton, N.J.: Princeton Univ. Press, 1981), article "Whig History," pp. 445–446.

[7] *Ibid.,* pp. 213, 214; and Thomas S. Kuhn, *The Structure of Scientific Revolutions* (Chicago: Univ. Chicago Press, 1962; 2nd ed., 1970).

In 1968 Kuhn defined the "maxims of the new internal historiography" which had for the previous thirty years "increasingly guided the best interpretive scholarship in the history of science":

> Insofar as possible (it is never entirely so, nor could history be written if it were), the historian should set aside the science that he knows. His science should be learned from the textbooks and journals of the period he studies. . . . Dealing with innovators, the historian should try to think as they did. . . . [T]he historian should ask what his subject thought he had discovered and what he took the basis of that discovery to be. And in this process of reconstruction the historian should pay particular attention to his subject's apparent errors, not for their own sake but because they reveal far more of the mind at work than do the passages in which a scientist seems to record a result or an argument that modern science still retains.[8]

Kuhn is himself a former scientist, with a Ph.D. in physics and at least two research publications in solid-state theory.[9] Yet he seems to imply that this background is a handicap in understanding the physics of earlier centuries. But without this background, how would he distinguish "his subject's apparent errors" from what "modern science still retains"?

Other historians of science shared Kuhn's view. Robert Fox, now a professor at Oxford, wrote in 1970:

> A knowledge of science, which is most frequently cited as a fundamental requirement for the historian of science, is in reality no more than an auxiliary tool in a study whose basic skills are those of the historian. . . . [F]or most work in the history of science, whether at the undergraduate or postgraduate level, [a background in science] is emphatically not essential; in fact, as I shall argue, such a background can even be positively harmful.[10]

There is a curious echo here of the claim by Robert Livingston Schuyler, criticizing present-minded historians:

> I submit that there is no reason at all why the historian, considered as such, should be concerned with what is going on around him. If, for example, a historian of the medieval English parliament knew nothing of the history of parliament since, say, 1689, would he for that reason be in a less favorable position, for investigating its beginnings than if his head were full of nineteenth-century reform bills and the present-day cabinet system? I think not. Perhaps he would be in a more favorable position, for he would be less likely to be led astray by modern developments.[11]

Butterfield's demand that the historian study the past for its own sake without reference to the present has proved hard to satisfy. In fact Butterfield himself did not

[8] Thomas S. Kuhn, "History of Science," *International Encyclopedia of the Social Sciences* (New York: Macmillan, 1968), Vol. 14, pp. 74–83, on pp. 76–77.

[9] Thomas S. Kuhn, "An Application of the W. K. B. Method to the Cohesive Energy of Univalent Metals," *Physical Review,* 1950, series 2, *79*:515–519; and Kuhn and J. H. Van Vleck, "A Simplified Method of Computing the Cohesive Energies of Monovalent Metals," *ibid.,* pp. 382–388.

[10] Robert Fox, "The History of Science," in *History: An Introduction for the Intending Student,* ed. H. J. Perkin (London: Routledge & Kegan Paul, 1970), pp. 173–186, on p. 175.

[11] Robert Livingstone Schuyler, "Some Historical Idols," *Political Science Quarterly,* 1932, *47:* 1–18, on p. 16. I owe this reference to Peter Novick, *That Noble Dream: The "Objectivity Question" and the American Historical Profession* (New York: Cambridge Univ. Press, 1988), p. 273, who says Schuyler "had been much impressed" with Butterfield's 1931 book.

consistently obey it. In 1944 he admitted that the whig interpretation is here to stay, and it is futile to go on attacking it:

> Those who, perhaps in the misguided austerity of youth, wish to drive out that whig interpretation, (that particular thesis which controls our abridgment of English history,) are sweeping a room which humanly speaking cannot long remain empty. They are opening the door for seven devils which, precisely because they are newcomers, are bound to be worse than the first. We, on the other hand, will not dream of wishing it away, but will rejoice in an interpretation of the past which has grown up with us, has grown up with the history itself, and has helped to make the history. . . . [W]e will celebrate this whig inheritance of ours with a robust but regulated pride.[12]

As for history of science, it has been noted by several readers that Butterfield's own book on the history of modern science is as whiggish as anything from the pen of George Sarton. While stressing the importance of understanding earlier theories and the context in which they were replaced by modern ones, Butterfield insists that the events of the 1940s helped historians to see the true significance of the Scientific Revolution of the seventeenth century more clearly than was possible at the beginning of the twentieth century.[13]

Although general historians and historians of science uncritically accepted Butterfield's original thesis for three or four decades, it gradually became clear that it suffers from severe defects. These may be summarized under two headings: first, it is impossible in principle to write a nonpresentist history of anything; second, nonpresentist history is not even a desirable ideal toward which to strive. Moreover, it seems perverse to deny that science *has,* in most respects, made substantial progress in understanding how nature works, whatever one may think of the uses to which that understanding has been put.

Before discussing these objections, we must decide: What is "nonwhiggish history" and what should we call it? Butterfield asserted that "there is not anything worth the name of 'the tory interpretation of English history,'"[14] although in the history of science one might give the label "tory interpretation" to the view that the quality of science has declined since a putative golden age (see, e.g., the writings of Clifford Truesdell). Studying the past for its own sake used to be called *antiquarianism,* but that word has acquired a negative connotation. Similarly, the term *historicism* has been so abused that it no longer has a generally accepted coherent meaning. Edward Harrison suggests that the extreme antiwhig historian should be called a "prig," denoting the "narrow-minded superiority" of those who make a virtue of their ignorance about modern science.[15]

For lack of a better word, *contextualism* seems to have emerged as a positive label for the doctrine that one should study the ideas and theories of a period in terms of the scientific knowledge and general culture of that period, without regard to what came afterwards. (Some historians would also limit their context to a local

---

[12] Herbert Butterfield, *The Englishman and His History* (Cambridge: Cambridge Univ. Press, 1994), pp. 3–4.
[13] Herbert Butterfield, *The Origins of Modern Science (1300–1800)* (1949), rev. ed. (New York: Macmillan, 1957), p. 148.
[14] Butterfield, *Englishman* (cit. n. 12), p. 2.
[15] Edward Harrison, "Whigs, Prigs, and Historians of Science," *Nature,* 1987, *329*:213–224. See comments by S. C. McCluskey and P. J. Bowler, "Historians, Whigs, and Progress," *Nature,* 1987, *330*:598.

community). Progress in historiography is to be measured by the extent to which the historian is liberated from "constraints imposed on inquiry by presuppositions about the past" stemming from knowledge of the present.[16] The contents of the new journal *Science in Context* may be taken to exemplify one currently popular version of contextualist history of science.[17]

It is also essential to distinguish between whiggism and the more general term *present-mindedness,* or *presentism* for short. The present-minded historian asks *questions* about the past inspired by concerns of the present; the whig historian gives *answers* that are distorted by those concerns. One can reject whiggism without completely rejecting presentism.[18] Thus it is presentist to ask why Gregor Mendel's work was ignored by his contemporaries (a historian writing before 1899 would have had no reason to ask that); one can see in the historiography of genetics a shift from whiggish to nonwhiggish answers to that question, yet the question itself is still considered legitimate.

Let us reserve the term *priggism* for the view that the historian should be exclusively contextual. The prig historian would insist that if a crucial theorem in Isaac Newton's *Principia* rests on a mathematically fallacious derivation but no one caught the error during Newton's lifetime, it is improper to point that out now; if Robert Boyle's conclusion about the transmission of sound at low pressures is vitiated by his misunderstanding of the physics of his vacuum apparatus, it is unacceptably whiggish to say so; and if Mendel's contemporaries ignored him, so should we! All that matters, for the prig historian, is what scientists thought to be true or important at the time. There are probably few if any prig historians of science; this is simply the position implied by taking seriously some of the more extreme antiwhig statements.

The first objection can then be stated thus: There is no unique and consistent way to define the context of an event in the history of science, just as there is no unique and consistent way to define the context of a text in literary analysis.[19] The choice of what context to consider has to be made by the individual historian, and there is no objective basis for preferring one choice to another. Contextualism assumes there is an objective past that can be uncovered by the historian, and an objective historical description that, once properly written, will not have to be rewritten by future historians.[20] But since the dogma that events in the present have an objective existence

---

[16] Kieran Egan, "Progress in Historiography," *Clio,* 1979, *8*(2):195–228, on p. 221.

[17] See the review of *Science in Context* by John Hendry, *Brit. J. Hist. Sci.,* 1989, *22*:102–105. Hendry also states that sociologically grounded works such as those of Martin Rudwick, Andrew Pickering, and especially Simon Schaffer and Steven Shapin, at present constitute the mainstream of the history of science.

[18] In developing this distinction (though these authors do not define terms quite as I do), I have been influenced by the arguments of David Hull, "In Defense of Presentism," *History and Theory,* 1979, *18*:1–15; Loren Graham, "Why Can't History Dance Contemporary Ballet? or, Whig History and the Evils of Contemporary Dance," *Science, Technology & Human Values,* 1981, *34*:3–6; Helge Kragh, *An Introduction to the Historiography of Science* (New York: Cambridge Univ. Press, 1987); Adrian Wilson and T. G. Ashplant, "Whig History and Present-Centred History," *Historical Journal,* 1988, *31*:1–16; Ashplant and Wilson, "Present-centred History and the Problem of Historical Knowledge," *ibid.,* pp. 253–274; and Ernst Mayr, "When Is Historiography Whiggish?" *Journal of the History of Ideas,* 1990, *51*:301–309.

[19] See David Harlan, "Intellectual History and the Return of Literature," *American Historical Review,* 1989, *94*:581–609.

[20] This is not a logical consequence of contextualism, and indeed one could just as well argue, especially in view of the diversity of contexts invoked by historians, that contextualism problematizes the notion of a unique objective past (I owe this point to George Stocking). Yet almost every individ-

no transcendence

independent of the observer has been resoundingly refuted (at least for events involving individual atoms) by modern physics, *a fortiori* the objective existence of events in the past, which in any case can be investigated only through relics surviving in the present, cannot be taken for granted. It is the observer (including the historian) who confers reality on the past as well as the present.[21]

For readers not convinced by the argument from physics, a linguistic proof may be more persuasive. Gary Hardcastle uses William V. O. Quine's thesis of the indeterminacy of translation to show that since any historical interpretation of a document or event involves "translation," based on some presentist assumption, there is no single correct answer to historical questions about science.[22]

Many commentators on the whig interpretation have pointed out that it is impossible to avoid presentism completely, since the very selection of topics to be studied is based on some present-minded preconception. Why do many more historians choose to study the history of astronomy than of astrology? Of Newton's theory than of Descartes's? Of Charles Darwin's theory than of Jean-Baptiste de Lamarck's? Having made a present-minded choice of topic, and of questions to ask about that topic, one can still try to avoid whiggish answers.

It is very difficult to carry through a complete contextualist criticism of a whiggish history without at some point invoking whiggish arguments. A recent example is David Leveson's indictment, on grounds of whiggism, of Claude Allègre's book on the development of plate tectonics. In discussing Allègre's evaluation of objections to Alfred Wegener's theory in 1926, Leveson complains that Allègre dismisses several objections which "are still considered valid. No one now proposes that sial drifts through sima, or that in doing so mountains are thrown up. Pleistocene moraines are not correlated across the Atlantic."[23]

A more interesting example is Kuhn's book on the origin of quantum theory.[24] He attacks the widely held view that Max Planck introduced the quantum hypothesis in 1900, arguing that this view arose from reading Planck's paper in the light of later developments. If we read the 1900 paper in the context of research by Planck and others before 1900, we can see that he did not really postulate a physical discontinuity in energy in that paper. But to make his argument plausible, Kuhn presents a detailed technical analysis that relies in part on twentieth-century understandings of statistical concepts, including a critique of Ludwig Boltzmann's derivations that uses standards of rigor rarely if ever applied by physicists in the nineteenth century.[24] Kuhn's book is a brilliantly successful contribution to the history of science, somewhat present-minded but not whiggish, and very much dependent on his own

---

ual historian tends to write as if the chosen context is unproblematic. This is noted explicitly by Novick, *That Noble Dream* (cit. n. 11).

[21] John Archibald Wheeler, "Frontiers of Time," in *Problems in the Foundations of Physics,* ed. N. Toraldo di Francia (Amsterdam: North-Holland, 1979), pp. 395–497.

[22] Gary L. Hardcastle, "Presentism and the Indeterminacy of Translation," *Studies in History and Philosophy of Science,* 1991, 22:321–345.

[23] David J. Leveson, "Whiggism and Its Sources in Allègre's *The Behavior of the Earth," Earth Sciences History,* 1991, 10:29–37, on p. 30; and Claude Allègre, *The Behavior of the Earth: Continental and Seafloor Mobility* (Cambridge, Mass.: Harvard Univ. Press, 1988). See also D. R. Oldroyd, "Sir Archibald Geikie (1835–1924), Geologist, Romantic Aesthete, and Historian of Geology: The Problem of Whig Historiography of Science," *Annals of Science,* 1980, 37:441–462.

[24] T. S. Kuhn, *Black-Body Theory and the Quantum Discontinuity, 1894–1912* (New York: Oxford Univ. Press, 1978), pp. 43–70, 100–101, 107.

expertise as a theoretical physicist. Ironically, it has received little recognition from other historians of science, in part because of the influence of Kuhn's own historiographic writings, which helped to make this kind of technical analysis unfashionable.

I turn to the second objection: presentism is not only inescapable but may even be desirable. Here we can begin with Butterfield's view: "History is not the study of origins; rather it is the analysis of all the mediations by which the past was turned into our present." Butterfield did not object to asking present-minded questions like "How did religious liberty arise" but rather to the assumption that one should try to identify and praise a particular person or persons in answer to the question, "To whom do we owe our religious liberty?" So he does not really disagree with A. Rupert Hall, who argues that questions of the former kind, which are now considered whiggish, are perfectly legitimate ones for the historian to consider:

> The most obvious of all historical questions is: "How did we arrive at the condition we are now in?" To argue that historical study cannot answer that question with finality [is] quite true but . . . also irrelevant if history is to be written. The question is put, and the questioner will find an answer somewhere; if academic historians are silent he or she will seek an answer in other ways. For if there is an academic supply of historical, literary and philosophical reading, there is also a non-academic or general demand for historical, literary and philosophical writing. . . . [I]t is surely deplorable for the intellectual health of a society if the two are totally out of step, and especially if the claims for the intellectual purity, rigour and rectitude of the academic supply are elevated to such extremes that the general demand goes entirely unrecognized by those best qualified to satisfy it. Academia does not exist solely for the sake of delighting and gratifying itself.[25]

But there are also academic justifications for abandoning extreme contextualism or priggism. Some historians have always accepted the thesis that the significance of a person or event depends at least partly on how it affects and is perceived by later generations. Some of the most exciting historical research in recent years started by taking a second look at the past with consciousness raised by late-twentieth-century sensitivity to the concerns of women, minorities, and the "common people." In particular, historians of science have asked fascinating questions about the "gender" of science, completely overlooked by earlier contextual studies.[26]

Advocates of "postmodernism" suggest that contextualism has become a sterile dogma, and that literary analysts, anthropologists, or hermeneutic historians like Hans-Georg Gadamer offer a more enlightened conception, involving the active participation of the historian in creating history. Michel Foucault insisted that "historical descriptions are necessarily ordered by the present state of knowledge." A typical postmodernist formulation is that of F. R. Ankersmit: "The essence of the past is not, or does not lie in, the essence of the past. It is the scraps, the slips of the tongue, the *Fehlleistungen* of the past, the rare moments when the past "let itself go," where we discover what is really of importance to us." According to Ankersmit, social history is still futilely seeking the essence of history, looking for the trunk of the

---

[25] Butterfield, *Whig Interpretation of History* (cit. n. 6), on pp. 46–47; and A. Rupert Hall, "On Whiggism," *History of Science,* 1983, *21*:45–59, on p. 54.

[26] See, e.g., Carolyn Merchant, "*Isis'* Consciousness Raised," *Isis,* 1982, *73*:398–409; Evelyn Fox Keller, *Reflections on Gender and Science* (New Haven: Yale Univ. Press, 1985); and Londa Schiebinger, *The Mind Has No Sex? Women in the Origins of Modern Science* (Cambridge, Mass.: Harvard Univ. Press, 1989).

tree; the postmodernist looks instead for the leaves—even though, now that it is autumn for Western historiography, the leaves are blowing off the tree. Their importance is not the place they once had on the tree but "the pattern we can form from them *now*." Thus we should "let go" of the historical context. "History here is no longer the reconstruction of what has happened to us in the various phases of our lives, but a continuous playing with the memory of this. The memory has priority over what is remembered. . . . The time has come when we should *think* about the past, rather than *investigate* it." [27]

This extravagant thesis provoked a revealing reply from Perez Zagorin. Postmodernist history must be rejected because it abandons even the pretense of objectivity; it fails to retain history's distinction between fact and fiction: "The historical work presents itself as consisting, to a great degree, of facts and true or probable statements about the past." [28]

I would not argue that postmodernism offers a satisfactory or even a coherent alternative to contextualism. Indeed, I have no strong objection to contextualism itself as one approach among several. I simply reject its claims to a monopoly over historical writing. The recent criticisms of contextualism do not show that it should be abandoned, but may serve the useful function of forcing historians of science to rethink and reformulate their assumptions, even if for no other reason than to avoid appearing hopelessly out of date to their colleagues in the humanities.

Liberation from the contextualist straitjacket could also allow historians of science to remedy one of the major shortcomings of our profession: failure to meet the widespread demand for broad synthetic overviews of the subject suitable for students and the public. This failure might be blamed on Butterfield's call for detailed studies of special problems as the only legitimate form of professional historical writing and his denigration of surveys as misleading "abridgments"—although he himself produced the successful survey of the history of science noted above. [29] To adapt Albert Einstein's famous advice: If you want to find out what methods historians use, don't listen to their words, fix your attention on their deeds! [30]

Finally, if historians of science can get over their fear of presentism, they could participate more freely in the discussion of contemporary problems, where politicians and others pontificate about the "lessons of history." Forman notes that Hunter Dupree, a highly respected historian of American science, was not reluctant to offer history-based advice on science policy. Lewis Pyenson reminds us that when our discipline was founded, one of its major goals was supposed to be "clarity to act in the present on the basis of an understanding of the past." In particular, Loren Graham suggested that historians of science could play a useful role on advisory boards

[27] Michel Foucault, *The Archaeology of Knowledge* (New York: Pantheon, 1972), on p. 5; and F. R. Ankersmit, "Historiography and Postmodernism," *Hist. Theory,* 1989, *28:*137–153, on pp. 148, 150, 152. See also Dominick LaCapra, *Rethinking Intellectual History: Texts, Contexts, Language* (Ithaca, N.Y.: Cornell Univ. Press, 1983); Harlan, "Intellectual History" (cit. n. 19); and Gabrielle M. Spiegel, "History, Historicism, and the Social Logic of the Text in the Middle Ages," *Speculum,* 1990, *65:*59–86.

[28] Perez Zagorin, "Historiography and Postmodernism: Reconsiderations," *Hist. Theory,* 1990, *29:* 263–274 on p. 272. See also Ankersmit, "Reply to Professor Zagorin," *ibid.,* pp. 275–296.

[29] Butterfield, *Whig Interpretation of History* (cit. n. 6); the survey is Butterfield, *Origins* (cit. n. 13). See also Hall, "On Whiggism" (cit. n. 25); and C Hakfoort, "The Missing Syntheses in the Historiography of Science," *Hist. Sci.,* 1991, *29:*207–216.

[30] Albert Einstein, *On the Method of Theoretical Physics* (Oxford: Clarendon Press, 1933).

such as those dealing with medical ethics. In fact a historian, Nicholas Steneck, was recently appointed chair of a Public Health Service Advisory Committee on Scientific Integrity.[31]

II. HISTORY OF SCIENCE OR HISTORY OF SCIENTISTS?

In 1966 Hunter Dupree announced to the American Historical Association that the field of the history of American science had "found itself." It had done so, he claimed, because historians, following the advice of Arthur M. Schlesinger, Sr., had recognized the central role of science in recent American history and had refused to let their ignorance of science prevent them from writing about it. Rejecting Sarton's demand that historians should be "deeply familiar with at least one branch of today's science" and "should have a more superficial acquaintance with various other branches," Dupree argued that they could dispense with such scientific training, since they are "not interested in the subject matter of the sciences per se . . . the history of science properly does not concern itself with the things of science: the plants, the animals, the molecules, the atoms, the ether, the quanta, or even the law or the equation." Their subject was the scientists and their relation to society: "The only object of study in the history of science is *Homo sapiens,* and since a scientist without communication is hard to conceive of, it is *Homo sapiens* in a social context that is the sole object of the historian's study of science."[32]

When Dupree said he was "not interested in the subject matter of the sciences," he meant not *professionally concerned* in his capacity as a historian; many of his works show that he does have an interest, in the sense of *curiosity,* in scientific matters. But some younger historians are "not interested" in a more literal sense— and not only indifferent but hostile. David Knight says that he and his colleagues are "tired of meeting the assumption that they must be interested in how things turned out; as though the student of the Austro-Hungarian Empire should necessarily be concerned about local government in present-day Budapest." For them, science is no longer the heroic enterprise it had seemed in the nineteenth century; now it has an "ambiguous" reputation. In a culture where there is widespread "alarm and unease about science, . . . there is no reason why its historian should also be its apologist."[33] Skepticism about the moral stature (and epistemological authority) of science pervades the comments of a dozen British historians invited to address the question "What Is the History of Science?" in the magazine *History Today.*[34]

Is alienation from science a prerequisite for objective study of its history? Anglo-American historians of science will not usually say that in print, but a Japanese

[31] Paul Forman, "1990 Sarton Medal Citation," *Isis,* 1991, *82*:281–283; Lewis Pyenson, "What Is the Good of History of Science?" *Hist. Sci.,* 1989, *27*:353–389; Graham, "Why Can't History Dance Contemporary Ballet?" (cit. n. 18); and David L. Wheeler, "NIH Director Seeks Changes in Office of Scientific Integrity," *Chronicle of Higher Education,* 24 July 1991, pp. A1, A6.

[32] This paragraph is adapted from an earlier article: Stephen G. Brush, "Looking Up: The Rise of Astronomy in America," *American Studies,* 1979, *20*(2):41–61. For the Dupree and Sarton quotations see A. Hunter Dupree, "The History of American Science—A Field Finds Itself," *Amer. Hist. Rev.,* 1966, *71*:863–874, on pp. 865, 869. For Schlesinger's advice see Schlesinger, "An American Historian Looks at Science and Technology," *Isis,* 1946, *36*:162–166.

[33] David M. Knight, "The History of Science in Britain: A Personal View." *Zeitschrift für Allgemeine Wissenschaftstheorie,* 1984, *15*:343–353, on p. 344.

[34] See, *e.g.,* Roger Cooter, "What Is the History of Science?" *History Today,* April 1985, No. 35, pp. 32–33, and following articles.

scholar, Shigeru Nakayama, has frankly admitted that for him and his colleagues history of science is "a subject for the frustrated." A person who is content with the way science is done may not be "a good critical historian of science" but will only write "a self-congratulatory narrative" or become "a dull, bureaucratic archivist"; those who reject the norms of the scientific community can borrow the authority of history to support their critique of that community.[35]

But this creates a dilemma for a historian who needs the cooperation of scientists. According to Joan Bromberg, "historians of modern physics have pusillanimity as an occupational hazard. We depend upon the good graces of the scientists we treat for our raw material—the interviews, access to archival and personal collections, even, occasionally, our funding. The temptation to treat scientists with kid gloves is enormous."[36]

This temptation is resisted by Paul Forman, who proclaims independence from science as crucial for the historian. Forman argues that scientists relinquish moral responsibility for their work by embracing the "transcendence" of science, but historians must not do so. The historian should not accept the scientist's judgment as to what is true or important, based on appeal to a transcendent reality, but instead should undertake an independent analysis based on "mundane factors and human actors." When scientists write history, they celebrate the success of science, but this "always involves some degree of mendacity" which "must inevitably prove damaging to historiographic practice." Since, Forman asserts, scientific knowledge is socially constructed, the historian "must focus either on social problems of science or on science as a social problem."[37]

Forman had earlier carried his attack on science-history into the enemy camp with a slashing critique of Jagdish Mehra's work in the widely read journal *Science*. Reviewing the first four volumes of *Historical Development of Quantum Theory*, written by Mehra with Helmut Rechenberg, Forman accused Mehra of "intellectual poverty, pompous pretension, depreciation of the quantity and significance of the extant historical writing in the field." Like those "without experience in historical research," Mehra (according to Forman) overestimates the value of recollections of the major actors. He assumes "that only one who was there . . . could know the true but secret history of his field." The historian, in contrast to the scientist, wants independence from the judgments of the participants. Forman faults Mehra and Rechenberg for using few of the many extant letters that could provide a more accurate record, for being uncritical of their sources and failing to give enough documentation for them, and for using the work of other historians without adequate credit.[38]

---

[35] Nakayama Shigeru, "The History of Science as a Subject for the Frustrated," in *Science and Society in Modern Japan: Selected Historical Sources,* ed. Nakayama *et al.* (Tokyo: Univ. Tokyo Press, 1974), pp. 3–16 (previously published as "The Externalist Orientation of Japanese Historians of Science," *Japanese Studies in the History of Science,* 1972, *11*:1–10). See also Morris Fraser Low, "The History of Japanese Physics and the Rise of the Scientist-Historian," review of *Proceedings of the Japan–USA Collaborative Workshops on the History of Particle Theory in Japan, 1935–1960: Japan–USA Collaboration, Second Phase, Historia Scientiarum,* 1989, *36*:117–120.

[36] Joan Bromberg, "When Physicists Served Fascism," *Science and Nature,* 1982, No. 5, pp. 143–145, on p. 145.

[37] Forman, "Independence, Not Transcendence" (cit. n. 4), p. 83, attributing the phrases and some of the arguments to Jerome Ravetz, Sal Restivo, and Robert Merton.

[38] Paul Forman, "A Venture in Writing History," *Science,* 1983, *220:*824–827, reviewing Jagdish Mehra and Helmut Rechenberg, *The Historical Development of Quantum Theory,* 5 vols. (New York: Springer-Verlag, 1982–1987).

Forman is not simply attacking a particular piece of science-history that he considers badly done; he makes it clear that he considers these shortcomings to be characteristic of *most* history written by scientists: "For scientists history is not the field upon which they wrestle for truth, but principally their field of celebration and self-congratulation. . . . History written in celebration of heroic ages and agents can scarcely avoid becoming propaganda pandering to the scientists' *amour propre*. To this danger Mehra has wholly succumbed." Forman concluded by scolding those who, just because they were "pleased and flattered" by the attention of Mehra and Rechenberg, supported their project—an insult which drew an indignant response from one of those supporters, Ilya Prigogine.[39]

Scientists have been more polite and less vociferous in their criticisms of non-scientists who write history of science. In 1950 Einstein told Robert Shankland: "Nearly all historians of science are philologists and do not comprehend what physicists are aiming at, how they thought and wrestled with their problems. Even most of the work on Galileo is poorly done."[40] At a 1967 conference of historians and physicists, all the physicists shook their heads in disagreement when a historian suggested that in the 1930s social factors rather than scientific merit determined access to accelerators.[41] Commenting on works by historians, scientists occasionally lament the absence or inaccuracy of the description of science and the (to them) inappropriate emphasis on personalities and social and institutional history. An engineer put it more strongly: reading a biography that fails to explain its subject's technical achievements is like "a biography of Babe Ruth written for people who have never seen a game and who aren't interested in learning much about the rules."[42]

The strongest criticism of writings by nonscientist historians has come from other nonscientist historians, especially Charles Gillispie. Delivering the Sarton Lecture at the annual meeting of AAAS in 1980, Gillispie questioned the profession's shift away from analysis of the substance of science toward discussion of cultural influences. The text of this lecture has not been published but Gillispie supplied the following abstract:

[39] Forman, "Venture"; and Ilya Prigogine, "History of Quantum Theory," *Science*, 1983, *221*:604. Prigogine did review the Mehra-Rechenberg books quite favorably, in *Foundations of Physics*, 1984, *14*:275–277, and 1987, *17*:1131–1136. In his second review he remarked that they show "the grand forward march" of quantum mechanics and inform the reader about "a turning point in the history of mankind." Another scientist also praised the work and was especially pleased by the use of interviews, which had been turned into "accounts that ring very true"; see Nicholas Kemmer, "Jump in Quantum History," *Nature*, 1988, *332*:745–746.

[40] R. S. Shankland, "Conversations with Albert Einstein" *American Journal of Physics*, 1963, *31*: 47–57.

[41] Charles Weiner, ed., *Exploring the History of Nuclear Physics* (New York: American Institute of Physics, 1972), p. 47.

[42] M. Granger Morgan, "The Wizard of Schenectady: An Unresolved Portrait," *American Scientist*, 1993, *81*:182–183, on p. 182. Cf. Eugene S. Rochow, letter to the editor, *Isis*, 1989, *80*:664; A. G. W. Cameron, review of *Nucleus: The History of Atomic Energy of Canada Limited*, by Robert Bothwell, *Physics Today*, 1989, *42*(5):78–79; and G. W. Wetherill, "Asteroid Paradox," *Science*, 1990, *247*: 1386–1387. Historians of science may resent criticism of their work on such grounds: see, e.g., Geoffrey Cantor, letter to the editor, *British Society for the History of Science Newsletter*, Sept. 1991, pp. 30–31. For similar complaints by nonscientist historians see "Science in Medicine," *Osiris*, 1985, *1*:37–58, on pp. 37–38, 48–49; Leonard Wilson, "Medical History without Medicine," *Journal of the History of Medicine*, 1980, *35*:5–7; I. B. Cohen, review of *The Launching of American Science, 1846–1876*, by Robert Bruce, *Nature*, 1987, *329*:209–210; R. W. Home, review of *Entstehung des modernen Systems wissenschaftlicher Disziplinen: Physik in Deutschland, 1740–1890*, by Rudolf Stichweh, *Social Studies of Science*, 1990, *20*:761–763; and Alan E. Shapiro, review of *Reappraisals*

In the last decade scientists have been subjected to a barrage of summonses to every sort of responsibility—ethical, environmental, economic, technological, social, political, military, and whatnot. In writings on the history of science, emphasis on the effects of science has produced such a shift in discourse that a large proportion of scholarship is now addressed to institutional and social factors accompanying the development of science rather than to technical and intellectual context. The change in fashion is in keeping with a widespread politicization and socialization of historical scholarship in general, and indeed of all scholarship—whether for good or for ill, the future may reveal more clearly than the present. In any case, the picture of science is drawn in political and social brush strokes. It conveys little of what science has achieved in the way of knowledge or even technique, and scientists are treated as actors in the political and social process like anybody else. The Sarton lecturer has no doubt that there is much to be learned from all this, but does sometimes feel concerned lest matters go too far and the baby drown in the bath. The lecture will be a modest plea to scientists to pay attention to what historians—and other social scientists—are making of their enterprise, and will venture to urge upon them yet another responsibility: that of exercising a measure of vigilance, at least over the references to technical matters that even the most externally minded commentators cannot altogether avoid, and perhaps more largely over the impression that is conveyed to the public of the experience of being a scientist.[43]

The lecture was reported in *Science* under the inflammatory headline, "History of Science Losing Its Science,'" with the opening sentence: "Once a highly respected field that focused on the conceptual evolution of scientific ideas, the history of science is losing its grip on science, leaning heavily on social history, and dabbling with shoddy scholarship." That report provoked a letter from the historian Robert Kohler, who defended the standards of the profession as being different from those of previous generations but not inferior; he did not directly address the issue of competence or interest of historians in dealing with the technical aspects of science.[44]

Another historian, Nathan Reingold, discussed the controversy at greater length a year later, expressing his surprise at "the firm support of a few historians in private conversations" for Gillispie's position but insisting that "a majority of the teachers and graduate students in the field have scientific backgrounds which inevitably influence their work." While conceding that the proportion of internalist work in history of science might have declined from more than 80 percent in the period 1955–1960 to less than two thirds, he suggested that part of the increase in nontechnical writings was generated by economists, sociologists, philosophers, and so forth, rather than by professional historians of science. More significantly, after describing the "emerging consensus in the history of science" as reflecting "influences from history, the social sciences, and the philosophy of science," Reingold cited as the single best example of this consensus a book, *Scientists in Power,* written by Spencer Weart—a historian of science who took a doctorate in astronomy.[45]

---

of the Scientific Revolution, ed. David C. Lindberg and Robert S. Westman, *Science,* 1990, *250:* 1600–1601.

[43] Charles C. Gillispie, abstract for "Is the Inwardness of Science Extraneous to Its History?" Sarton Lecture at Annual Meeting of American Association for the Advancement of Science, 1980 (unpublished).

[44] William J. Broad, "History of Science Losing Its Science," *Science,* 1980, *207:*389; and Robert Kohler, "History of Science: Perceptions," *Science,* 1980, *207:*934–935. See also Gillispie, "History of Science: Perceptions," *ibid.,* p. 934.

[45] Nathan Reingold, "Science, Scientists, and Historians of Science," *Hist. Sci.,* 1981, *19:*274–283, quoting from pp. 274, 276, 280.

What Einstein wanted from the history of science, and what at least one reader of Reingold's paper found missing there, was an understanding of "the intense desire of research scientists to produce new basic discoveries"—an "intense excitement in the workings of nature" which takes priority over any future consequences of the discoveries.[46] That does not mean that historians must approve of the values of the scientist or share in what Forman calls the "transcendence" of science. It does mean that there is a place in our discipline for *some* historians who have enough personal experience in scientific research to describe what it feels like. As Gerald Holton interprets Einstein's statement, we need "a particular kind of historical sense, one that largely intuits how a scientist may have proceeded, even in the absence of 'the real facts' about the creative phase."[47]

III. DISCOVERY VERSUS CONSTRUCTION

The transcendence of scientific research relies on the implicit assumption that one is objectively uncovering true facts about the world, not just making them up. The independence of historical research includes the freedom to argue that facts as well as theoretical concepts are subjectively constructed with the help of philosophical preconceptions, social interests, and rhetorical devices. Scientists, like historians or anyone else, may be dismayed if such "construction" arguments undermine the credibility of their work. But it is also scientists, acting as or in cooperation with historians, who have a great deal to say about the validity of construction arguments.

For an earlier generation of historians of science, the most famous construction argument was Alexandre Koyré's claim that Galileo did not actually perform the experiments he mentioned in his dialogues, but instead described how they *must* come out on the basis of his Platonic presuppositions about how nature works. Koyré supported his claim by asserting that if Galileo had really done those experiments with the apparatus available to him, he would have obtained different results, or at least results not as accurate as he stated. But Koyré did not do the experiments either, relying on his own presuppositions about how nature works as well as about how Galileo worked. In the 1960s the discussion moved to a different plane when Thomas Settle and other historians of science with competence in physics tried to reproduce these experiments with the kinds of apparatus used in the seventeenth century. As a result of this contribution from science history, we now have a much better understanding of the role of philosophical preconceptions in Galileo's work, and the extent to which his results might have been constructed rather than discovered—apparently much less than Koyré thought.[48]

Science history has also played a role in the historiography of John Dalton's chem-

---

[46] M. A. B. Whitaker, "Science, Scientists, and History of Science," *Hist. Sci.,* 1984, *22*:421–424.

[47] Gerald Holton, *Thematic Origins of Scientific Thought: Kepler to Einstein,* rev. ed. (Cambridge, Mass.: Harvard Univ. Press, 1988), p. 346.

[48] Alexandre Koyré, *Études galiléennes* (Paris: Hermann, 1939); Koyré, *Metaphysics and Measurement* (Cambridge, Mass.: Harvard Univ. Press, 1968); Thomas Settle, "An Experiment in the History of Science," *Science,* 1961, *133*:19–23; Settle, "Galileo and Early Experimentation," in *Springs of Scientific Creativity,* ed. R. Aris *et al.* (Minneapolis: Univ. Minnesota Press, 1983), pp. 3–20; James MacLachlan, "A Test of an 'Imaginary' Experiment of Galileo's," *Isis,* 1973, *64*:374–379; MacLachlan, "Galileo's Experiments with Pendulums: Real and Imaginary," *Ann. Sci.,* 1976, *33*:173–185; and Stillman Drake, "Galileo's Experimental Confirmation of Horizontal Inertia: Unpublished Manuscripts (Galileo Gleanings XXII)," *Isis,* 1973, *64*:291–305.

ical atomic theory. J. R. Partington and Leonard Nash both tried to replicate one of Dalton's experiments; both concluded that since the results were not as reported by Dalton, they must have been adjusted later to illustrate the theory rather than providing the inspiration for it as Dalton claimed.[49] The issue here is not so much whether Dalton really constructed his experimental facts from his theory, but whether that issue can even be discussed by an historian who does not have some competence in chemistry.

Recently it has been suggested that Robert Brown did not actually observe the kind of particle-movement now named after him.[50] Again, the technical expertise of the scientist is needed to test this historical hypothesis.

It is taken for granted that professional historians can read original sources in the language in which they are written. But what about mathematics, the language of science? Ivor Grattan-Guinness points out that mathematics is simply ignored in most current writing on the history of science.[51] By default, historians of science have relinquished to practitioners not only the history of mathematics but topics like the development of general relativity, even though, as John Stachel puts it, "the skills of trained historians of science are badly needed, not least in the formulation of the right questions to be answered by future research."[52]

Even in the history of biology some mathematical competence is needed. A vexing question in Mendel scholarship is whether the venerable abbé "cooked" the data he reported in support of his theory. The question is asked only because a scientist, R. A. Fisher, pointed out that the data are, from a statistical viewpoint, too good to be true. Obviously a mathematician-historian like B. L. van der Waerden will have an advantage over most historians of biology in discussing this question.[53]

A more general problem for any attempt to write a social history of science is that social influences on the detailed process of research can be discerned only if one understands the process itself at a technical level.[54] The most successful and convincing studies of the social aspects of science are precisely those whose authors are willing to immerse themselves in the life of the laboratory, not as an anthropologist visiting from an alien culture but as a participant-observer. Whether or not the historian has originally been trained as a scientist, she or he needs to acquire some knowledge of how research was done by the scientists being studied. Those who lack this knowledge can still do useful work on other aspects of the history of science (as Schlesinger and Dupree argue) but cannot claim to provide a *complete* description of the scientific enterprise.

[49] J. R. Partington, "The Origins of the Atomic Theory." *Ann. Sci.*, 1939, *4*:245–281; and Leonard K. Nash, "The Origin of Dalton's Chemical Atomic Theory," *Isis*, 1956, *47*:101–116.

[50] Daniel Deutsch, "Did Robert Brown Observe Brownian Motion: Probably Not," *Bulletin of the American Physical Society*, 1991, *36*:1374; and John Rennie, "A Small Disturbance," *Scientific American*, 1991, *265*(2):20.

[51] I. Grattan-Guinness, "Does History of Science Treat of the History of Science? The Case of Mathematics," *Hist. Sci.*, 1990, *28*:149–173.

[52] John Stachel, introduction to *Einstein and the History of General Relativity*, ed. D. Howard and J. Stachel (Boston: Birkhäuser, 1989), pp. 1–4. See also B. Bertotti *et al.*, eds., *Modern Cosmology in Retrospect* (Cambridge: Cambridge Univ. Press), pp. xi–xx.

[53] R. A. Fisher, "Has Mendel's Work Been Rediscovered?" *Ann. Sci.*, 1936, *1*:115–137; Joan Fisher Box, *R. A. Fisher: The Life of a Scientist* (New York: Wiley, 1978), pp. 296–300; and B. L. van der Waerden, "Mendel's Experiments," *Centaurus*, 1968, *12*:275–288.

[54] F. L. Holmes, "The Fine Structure of Scientific Creativity," *Hist. Sci.*, 1981, *19*:60–70; and Holmes, *Lavoisier and the Chemistry of Life* (Madison: Univ. Wisconsin Press, 1985), pp. 501–502.

IV. CONCLUDING REMARKS

There is no doubt that much historical writing by scientists is whiggish or merely celebratory. Some scientists acknowledge this criticism but publish their memoirs anyway, leaving it to the reader to decide what has historical value.[55] Autobiographies, interviews, and efforts to preserve documents are all useful contributions that scientists can make to history without becoming historians themselves.[56] There is a vast amount of "amateur" writing on history of science which professional historians treat as evidence of how scientists perceive their own enterprise and how they justify it to the public, rather than as historical scholarship. But the issue here is the attitude of professional historians of science toward those scientists who have taken seriously the criticisms of science history noted above, and have made a conscious effort to appreciate (and to some extent adopt) the methods and attitudes of the historian. The ultimate justification for the participation of scientists in writing the history of science must rest on the quality of their *best* work. That work should not be denigrated if it starts from presentist questions, employs skills and knowledge from modern science, and ignores the current fad for social and institutional history. Nor should one overlook the fact that some historians with solid credentials in science have become leaders in social and contextual analysis.

In the past, science historians like Partington, Pierre Duhem, and Joseph Needham pioneered in mapping large regions of our subject—chemistry, the Middle Ages, and China, respectively. Without their efforts to build on, our knowledge of those regions would now be much less advanced than it is.

To assess the more recent contributions of historians who have earned doctorates in the sciences, one can use (with some caution) the membership directory of the History of Science Society and the indexes to its journals. Anyone may join the History of Science Society (by paying the dues), but only those papers that satisfy the scholarly criteria of the discipline can get published in *Isis* or *Osiris*. For example, among those who hold a doctorate in physics we find the following major contributors to the history of physics: Geoffrey Cantor, Allan Franklin, Peter Galison, Gerald Holton, Martin J. Klein, Thomas S. Kuhn, Arthur I. Miller, Andy Pickering, Sylvan S. Schweber, Daniel Siegel, Roger Stuewer, and M. Norton Wise. This list could easily be enlarged by including others with doctorates in physics who did not indicate that in the directory entry, and several who have published monographs or articles in other history of science journals.[57] But even as it stands, this list already includes the authors of a substantial portion of the best work (by any standards) in the history of modern physics; their publications at the same time display a striking diversity of styles and approaches. Probably a similar statement could be made by someone familiar with recent scholarship in the history of biology, geology, or any other science.

I would not argue that science historians who have published in history of science journals share any single historiographic perspective, or even that all of them make

---

[55] Walter H. Stockmayer, "When Polymer Science Looked Easy," *Annual Review of Physical Chemistry*, 1984, *35*:1–21.

[56] E. N. Hiebert, review of *Fifty Years of Neutron Diffraction*, ed. G. E. Bacon, *Archives Internationales d'Histoire des Science*, 1990, *40*:129–130.

[57] P. Thomas Carroll, ed., *Guide to the History of Science*, 8th ed. (Philadelphia: History of Science Society, 1992).

also PhD history of science
galison
wise

use of their scientific training in their historical research. But at least some of them can remedy one of the shortcomings of contextualism as it is often practiced: I mean the tendency to focus so strongly on what happened at one particular time, and on the events *before* that time, that one overlooks the impact of the events on the *subsequent* development of science. Daniel Siegel writes, in connection with the analysis of Maxwell's electromagnetic theory:

> The fantasy has been entertained of having the history of science written by individuals who do not know the modern theories, and who will therefore be able to enter into the spirit of past theories unprejudiced. Such individuals, however, would be ill situated to interpret the theories of the past for audiences acquainted with the modern theories, and would be ill equipped to discern patterns of development leading from past to present. It is just as well, then, that there are historians of science who are conversant with modern science.[58]

Siegel's goal is to understand Maxwell's theory in its nineteenth-century context, but in a more detailed way that can be connected with later developments.

To summarize: scientists have much to contribute to the history of science, and there are certain kinds of important questions that can be discussed only by those who have considerable technical background. Holmes has pointed out that "context" by itself is only a background that situates a foreground:

> The study of what is variously referred to as the "intellectual" history of science, the "internal dynamic," or the "cognitive" side of intellectual development is as fresh and new, as underdeveloped, as urgently in need of more concentrated, penetrating analysis as is the study of the "social dimension." . . . [T]he study of these subjects should remain at the heart of the discipline of the history of science; for it is only through a profound understanding of these subjects that we can know *what it is* that the various contexts surround.[59]

In this enterprise those scientists who are willing to learn historical methods and study original sources have a continuing and essential role to play.

---

[58] Daniel Siegel, "The Origin of the Displacement Current," *Historical Studies in the Physical and Biological Sciences,* 1986, *17*:99–146, on p. 101.
[59] Holmes, "Fine Structure" (cit. n. 54), on p. 60.

Bibliography
A Guide to the Life Sciences

I. BIBLIOGRAPHICAL, BIOGRAPHICAL, AND AUTOBIOGRAPHICAL RESOURCES

Allen, Garland E. "Life Sciences in the Twentieth Century." In *Teaching in the History of Science: Resources & Strategies*. Philadelphia: History of Science Society, 1989, pp. 13–22.

Bearman, David; John T. Edsall, editors. *Archival Sources for the History of Biochemistry and Molecular Biology: A Reference Guide and Report*. Philadelphia: American Philosophical Society, 1980.

Dictionary of Scientific Biography. Edited by Charles C. Gillispie. 16 vols. New York: Charles Scribner's Sons, 1970–1980. Supplement II. Edited by Frederic L. Holmes. 2 vols. 1990.

Edsall, John T.; David Bearman. "Historical Records of Scientific Activity: The Survey of Sources for the History of Biochemistry and Molecular Biology." *Proceedings of the American Philosophical Society*, 1979, *123*:279–292.

The Excitement and Fascination of Science: Reflections by Eminent Scientists. 2 vols. Palo Alto, Calif.: Annual Reviews, 1965, 1978.

Fox, Daniel M.; Marcia Meldrum; Ira Rezak, editors. *Nobel Laureates in Medicine or Physiology: A Biographical Dictionary*. New York: Garland Publishing, 1990.

Fruton, Joseph S. *A Bio-Bibliography for the History of the Biochemical Sciences since 1800*. Philadelphia: American Philosophical Society, 1990.

Glass, Bentley, editor. *A Guide to the Genetics Collections of the American Philosophical Society*. Philadelphia: American Philosophical Society Library Publication Number 13, 1988.

Kay, Lily E., editor. *Molecules, Cells, and Life: An Annotated Bibliography of Manuscript Sources on Physiology, Biochemistry and Biophysics, 1900–1960, in the Library of the American Philosophical Society*. American Philosophical Society Library Publication, Number 14. Philadelphia, 1989.

McGraw-Hill Modern Scientists and Engineers. 3 vols. New York: McGraw-Hill, 1980.

Nobel Prize Winners: An H. W. Wilson Biographical Dictionary. Edited by Tyler Wasson. New York: H. W. Wilson, 1987.

II. MODERN LIFE SCIENCES: GENERAL

Adams, Mark B., editor. *The Wellborn Science: Eugenics in Germany, France, Brazil, and Russia*. New York: Oxford University Press, 1990.

Allen, Garland E. *Life Science in the Twentieth Century*. New York: Wiley, 1975; Cambridge University Press, 1978.

Benison, Saul; A. Clifford Barger; Elin L. Wolfe. *Walter B. Cannon: The Life and Times of a Young Scientist*. Cambridge: Harvard University Press, 1987.

Benson, Keith R.; Jane Maienschein; Ronald Rainger, editors. *The Expansion of American Biology*. New Brunswick, N.J.: Rutgers University Press, 1991.

Bertalanffy, Ludwig von. *Modern Theories of Development*. Translated by J. H. Woodger. New York: Harper, 1962.

Bliss, Michael. *The Discovery of Insulin*. Chicago: University of Chicago Press, 1982.

Borell, Merriley. "Organotherapy and the Emergence of Reproductive Endocrinology." *Journal of the History of Biology*, 1985, *18*:1–30.

———. "Extending the Senses: The Graphic Method." *Medical Heritage*, 1986, *2*:114–121.

————. "Instruments and an Independent Physiology: The Harvard Physiological Laboratory, 1871–1906." In *Physiology in the American Context,* ed. Geison (q.v.), pp. 293–321.

————. "Instrumentation and the Rise of Modern Physiology." *Science & Technology Studies,* 1987, *5:*53–62.

————. *The Biological Sciences in the Twentieth Century.* Album of Science, 5. New York: Charles Scribner's Sons, 1989.

Brown, E. Richard. *Rockefeller Medicine Men: Medicine and Capitalism in America.* Berkeley: University of California Press, 1979.

Clarke, Adele. "Research Materials and Reproductive Science in the United States, 1910–1940." In *Physiology in the American Context,* ed. Geison (q.v.), pp. 323–350.

Clarke, Adele; Joan Fujimura, editors. *The Right Tools for the Job: At Work in Twentieth-Century Life Sciences.* Princeton: Princeton University Press, 1992.

Clause, Bonnie T. "The Wistar Rat as a Right Choice: Establishing Mammalian Standards and the Ideal of a Standardized Mammal." *Journal of the History of Biology,* 1993, *26:*329–350.

Coleman, William; Frederic L. Holmes, editors. *The Investigative Enterprise: Experimental Physiology in Nineteenth-Century Medicine.* Berkeley: University of California Press, 1988.

Cross, Stephen J.; William R. Albury. "Walter B. Cannon. L. J. Henderson, and the Organic Analogy." *Osiris,* 1987, *3:*165–192.

Ettling, John. *The Germ of Laziness: Rockefeller Philanthropy and Public Health in the New South.* Cambridge, Mass.: Harvard University Press, 1981.

Fleming, Donald. "Walter B. Cannon and Homeostasis." *Social Research,* 1984, *51:*609–640.

Flexner, Abraham. *Medical Education in the United States and Canada.* New York: Carnegie Foundation, 1910.

Geison, Gerald L. "'Divided We Stand': Physiologists and Clinicians in the American Context." In *The Therapeutic Revolution,* ed. Vogel and Rosenberg (q.v.), pp. 69–90.

————, editor. *Physiology in the American Context, 1850–1940.* Bethesda, Md.: American Physiological Society, 1987.

Gerard, Ralph W. *Mirror to Physiology: A Self-Survey of Physiological Science.* Washington, D.C.: American Physiological Society, 1958.

Hacking, Ian. "On the Stability of the Laboratory Sciences." *Journal of Philosophy,* 1988, *85:*507–514.

Hall, Diana Long. "Biology, Sex Hormones, and Sexism in the 1920s." *Philosophical Forum,* 1974, *5:*81–96.

————. "Physiological Identity of American Sex Researchers between the Two World Wars." In *Physiology in the American Context,* ed. Geison (q.v.), pp. 263–268.

Haraway, Donna J. *Crystals, Fabrics, and Fields: Metaphors of Organicism in Twentieth-Century Developmental Biology.* New Haven: Yale University Press, 1976.

Holmes, Frederic L. *Between Biology and Medicine: The Formation of Intermediary Metabolism.* Berkeley: Office for History of Science and Technology, University of California, 1992.

————. *Claude Bernard and Animal Chemistry.* Cambridge, Mass.: Harvard University Press, 1974.

————. *Hans Krebs: The Formation of a Scientific Life, 1900–1933.* Vol. I. New York: Oxford University Press, 1991.

————. "Manometers, Tissue Slices, and Intermediary Metabolism." In *The Right Tools for the Job,* ed. Clarke and Fujimura (q.v.), pp. 151–171.

Hughes, Arthur F. "A History of Endocrinology." *Journal of the History of Medicine and Allied Sciences,* 1977, *32:*292–313.

Kingsland, Sharon. "The Battling Botanist: Daniel Trembly MacDougal, Mutation Theory, and the Rise of Experimental Evolutionary Biology in America, 1900–1912." *Isis,* 1991, *82:*479–509.

Kohler, Robert E. *Partners in Science: Foundations and Natural Scientists, 1900–1945.* Chicago: University of Chicago Press, 1991.

Latour, Bruno; Steve Woolgar. *Laboratory Life: The Construction of Scientific Facts.* Princeton: Princeton University Press, 1987.

Ludmerer, Kenneth M. *Learning to Heal: The Development of American Medical Education.* New York: Basic Books, 1985.

Lynch, Michael. "Sacrifice and the Transformation of the Animal Body into a Scientific Object: Laboratory Culture and Ritual Practice in the Neurosciences." *Social Studies of Science,* 1988, *18:*265–289.

Magner, Lois N. *A History of the Life Sciences.* New York: Marcel Dekker, 1979.

Maienschein, Jane. "History of Biology." *Osiris,* 1985, *1:*147–162.

————, editor. *Defining Biology: Lectures from the 1890s.* Cambridge: Harvard University Press, 1986.

Maienschein, Jane; Ronald Rainger; Keith R. Benson. "Special Section on American Morphology at the Turn of the Century." *Journal of the History of Biology,* 1981, *14:*83–191.

Manning, Keith R. *Black Apollo of Science: The Life of Ernest Everett Just.* New York: Oxford University Press, 1983.

Marks, Harry. "Notes from the Underground: The Social Organization of Therapeutic Research." In *Grand Rounds: One Hundred Years of Internal Medicine,* ed. R. C. Maulitz and D. L. Hall (Philadelphia: University of Pennsylvania Press, 1986), pp. 297–338.

Mayr, Ernst. *The Growth of Biological Thought.* Cambridge, Mass.: Harvard University Press, 1982.

Medawar, Peter B. *The Threat and the Glory: Reflections on Science and Scientists.* New York: HarperCollins Publishers, 1990.

Oppenheimer, Jane. *Essays in the History of Embryology and Biology.* Cambridge, Mass.: MIT Press, 1967.

Pauly, Philip J. *Controlling Life: Jacques Loeb and the Engineering Ideal in Biology.* New York: Oxford University Press, 1987.

Proctor, Robert N. *Racial Hygiene: Medicine under the Nazis.* Cambridge, Mass.: Harvard University Press, 1988.

Rainger, Ronald; Keith R. Benson; Jane Maienschein, editors. *The American Development of Biology.* Philadelphia: University of Pennsylvania Press, 1988.

Rosenberg, Charles E. *No Other Gods: On Science and American Social Thought.* Baltimore: Johns Hopkins University Press, 1976.

Rothschuh, Karl E. *History of Physiology.* Translated by Guenther B. Risse. Huntington, N.Y.: R.E. Krieger, 1973.

Ruse, Michael; P. Taylor. "Special Issue on Pictorial Representation in Biology." *Biology and Philosophy,* 1991, *6:*125–294.

Russett, Cynthia E. *The Concept of Equilibrium in American Social Thought.* New Haven: Yale University Press, 1968.

Servos, John. *Physical Chemistry from Ostwald to Pauling: The Making of a Science in America.* Princeton: Princeton University Press, 1990.

Shapin, Steven. "The Invisible Technician." *American Scientist,* 1989, *77:*554–563.

Silverstein, Arthur M. *A History of Immunology.* New York: Academic Press, 1989.

Starr, Paul. *The Social Transformation of American Medicine.* New York: Basic Books, 1982.

Studer, Kenneth E.; Daryl E. Chubin. *The Cancer Mission: Social Contexts of Biological Research.* Beverly Hills, Calif: Sage Publications, 1980.

Vogel, Morris J.; Charles E. Rosenberg, editors. *The Therapeutic Revolution: Essays in the Social History of American Medicine.* Philadelphia: University of Pennsylvania Press, 1979.

Warner, John Harley. "Science in Medicine." *Osiris,* 1985, *1:*37–58.

Wilson, J. W. "Biology Attains Maturity in the Nineteenth Century." In *Critical Problems in the History of Science,* ed. Marshall Clagett (Madison: University of Wisconsin Press, 1959), pp. 401–408.

III. GENETICS, MOLECULAR BIOLOGY, VIROLOGY

Abir-Am, Pnina G. "The Discourse of Physical Power and Biological Knowledge in the 1930's: A Reappraisal of the Rockefeller Foundation's 'Policy' in Molecular Biology." *Social Studies of Science,* 1982, *12:*341–382.

————. "Essay Review: How Scientists View Their Heroes: Some Remarks on the Mechanism of Myth Construction." *Journal of the History of Biology,* 1982, *15:*281–315.

————. "The Biotheoretical Gatherings—Transdisciplinary Authority and the Incipient Legitimation of Molecular Biology in the 1930's: New Perspective on the Historical Sociology of Science." *History of Science,* 1987, *25:*1–70.

Allen, Garland E. "The Introduction of *Drosophila* into the Study of Heredity and Evolution: 1900–1910. In *Science in America since 1920,* ed. Nathan Reingold (New York: Science History Publishers, 1976), pp. 266–277.

————. *Thomas Hunt Morgan: The Man and His Science.* Princeton: Princeton University Press, 1978.

————. "T. H. Morgan and the Split between Embryology and Genetics, 1910–1935." In *A History of Embryology,* ed. T. J. Horder, J. A. Witkowski, and C. C. Wylie (Cambridge: Cambridge University Press, 1986), pp. 113–144.

Angier, Natalie. *Natural Obsessions: The Search for the Oncogene.* Boston: Houghton Mifflin, 1988.

Austoker, Joan. *A History of the Imperial Cancer Research Fund, 1902–1986.* Oxford: Oxford University Press, 1988.

Bang, Frederick B. "History of Tissue Culture at Johns Hopkins." *Bulletin of the History of Medicine,* 1977, *51:*516–537.

Bertran, J.; R. C. Lewontin. "The Political Economy of Hybrid Corn." *Monthly Review,* 1986, *38:*35–47.

Bishop, J. Michael. "Oncogenes." *Scientific American,* 1982, *246:*81–92.

Bishop, Jerry E.; Michael Waldholz. *Genome: The Story of the Most Astonishing Scientific Adventure of our Time—The Attempt to Map All the Genes in the Human Body.* New York: Simon & Schuster, 1990.

Boyd, T. E. Memo to Basil O'Connor, "Contributions to Science in the Field of Poliomyelitis," March of Dimes Birth Defects Foundation Archives, White Plains, New York, [1956].

Brannigan, A. "The Reification of Mendel." *Social Studies of Science,* 1979, *9:*423–454.

Bulloch, William. *The History of Bacteriology.* London: Oxford University Press, 1938, 1960.

Burian, Richard M.; Jean Gayon; Doris Zallen. "The Singular Fate of Genetics in the History of French Biology, 1900–1940." *Journal of the History of Biology,* 1988, *21:*357–402.

Cairns, John; Gunther Stent; James D. Watson, editors. *Phage and the Origins of Molecular Biology.* Cold Spring Harbor, N.Y.: Cold Spring Harbor Laboratory of Quantitative Biology, 1966.

Cambrosio, Antonio; Peter Keating. "'Going Monoclonal': Art, Science, and Magic in the Day-to-Day Use of Hybridoma Technology." *Social Problems,* 1988, *35:*244–260.

Carlson, Elof Axel. *The Gene: A Critical History.* Philadelphia: Saunders, 1966.

————. *Genes, Radiation and Society: The Life and Work of H. J. Muller.* Ithaca, N.Y.: Cornell University Press, 1981.

Chargaff, Erwin. "A Quick Climb up Mount Olympus." *Science,* 1968, *159:*1448–1449.

————. *Heraclitean Fire: Sketches from a Life before Nature.* New York: Rockefeller University Press, 1978.

Cohen, S. S. "The Biochemical Origins of Molecular Biology (Introduction)." *Trends in Biochemical Sciences,* 1984, *9:*334–336.

Corner, George W. *A History of the Rockefeller Institute, 1901–1953: Origins and Growth.* New York: Rockefeller Institute Press, 1964.

Crick, Francis. *What Mad Pursuit: A Personal View of Scientific Discovery.* New York: Basic Books, 1988.

Davis, Joel. *Mapping the Code: The Human Genome Project and the Choices of Modern Science.* New York: Wiley, 1990.

Dubos, René. *The Professor, the Institute, and DNA: Oswald T. Avery, His Life and Scientific Achievements.* New York: Rockefeller University Press, 1976.

Dulbecco, Renato. *Aventurier du vivant.* Paris: Plon, 1989.

————. "Oncogenic Viruses: The Last Twelve Years." In *Tumor Viruses.* Cold Spring Harbor Symposia on Quantitative Biology, 29. 2 pts. Cold Spring Harbor, N.Y.: Cold Spring Harbor Laboratory, 1975, pt. 1, pp. 1–7.

Enders, John F.; Frederic Robbins; Thomas H. Weller. "The Cultivation of the Poliomyelitis Viruses in Tissue Culture." Nobel Lecture, 11 Dec. 1954. In *Nobel Lectures . . . Physiology or Medicine, 1942–1962* (Amsterdam: Elsevier, 1964), pp. 448–467.

Fischer, Ernst Peter; Carol Lipson. *Thinking about Science: Max Delbrück and the Origins of Molecular Biology.* New York: Norton, 1988.

Fischer, Jean-Louis; William H. Schneider, editors. *Histoire de la génétique: Pratiques, techniques, et théories.* Paris: A.R.P.E.M./Editions Sciences en Situation, 1990.

Fitzgerald, Deborah. *The Business of Breeding: Hybrid Corn in Illinois, 1890–1940.* Ithaca, N.Y.: Cornell University Press, 1990.

Fleming, Donald. "Emigré Physicists and the Biological Revolution." In *The Intellectual Migration: Europe and America, 1930–1960,* ed. Donald Fleming and Bernard Bailyn (Cambridge: Harvard University Press, 1969), pp. 152–189.

Foster, H. L. "The History of Commercial Production of Laboratory Rodents." *Laboratory Animal Science,* 1980, *30*:793–798.

Foster, William D. *A History of Medical Bacteriology and Immunology.* London: William Heinemann Medical Books, 1970.

Fruton, Joseph S. *Molecules and Life: Historical Essays on the Interplay of Chemistry and Biology.* New York: Wiley, 1972.

———. *Contrasts in Scientific Style: Research Groups in the Chemical and Biochemical Sciences.* Philadelphia: American Philosophical Society, 1990.

Gallo, Robert. *Virus Hunting: AIDS, Cancer, and the Human Retrovirus, A Story of Scientific Discovery.* New York: Basic Books, 1991.

Galperin, Charles. "Le bactériophage, la lysogénie et son déterminisme génétique." *History and Philosophy of the Life Sciences,* 1987, *9*:175–224.

Gaudillière, Jean-Paul. *Biologie moleculaire et biologistes dans les années soixante: La naissance d'une discipline—Le cas français.* Dissertation, Doctorat d'Histoire des Sciences, Université Paris VII, 1991.

———. "J. Monod, S. Spiegelman et l'adaptation enzymatique: Programmes de recherche, cultures locales et traditions disciplinaires." *History and Philosophy of the Life Sciences,* 1992, *14*:29–78.

———. "Entre laboratoire et hôpital: biochimistes et biomédecine dans l'après guerre— Deux itinéraires." *Sciences Sociales et Santé,* 1992, *10*:104–149.

———. "Molecular Biology in the French Tradition? Redefining Local Traditions and Disciplinary Patterns." *Journal of the History of Biology,* 1993, *26*:473–498.

———. "Oncogenes as Metaphors for Human Cancer: Articulating Laboratory Practices and Medical Demands." In *Medicine and Change: Historical and Sociological Studies of Medical Innovation,* ed. Ilana Löwy (Colloque Inserm, 220) (London: John Libbey, 1993), pp. 213–248.

———. "Cancer between Heredity and Contagion: About the Production of Mice and Viruses." In *Human Genetics,* ed. Everett Mendelsohn (Sociology of Science Yearbook, 1995, in press).

Gilbert, Scott F. "The Embryological Origins of the Gene Theory." *Journal of the History of Biology,* 1978, *11*:307–351.

———. "Intellectual Traditions in the Life Sciences: Molecular Biology and Biochemistry." *Perspectives in Biology and Medicine,* 1982, *26*:151–162.

Glass, Bentley. "A Century of Biochemical Genetics." *Proceedings of the American Philosophical Society,* 1968, *109*:227–236.

Grmek, Mirko D. *History of AIDS: Emergence and Origin of a Modern Pandemic.* Translated by Russell C. Maulitz and Jacalyn Duffin. Princeton: Princeton University Press, 1990.

Gross, Ludwik. *Oncogenic Viruses.* 2nd ed. Oxford: Pergamon Press, 1970.

Hall, Stephen S. *Invisible Frontiers: The Race to Synthesize a Human Gene.* New York: Atlantic Monthly Press, 1987.

Hsu, T. C. *Human and Mammalian Cytogenetics: An Historical Perspective.* New York: Springer-Verlag, 1979.

Huebner, Robert J.; George J. Todaro. "Oncogenes of RNA Tumor Viruses as Determinants of Cancer." *Proceedings of the National Academy of Sciences,* 1969, *64*:1087–1088.

Hughes, Sally Smith. *The Virus: A History of the Concept.* New York: Science History Publications, 1977.

Jacob, François. *The Statue Within: An Autobiography.* New York: Basic Books, 1988.

Judson, Horace F. *The Eighth Day of Creation: The Makers of the Revolution in Biology.* New York: Simon & Schuster, 1979.

Kay, Lily E. "Conceptual Models and Analytical Tools: The Biology of Physicist Max Delbrück." *Journal of the History of Biology,* 1985, *18:*207–246.

———. "Cooperative Individualism and the Growth of Molecular Biology at the California Institute of Technology, 1928–1953." Ph.D. dissertation, Johns Hopkins University, 1987.

———. "Laboratory Technology and Biological Knowledge: The Tiselius Electrophoresis Apparatus, 1930–1945." *History and Philosophy of the Life Sciences,* 1988, *10:*51–72.

———. *The Molecular Vision of Life: Caltech, the Rockefeller Foundation, and the Rise of the New Biology.* New York: Oxford University Press, 1991.

Keller, Evelyn Fox. *A Feeling for the Organism: The Life and Work of Barbara McClintock.* New York: W. H. Freeman, 1983.

———. "Physics and the Emergence of Molecular Biology: A History of Cognitive and Political Synergy." *Journal of the History of Biology,* 1990, *23:*389–409.

Kevles, Daniel J. "Genetics in the United States and Great Britain, 1890–1930: A Review with Speculations." In *Biology, Medicine, and Society, 1840–1930,* ed. Charles L. Webster (Cambridge: Cambridge University Press, 1981), pp. 193–215.

———. *In the Name of Eugenics: Genetics and the Uses of Human Heredity.* New York: Alfred A. Knopf, 1985.

———. "Pursuing the Unpopular: A History of Courage, Viruses, and Cancer." In *Hidden Histories of Science,* ed. Robert Silvers (New York: New York Review of Books, 1995), pp. 69–112.

———. "Renato Dulbecco and the New Animal Virology: Medicine, Methods, and Molecules." *Journal of the History of Biology,* 1993, *26:*409–442.

Kevles, Daniel J.; Leroy Hood, editors. *The Code of Codes: Scientific and Social Issues in the Human Genome Project.* Cambridge, Mass.: Harvard University Press, 1992.

Kimmelman, Barbara. "The American Breeders' Association: Genetics and Eugenics in an Agricultural Context, 1903–1913." *Social Studies of Science,* 1983, *13:*163–204.

Klein, George. *The Atheist and the Holy City: Encounters and Reflections.* Translated by Theodore and Ingrid Friedman. Cambridge, Mass.: MIT Press, 1990.

Kloppenburg, Jack R., Jr. *First the Seed: The Political Economy of Plant Biotechnology, 1492–2000.* Cambridge: Cambridge University Press, 1988.

Kohler, Robert E. *Lords of the Fly:* Drosophila *Genetics and the Experimental Life.* Chicago: University of Chicago Press, 1994.

———. "The Management of Science: The Experience of Warren Weaver and the Rockefeller Foundation Programme in Molecular Biology." *Minerva,* 1976, *14:*279–306.

———. *From Medical Chemistry to Biochemistry.* Cambridge: Cambridge University Press, 1982.

Kornberg, Arthur. *For the Love of Enzymes: The Odyssey of a Biochemist.* Cambridge, Mass.: Harvard University Press, 1989.

Levine, Arnold J. *Viruses.* New York: Scientific American Library, 1992.

Lewontin, Richard C.; J. Bertran. "Technology, Research, and the Penetration of Capital: The Case of U.S. Agriculture." *Monthly Review,* 1986, *38:*21–34.

Lipmann, Fritz. *Wanderings of a Biochemist.* New York: Wiley-Interscience, 1971.

Luria, Salvador E. *A Slot Machine, A Broken Test Tube: An Autobiography.* New York: Harper & Row, 1984.

Lwoff, André; Agnès Ullmann. *Les origines de la biologie moléculaire: Un hommage à Jacques Monod.* Paris-Montréal: Études vivantes, 1980.

Mazumdar, Pauline M., editor. *Immunology, 1930–1980: Essays on the History of Immunology.* Toronto: Wall & Thompson, 1989.

McCarty, Maclyn. *The Transforming Principle: Discovering that Genes Are Made of DNA.* New York: W. W. Norton, 1985.

Morange, Michel, editor. *L'Institut Pasteur: Contributions à son histoire.* Paris: Éditions La Decouverte, 1991.

———. *Histoire de la biologie moléculaire.* Paris: Éditions La Decouverte, 1994.

Moulin, Anne Marie. *Le dernier langage de la médecine: Histoire de l'immunologie de Pasteur au Sida.* Paris: Presses Universitaires de France, 1991.

Olby, Robert C. "Francis Crick, DNA and the Central Dogma." *Daedalus,* 1970, *99:*938–987.

———. "Schrodinger's Problem: What Is Life?" *Journal of the History of Biology,* 1971, *4:*119–148.

———. *The Path to the Double Helix.* Seattle: University of Washington Press, 1974.

Patterson, James T. *The Dread Disease: Cancer and Modern American Culture.* Cambridge, Mass.: Harvard University Press, 1987.

Pauling, Linus. "Fifty Years of Progress in Structural Chemistry and Molecular Biology." *Daedalus,* 1970, *99:*998–1014.

Pinell, Patrice. *Naissance d'un fléau: Histoire de la lutte contre le cancer en France (1900–1940).* Paris: Éditions Métailié, 1992.

Provine, William B. *Sewall Wright and Evolutionary Biology.* Chicago: University of Chicago Press, 1986.

Radetsky, Peter. *The Invisible Invaders: The Story of the Emerging Age of Viruses.* New York: Little, Brown, 1991.

Rosenberg, Steven A.; John M. Barry. *The Transformed Cell: Unlocking the Mysteries of Cancer.* New York: G. P. Putnam's Sons, 1992.

Rubin, H. "Quantitative Tumor Virology." In *Phage and the Origins of Molecular Biology,* ed. Cairns, Stent, and Watson (q.v., in Sect. II), pp. 287–300.

Sapp, Jan. *Beyond the Gene: Cytoplasmic Inheritance and the Struggle for Authority in Genetics.* New York: Oxford University Press, 1987.

———. *Where the Truth Lies: Franz Moewus and the Origins of Molecular Biology.* Cambridge: Cambridge University Press, 1990.

Sayre, Anne. *Rosalind Franklin and DNA.* New York: Norton, 1975.

Schaffner, Kenneth. "The Peripherality of Reductionism in the Development of Molecular Biology." *Journal of the History of Biology,* 1974, *7:*111–139.

Shapiro, Robert. *The Human Blueprint: The Race to Unlock the Secrets of Our Genetic Script.* New York: St. Martin's Press, 1991.

Shimkin, Michael B. *Contrary to Nature: . . . the Development of Knowledge concerning Cancer.* DHEW Pub. No. (NIH) 76-720. Washington, D.C.: U.S. Department of Health, Education, and Welfare, 1977.

Smith, Jane S. *Patenting the Sun.* New York: William Morrow, 1990.

Stent, Gunther, "DNA." *Daedalus,* 1970, *99:*909–937.

Vogt, Marguerite; Renato Dulbecco. "Virus-Cell Interaction with a Tumor-Producing Virus." *Proceedings of the National Academy of Sciences,* 1960, *46:*365–370.

Waterson, Alan P.; Lise Wilkinson. *An Introduction to the History of Virology.* Cambridge: Cambridge University Press, 1978.

Watson, James D. "Foreword." *Viral Oncogenes.* Cold Spring Harbor Symposia on Quantitative Biology, 44. Cold Spring Harbor, N.Y.: Cold Spring Harbor Laboratory, 1980.

———. *The Double Helix.* New York: Athenaeum, 1968.

Watson, James D.; Francis Crick. "Molecular Structure of Nucleic Acids." *Nature,* 1953, *171:*737–738.

Wingerson, Lois. *Mapping Our Genes: The Genome Project and the Future of Medicine.* New York: Dutton, 1990.

Yoxen, E. "Giving Life a New Meaning: The Rise of the Molecular Biology Establishment." In *Scientific Establishments and Hierarchies,* ed. Norbert Elias, Herminio Martins, and Richard Whitley (Dordrecht: Reidel, 1982), pp. 123–143.

IV. NEUROBIOLOGY

Babkin, Boris Petrovich. *Pavlov: A Biography.* Chicago: University of Chicago Press, 1949.

Bennett, M. V. L. "Nicked by Occam's Razor: Unitarianism in the Investigation of Synaptic Transmission." *Biological Bulletin,* 1985, *168:*159–167.

Billings, Susan. "Concepts of Nerve Fiber Development, 1839–1930." *Journal of the History of Biology,* 1971, *4:*275–306.

Brazier, Mary A. B. "The Historical Development of Neurophysiology." In *Handbook of Physiology*, ed. J. Field, H. W. Magoon, and V. E. Hall (Baltimore: Waverly, 1960), pp. 1–57.

Cannon, Walter B. "Organization for Physiological Homeostasis." *Physiological Reviews*, 1929, *9*:399–431.

———. "The Autonomic Nervous System, and Interpretation." *The Lancet*, 1930, *1*:1109–1115.

———. *The Wisdom of the Body*. New York: Norton, 1932.

Clarke, Edwin; Charles D. O'Malley. *The Human Brain and Spinal Cord: A Historical Study Illustrated by Writings from Antiquity to the Twentieth Century*. Berkeley: University of California Press, 1968.

Clarke, Edwin; Louise S. Jacyna. *Nineteenth-Century Origins of Neuroscientific Concepts*. Berkeley: University of California Press, 1987.

Frank, Robert G., Jr. "The A. V. Hill Papers at Churchill College, Cambridge." *Journal of the History of Biology*, 1978, *11*:211–214.

———. "The Joseph Erlanger Collection at Washington University School of Medicine, St. Louis." *Journal of the History of Biology*, 1979, *12*:193–201.

———. "Instruments, Nerve Action, and the All-or-None Principle." *Osiris*, 1994, *9*:208–235.

———. "The Telltale Heart: Physiological Instruments, Graphic Methods, and Clinical Hopes, 1854–1914." In *The Investigative Enterprise*, ed. Coleman and Holmes (q.v., in Sect. II), pp. 211–290.

Frank, Robert G., Jr.; Judith H. Goetzel. "The Alexander Forbes Papers." *Journal of the History of Biology*, 1978, *11*:429–435.

Frank, Robert G., Jr.; L. H. Marshall; H. W. Magoun. "The Neurosciences." In *Advances in American Medicine: Essays at the Bicentennial*. New York: Josiah Macy, Jr., Foundation, 1976.

Fruton, Joseph S. *A Skeptical Biochemist*. Cambridge: Harvard University Press, 1992.

Geddes, L. A. "A Short History of the Electrical Stimulation of Excitable Tissue, Including Electrotherapeutic Applications." *The Physiologist*, 1984, supplement *27*:1–47.

Geison, Gerald L. "Keith Lucas." In *Dictionary of Scientific Biography* (q.v., in Sect. I).

———. *Michael Foster and the Cambridge School of Physiology: The Scientific Enterprise in Victorian Society*. Princeton: Princeton University Press, 1978.

Greenspan, Ralph J. "The Emergence of Neurogenetics." *The Neurosciences*, 1990, *2*:145–157.

Harrington, Anne. *Medicine, Mind, and the Double Brain: A Study in Nineteenth-Century Thought*. Princeton: Princeton University Press, 1987.

Howell, Joel. "Cardiac Physiology and Clinical Medicine? Two Case Studies." In *Physiology in the American Context*, ed. Geison (q.v., in Sect. II), pp. 279–292.

Jung, R. "Some European Neuroscientists: A Personal Tribute." In *The Neurosciences*, ed. Worden, Swazey, and Adelman (q.v.), pp. 477–511.

Lenoir, Timothy. "Models and Instruments in the Development of Electrophysiology, 1845–1912." *Historical Studies in the Physical and Biological Sciences*, 1986, *17*:1–54.

Marshall, Louise. "The Fecundity of Aggregates: The Axonologists at Washington University, 1922–1942." *Perspectives in Biology and Medicine*, 1983, *26*:613–636.

———. "Instruments, Techniques, and Social Units in American Neurophysiology, 1870–1950." In *Physiology in the American Context*, ed. Geison (q.v., in Sect. II), pp. 351–369.

Ramon y Cajal, Santiago. *Recollections of My Life*. Philadelphia: American Philosophical Society, 1937.

Sherrington, Charles S. *The Integrative Action of the Nervous System*. New Haven: Yale University Press, 1907.

Star, Susan L. *Regions of the Mind: Brain Research and the Quest for Scientific Certainty*. Palo Alto, Calif.: Stanford University Press, 1989.

Swazey, Judith P. *Reflexes and Motor Integration: Sherrington's Concept of Integrative Action*. Cambridge, Mass.: Harvard University Press, 1969.

Swazey, Judith P.; Frederic G. Worden. "On the Nature of Research in Neuroscience." In *The Neurosciences*, ed. Worden, Swazey, and Adelman, pp. 569–587.

Worden, Frederic G.; Judith P. Swazey; George Adelman, editors. *The Neurosciences: Paths of Discovery.* Cambridge, Mass.: MIT Press, 1975.

Young, Robert M. *Mind, Brain and Adaptation in the Nineteenth Century: Cerebral Localization and Its Biological Context from Gall to Ferrier.* New York: Oxford University Press, 1990.

N.b.: Cross references to works in the same section of the bibliography use *(q.v.).* Cross references to works in other sections give the number of the section as well.

Notes on Contributors

Stephen G. Brush was a theoretical physicist before coming to the University of Maryland as a historian of science. He was president of the History of Science Society in 1990–1991. His three-volume work on theories of the origin and structure of the earth will be published in 1995–1996.

Lorraine Daston teaches at the University of Chicago, in the Department of History and the Committee on the Conceptual Foundations of Science. The author of *Classical Probability in the Enlightenment* (1988), she has written recently on curiosity in early modern science, the meanings of calculation in the Enlightenment, and forms of scientific objectivity in the nineteenth century.

Gerald Geison is Professor of History at Princeton University. With Frederic L. Holmes he edited *Research Schools: Historical Reappraisals,* Volume 8 of *Osiris* (1993). His new book, *The Private Science of Louis Pasteur,* will be published by Princeton University Press in the spring of 1995.

Evelyn Fox Keller took her Ph.D. in theoretical physics and is now Professor of History and Philosophy of Science in the Program in Science, Technology, and Society at MIT. She is the author of *A Feeling for the Organism: The Life and Work of Barbara McClintock* (1983), *Reflections on Gender and Science* (1985), and *Refiguring Life: Metaphors of Twentieth-Century Biology* (in press). Her current research is on the history of developmental biology.

Daniel J. Kevles is the Koepfli Professor of Humanities at the California Institute of Technology, where he heads the Program on Science, Ethics, and Public Policy. His work has ranged across the history of physics, virology and cancer research, eugenics and human genetics, environmentalism, and intellectual property in living organisms.

Sally Gregory Kohlstedt, a professor in the History of Science and Technology Program at the University of Minnesota, recently edited *The Origins of Natural Science in the United States: The Essays of George Brown Goode* (1991), and, with Roderick W. Home, *International Science and National Scientific Identity: Australia between Britain and America* (1991). She is currently studying the history of the nature study movement.

David C. Lindberg is Professor of History of Science at the University of Madison, Wisconsin, and president of the History of Science Society. His specialty is reflected in the title of his recent book, *The Beginnings of Western Science: The European Scientific Tradition in Philosophical, Religious, and Institutional Context, 600 B.C. to A.D. 1450* (1992).

Nakayama Shigeru is a professor in the Science and Technology Studies Center at Kanagawa University in Hiratsuka, Japan. Trained in astrophysics and in history of science, he is the author of six books in English and more than thirty in Japanese; his most readily accessible work is "Japanese Scientific Thought" in the *Dictionary of Scientific Biography*. His interests have shifted gradually from traditional East Asian sciences to science policy studies.

Nancy J. Nersessian is Professor of Cognitive Sciences and Coordinator of the Cognitive Science Program at Georgia Institute of Technology. She writes on the history of electromagnetic theory in the nineteenth and twentieth centuries and on conceptual change in science. She is at work on *Creativity and Conceptual Change: A Constructivist View of Science,* to be published by MIT Press.

Thomas Nickles is Professor of Philosophy at the University of Nevada, Reno. He is working on history of methodology, naturalized and socialized accounts of inquiry, and discovery and problem solving. He is the editor of *Scientific Discovery, Logic, and Rationality* and *Scientific Discovery: Case Studies* (both Reidel, 1980).

Joan L. Richards is Associate Professor of History at Brown University. She is the author of *Mathematical Visions* (1988), a study of non-Euclidean geometry in England. Her current project is a book about English interpretations of mathematical knowing in the first half of the nineteenth century.

John Harley Warner is Professor of History of Medicine and Life Sciences at Yale University. He is the author of *The Therapeutic Perspective: Medical Practice, Knowledge, and Identity in America, 1820–1885* (Harvard, 1986). He is now completing *Against the Spirit of System: The French Impulse in Nineteenth-Century Medicine.*

Index